PRINCIPLES OF MATHEMATICS

PRENTICE-HALL MATHEMATICS SERIES

PRENTICE-HALL INTERNATIONAL, INC., *London*
PRENTICE-HALL OF AUSTRALIA, PTY., LTD., *Sydney*
PRENTICE-HALL OF CANADA, LTD., *Toronto*
PRENTICE-HALL OF INDIA (PRIVATE) LTD., *New Delhi*
PRENTICE-HALL OF JAPAN, INC., *Tokyo*

PRINCIPLES OF MATHEMATICS

Professor of Mathematics
Louisiana State University

Prentice-Hall, Inc.,   Englewood Cliffs, New Jersey

A revision of *Freshman Mathematics* by Paul K. Rees

© 1959, 1965 by Prentice-Hall, Inc., Englewood Cliffs, N. J.

Second printing........October, 1965

Library of Congress Catalog Card Number 65-16592

Printed in the United States of America
C-70945

PREFACE

This book is intended primarily for non-science majors in the college of arts and science and for prospective and in-service elementary teachers; however, it can be used advantageously by those persons who need a refresher course in algebra, by those who need an algebraic background for work in mathematics of finance, and by those who want some insight into mathematics.

A student with one year of high school algebra can use this book as a text with profit; a student with less algebraic background might well benefit by its use since it contains a long chapter on arithmetic; furthermore, a student with more than one year of high school algebra might find such chapters as those on an introduction to sets, the trigonometric functions, right triangles, the laws of sines and cosines, polar coordinates, introduction to analytic geometry, annuities, statistics, permutations, combinations, probability, inequalities, and a glimpse at the Calculus to his liking and fitting his needs.

In addition to retaining all of the material of the first edition, the present book contains several new articles and chapters and a considerable bit of revising and repolishing as well as some modernizing. The new and rewritten material includes work on income tax, significant figures, functions and functional notation, calculations with numbers and angles, nominal and effective rates, equations of value, and use of determinants in solving systems of linear equation besides chapters on introduction to sets, polar coordinates, analytic geometry, inequalities, and Calculus.

A study of the contents will reveal that the book is made up about as follows: algebra 30 per cent; arithmetic, 20 per cent; trigonometry, 18 per cent; mathematics of finance, 9 per cent; Calculus, 8 per cent; other topics, 15 per cent.

A considerable bit of time, effort, and thought has been expended to make the worked-out examples illustrate the text and the problems in the exercise that follows. We hope and believe that the exercises are a normal lesson apart, but realize that conditions may be such that it is advisable to spend more than one lesson on some exercises. The problems are all new and are in groups of four similar ones of about the same level of difficulty. Because of this, an instructor can make a good assignment by giving each fourth problem without having to study through the list.

This edition contains some 3000 problems in 87 exercises as compared to about 2200 in 70 exercises for the former edition (published as *Freshman Mathematics*). Thus, the book furnishes enough material for six semester hours of work but can be used with a considerable degree of choice for a three semester hour course.

The author wishes to express his appreciation to Louisiana State University for granting a sabbatical leave during the spring of 1957 and another during the spring of 1964; much of the work on the two editions of this book was done during those periods.

PAUL K. REES

CONTENTS

1

ARITHMETIC

1.1. Mathematics

Experience in and out of the classroom has convinced me that mathematics is many things to many people: to many students it is simply one of the obstacles that must be overcome to obtain a degree; to most graduates of college or high school, it is a subject they wish they had studied more of and more diligently; to some of us, it is a constantly used tool or a language— a very concise language that makes exact and logical statements easier to form. To others, mathematics is a logical development made up of undefined terms, principles of logic, hypotheses, and, finally, conclusions that follow from the latter—conclusions that are merely logical consequences and make no claim concerning absolute truth or falsity. Finally, to mathematicians, our subject is not only a means of livelihood, but also a pleasant ruler in our way of living.

1.2. Graphical Representation of Positive Integers

We shall use the term *positive integer* to designate a number that is used in counting objects. Thus, 1, 2, 3, 4, ... are positive integers. It is desirable to be able to represent the positive integers graphically since such a representation enables us to visualize the procedures of addition and multiplication. In order to represent them graphically, we begin with a ray, mark its beginning point with 0, and mark off the positive integers at equal intervals in the order in which they come in counting as shown in Fig. 1-1. We must use a ray as distinguished from a line segment since there is no last positive integer as evidenced by assuming there is a last one and then obtaining a larger one by adding any nonzero integer to the assumed last one.

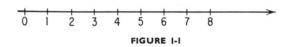

FIGURE 1-1

1.3. Addition of Positive Integers

If we use the representation along a line, we think of 3 as the number we arrive at by going 3 units along the ray from the starting point and of 5 as the number we arrive at by going 5 units from the starting point; furthermore, we think of 3 plus 5 as the number we arrive at by going 5 units beyond 3 and of 5 plus 3 as the number we arrive at by going 3 units beyond 5. The reader can readily verify the fact that we arrive at the same number, 8, regardless of the order in which 3 and 5 are added. In general, we state that the order in which two numbers are added is immaterial by saying that

Addition is commutative, or symbolically, $a + b = b + a$ for any two positive integers a and b.

In order to make another rule of operation on positive integers seem reasonable, we shall notice that we get the same results by adding 6 to the sum of 2 and 3 as indicated by $(2 + 3) + 6$ and by adding the sum of 3 and 6 to 2 as indicated by $2 + (3 + 6)$. By use of representation along a line, we see that $(2 + 3) + 6 = 5 + 6 = 11$ and $2 + (3 + 6) = 2 + 9 = 11$. This situation can be expressed in general by saying that

Addition is associative, or symbolically, $(a + b) + c = a + (b + c)$ for any three positive integers a, b, and c.

1.4. Multiplication of Positive Integers

We saw in the last two articles how to represent a positive integer graphically and how to add two or more such integers. We shall now consider the addition of a positive integer to itself any integral number of times. As a special case, $3 + 3 + 3 + 3 = (3 + 3) + (3 + 3) = 6 + 6 = 12$. If the number of threes to be added together increases, it becomes more and more a task to indicate the sum. Fortunately, there is another way of indicating the sum. It is done by telling how many times the number is to be used. Thus, $3 + 3 + 3 + 3$ is four threes, and we symbolize this by writing 4×3 or $4 \cdot 3$ or $(4)(3)$. This can be interpreted as going out three units from the point marked 0, then a second 3 units, then another 3 and finally a fourth 3 units. The reader can readily verify the fact that $(3)(4) = (4)(3)$. This is a special case of the more general statement that

Multiplication is commutative, *or symbolically,* $ab = ba$ *for any two positive integers* a *and* b.

We can readily verify the fact that we get the same product if we multiply the product of 2 and 3 by 5 and if we multiply 2 by the product of 3 and 5. This situation is expressed in general terms by saying that

Multiplication is associative, *or symbolically,* $(ab)c = a(bc)$ *for any three positive integers* a, b, *and* c.

There is one other fundamental assumption that is made and used in connection with addition and multiplication. It is that

Multiplication is distributive with respect to addition, *or symbolically,* $a(b + c) = ab + ac$ *for any three positive integers* a, b, *and* c.

EXAMPLES

1. By use of the representation of positive integers as points along a ray we see that $(2)(3)$ is represented by going out 3 units and then going out another 3 from there, and that $(3)(2)$ is obtained by going out 2 units, then another 2 and finally a third 2 units and 6 is obtained in each case; hence, $(2)(3) = (3)(2)$.

2. $(3 \times 4)5 = (12)5 = 60$ and $3(4 \times 5) = 3(20) = 60$; therefore, $(3 \times 4)5 = 3(4 \times 5)$.

3. $3(4 + 5) = 3 \times 4 + 3 \times 5 = 12 + 15 = 27$ and $3(9) = 27$.

1.5. Subtraction of Positive Integers

If one opens and then closes a door, things are as they were originally; thus, closing the door has offset opening it. In general, if an operation does something, then the operation which offsets or undoes that something is called the *inverse operation*.

Subtraction is the inverse of addition and is defined by the statement

$$a - b = x \quad \text{provided} \quad b + x = a.$$

Thus $a - b$ is x provided x is the number which must be added to b in order to obtain *a*. If we allow *b* to be equal to *a* and interpret subtraction in connection with points on a line as going in the direction opposite to that for addition, we see that *a* − *a* leads us to the starting point and we have a new number which we call *zero*. Furthermore, if we extend rays in both directions from the starting point as in Fig. 1-2 and allow *b* to be larger than *a*, then the point which represents *a* − *b* is to the left of the starting point and we have

FIGURE I-2

a second new type of number which we call *negative*. Thus $2 - 5 = -3$, since starting at 0, going 2 units to the right and then 5 to the left, we arrive at the point that is 3 units to the left of the starting point.

1.6. Division of Positive Integers

Division is the inverse of multiplication and is defined by the statement

$$a \div b = x \quad \text{provided} \quad b \times x = a, b \neq 0.$$

Thus $a \div b$ is *x* provided *x* is the number that must be multiplied by *b* in order to obtain *a*.

EXAMPLE

1. $12 \div 3$ is the number that 3 must be multiplied by in order to give 12 and is 4.

Since we can think of *ab* as $b + b + \cdots$ on to *a* terms for *a* and *b* positive integers, we see that *multiplication can be thought of as an addition process*; hence, *division can be thought of as a subtraction process* since that is the inverse of addition.

EXAMPLE

2. We can now think of $12 \div 3$ as the number of times 3 must be subtracted from 12 in order to exactly reach zero. Thus, $12 - 3 = 9$, $9 - 3 = 6$, $6 - 3 = 3$ and $3 - 3 = 0$. Therefore, $12 \div 3$ is 4, the number of times 3 must be subtracted from 12 to produce zero.

If the subtraction process as applied to two positive integers does not exactly produce zero, then the indicated division gives a number that is not a positive integer. It is the quotient of two integers and is called a *rational fraction*. Thus $17 \div 5$ does not lead to zero by the subtraction process and hence $71 \div 5 = \frac{17}{5}$ is a rational fraction. It can be put in another form by noticing that 5 goes into 17 three times and that 2 is left over, we then write $\frac{17}{5} = 3\frac{2}{5}$. The number left over in a division is called the *remainder*.

The positive and negative integers along with the positive and negative rational fractions and zero are called the *rational number system.*

There are real numbers that can not be expressed as the quotient of two integers. They are called *irrational numbers.*

1.7. Zero in the Division Process

If, in $a \div b = x$ provided $b \times x = a$, we let $b = 0$, we have $0 \times x = a$. If $a \neq 0$, read a is not equal to zero, the statement $0 \times x = a$ is clearly false since zero times any number is zero; consequently, *division by zero is not an admissible operation.* If, in $bx = a$, $a = 0$ and $b \neq 0$, we have $bx = 0$; hence, $x = 0$. Therefore, *zero divided by any nonzero number is zero.*

Perhaps a less formal look at zero in the division process would be in order. If we think of division in terms of the subtraction process, we see that

$$\frac{0}{a} = 0, \qquad a \neq 0,$$

since zero is the number of times that $a \neq 0$ must be subtracted from zero to obtain zero.

If we think of division as the inverse of multiplication, then a divided by zero can be equal to a number x only if x is the number of times that zero must be subtracted from a to obtain zero; there is no such number, since regardless of the number of times zero is subtracted from a we do not obtain zero. Consequently, as we did earlier in this section, we reach the conclusion that division by zero is not defined; hence, is not an admissible operation.

1.8. Operations with Positive and Negative Integers

We have seen how to perform the four fundamental operations on positive integers and shall now discuss performing them on a combination of positive and negative integers. If we think of the rational numbers as being represented along a line, and of plus and minus as indicating movement to the right and to the left, respectively, then $6 + (-2)$ would be performed by moving 6 units to the right from the starting point and then 2 units to the left. We thus arrive at 4. To find the sum indicated by $(-3) + (-5)$, we move 3 units to the left of the starting point and then 5 units further to the left; hence, arriving at -8. In general, the following procedure is followed in addition.

In order to add two numbers with the same sign, we omit the sign, find the sum of the new numbers and give it the common sign. In order to add two numbers with opposite signs, we omit the signs, find the difference of the new numbers and give it the sign of the one of greater size.

EXAMPLES

1. $3 + 7 = 10.$ **2.** $(-2) + (-4) = -6.$

3. $9 + (-3) = 6$ since 6 is the difference between 9 and 3 and the larger is positive.

4. $5 + (-14) = -9$ since 9 is the difference between 5 and 14 and the one of greater size is negative.

In order to subtract one number from another, we change the sign of the one that is being subtracted and then add, since addition and subtraction are inverse operations.

EXAMPLES

5. $3 - 7 = 3 + (-7) = -4.$ **6.** $5 - (-2) = 5 + (+2) = 7.$

In order to make the definitions concerning products seem reasonable, we shall consider several examples in connection with representation of numbers along a line.

EXAMPLES

7. $2(3)$ means that we are to go 3 units to the right of the starting point and then another 3; hence, $2(3) = +6.$

8. $2(-3)$ means that we are to go 3 units to the left of the starting point and then another 3 units in that direction; hence, $2(-3) = -6.$

9. If we assume that multiplication in connection with negative numbers is commutative, then $(-3)2 = -6$ as was $2(-3).$

10. If we think of $(-3)(-2)$ as $-1(3)(-2)$, we then see that $(-3)(-2)$ and $3(-2)$ can be thought of as being the same size but of opposite signs; hence, we say that $(-3)(-2) = +6$ since $(3)(-2) = -6.$

We shall now state the following rules for multiplication of signed numbers.

Multiplication of two numbers with the same sign gives a positive number.
Multiplication of two numbers with opposite signs gives a negative number.

We shall use the same rules of signs for division as for multiplication and have:

Division of two numbers with the same sign gives a positive number.
Division of two numbers with opposite signs gives a negative number.

EXAMPLES

11. $\dfrac{10}{-5} = -2.$ **12.** $\dfrac{-10}{-5} = 2.$

Exercise I.I

Locate the following points along a line.

1. 3, −2, 4.

2. −4, 5, −3.

3. −5, −7, 1.

4. 8, 2, −6.

Find the value of each of the following.

5. $2 + 5, 3 + 6$.

6. $-3 + (-4), -1 + (-2)$.

7. $3 + (-2), 3 - 2$.

8. $4 - (-3), 9 - (-5)$.

9. $2(5), 3(6)$.

10. $(-3)(-4), (-1)(-2)$.

11. $3(-2), (-3)(2)$.

12. $4(-3), (-9)(5)$.

13. $8 \div 4, 8 \div (-4)$.

14. $-10 \div 5, -10 \div (-5)$.

15. $0 \div 5, 0 \div (-5)$.

16. $3 \div 0, -2 \div 0$.

17. $(3 \times -2)4$.

18. $-5(2 \times 6)$.

19. $4(-3 \times -2)$.

20. $-7(5 \times -3)$.

21. $4(3 + 2)$.

22. $5(4 + 1)$.

23. $5(6 - 3)$.

24. $2(7 - 4)$.

25. $-7(-2 + 6)$.

26. $-3(8 - 5)$.

27. $-3(-2 - 5)$.

28. $-6(-1 - 4)$.

29. $54 + 6(-5 - 3)$.

30. $17 - 2(11 - 3)$.

31. $3(-3 + 8) - 11$.

32. $-4(7 - 3) + 17$.

I.9. Multiplication, Division, and Reduction of Fractions

In Art. 1.6, we defined a rational fraction as the indicated quotient of two integers and shall now continue our study of fractions. If the fraction is represented by a/b, then a is called the *dividend* or *numerator* of the fraction and b is called the *divisor* or *denominator*. The dividend and divisor are referred to collectively as the *members* of the fraction. We can think of the integers as fractions with a divisor of one.

We shall employ the following axiom from time to time:

For every nonzero real number a there is a real number $1/a$ such that $a(1/a) = 1$.

The number $1/a$ is called the *reciprocal* of a.

The product of the two fractions a/b and c/d is defined by the equation

(1)
$$\frac{a}{b} \times \frac{c}{d} = \frac{ac}{bd}.$$

This can be stated in words by saying that *the product of two fractions is a fraction whose numerator is the product of the numerators and whose denominator is the product of the denominators.*

EXAMPLE

1. The product of $\frac{2}{3}$ and $\frac{5}{7}$ is

$$\frac{2}{3} \times \frac{5}{7} = \frac{2 \times 5}{3 \times 7} = \frac{10}{21}.$$

If $d = c$ in (1), it becomes

(2)
$$\frac{a}{b} \times \frac{c}{c} = \frac{ac}{bc};$$

furthermore, $c/c = 1$ and we have

(3)
$$\frac{a}{b} \times \frac{c}{c} = \frac{a}{b} \times 1 = \frac{a}{b}.$$

Since the left members of (2) and (3) are equal, the right members are equal also, and we have

(4)
$$\frac{a}{b} = \frac{ac}{bc}.$$

If we read (4) from left to right and from right to left, we see that *if the numerator and denominator of a fraction are multiplied or divided by the same nonzero number, the value of the fraction is not changed.* This fact enables us to simplify a fraction by dividing the numerator and denominator by any common factor that they may have. Removing a common factor from the numerator and denominator of a fraction is called *reducing the fraction to lower terms.* We suggest the following procedure for reducing a product of fractions.

1. *Factor both members of each fraction.*
2. *Take out each factor that occurs in both a numerator and a denominator.*
3. *Write the product of the factors which remain in the numerators as the numerator of the product and the product of the factors which remain in the denominators as the denominator of the product.*

EXAMPLES

2. $\dfrac{6}{15} = \dfrac{2 \times 3}{5 \times 3} = \dfrac{2}{5}$ by dividing numerator and denominator by 3.

3. If we factor both numerator and denominator and then take out the common factors, we have

$$\frac{330}{105} = \frac{3 \times 5 \times 2 \times 11}{3 \times 5 \times 7} = \frac{22}{7}.$$

4. In order to change $\frac{7}{5}$ to a fraction with 15 as a denominator, we multiply both numerator and denominator by 3 and have

$$\frac{7}{5} = \frac{7 \times 3}{5 \times 3} = \frac{21}{15}.$$

5. If we factor each numerator and each denominator, we have

$$\frac{22}{6} \times \frac{5}{21} \times \frac{105}{10} = \frac{2 \times 11}{2 \times 3} \times \frac{5}{3 \times 7} \times \frac{3 \times 5 \times 7}{2 \times 5}$$

$$= \frac{11 \times 5}{3 \times 2}, \text{ taking out common factors,}$$

$$= \frac{55}{6}.$$

We shall now develope a procedure for dividing one fraction by another. We have

$$\frac{a}{b} \div \frac{c}{d} = \frac{ad}{bc} \div \frac{cd}{dc}, \text{ multiplying numerator and denominator by } \frac{d}{c},$$

$$= \frac{ad}{bc} \div 1 \text{ since } \frac{cd}{dc} = 1$$

$$= \frac{ad}{bc}.$$

Consequently, *in order to divide one fraction by another, we invert the one we are dividing by and multiply.*

EXAMPLES

6. $\dfrac{2}{3} \div \dfrac{5}{7} = \dfrac{2}{3} \times \dfrac{7}{5}$, inverting the denominator,

$$= \frac{14}{15}.$$

7. $\dfrac{35}{6} \div \dfrac{15}{12} = \dfrac{35}{6} \times \dfrac{12}{15}$, inverting the denominator,

$$= \dfrac{7 \times 5}{2 \times 3} \times \dfrac{2 \times 2 \times 3}{3 \times 5}, \text{ factoring,}$$

$$= \dfrac{7 \times 2}{3}, \text{ removing common factors,}$$

$$= \dfrac{14}{3} = 4\dfrac{2}{3}.$$

8. $\dfrac{21}{15} \times \dfrac{10}{55} \div \dfrac{26}{39} = \dfrac{3 \times 7}{3 \times 5} \times \dfrac{2 \times 5}{11 \times 5} \times \dfrac{3 \times 13}{2 \times 13}$, inverting and factoring,

$$= \dfrac{7 \times 3}{5 \times 11} = \dfrac{21}{55}.$$

Exercise 1.2

Change the fraction in each of Problems 1 through 8 to an equal fraction that has the number at the right as a denominator.

1. $\dfrac{2}{3}$, 6.

2. $\dfrac{3}{5}$, 25.

3. $\dfrac{4}{3}$, 15.

4. $\dfrac{6}{7}$, 28.

5. $\dfrac{4}{10}$, 5.

6. $\dfrac{10}{15}$, 3.

7. $\dfrac{21}{12}$, 4.

8. $\dfrac{30}{12}$, 2.

Change the fraction in each of Problems 9 through 16 to an equal fraction that has the number at the right as numerator.

9. $\dfrac{25}{35}$, 5.

10. $\dfrac{33}{11}$, 3.

11. $\dfrac{27}{18}$, 9.

12. $\dfrac{30}{21}$, 10.

13. $\dfrac{15}{42}$, 5.

14. $\dfrac{48}{16}$, 12.

15. $\dfrac{162}{36}$, 27.

16. $\dfrac{105}{42}$, 5.

Reduce the fraction in each of Problems 17 through 28 to lowest terms.

17. $\dfrac{39}{52}$.

18. $\dfrac{21}{35}$.

19. $\dfrac{25}{45}$.

20. $\dfrac{28}{70}$.

21. $\dfrac{80}{112}$.

22. $\dfrac{51}{68}$.

23. $\dfrac{64}{80}$.

24. $\dfrac{46}{115}$.

25. $\dfrac{42}{66}$.

26. $\dfrac{60}{105}$.

27. $\dfrac{56}{84}$.

28. $\dfrac{63}{147}$.

Perform the indicated operations in Problems 29 through 52.

29. $\dfrac{3}{4} \times \dfrac{7}{5}$.   **30.** $\dfrac{2}{9} \times \dfrac{11}{5}$.   **31.** $\dfrac{3}{10} \times \dfrac{13}{14}$.   **32.** $\dfrac{5}{8} \times \dfrac{7}{4}$.

33. $\dfrac{2}{7} \times \dfrac{14}{3}$.   **34.** $\dfrac{3}{8} \times \dfrac{4}{5}$.   **35.** $\dfrac{5}{12} \times \dfrac{6}{7}$.   **36.** $\dfrac{15}{9} \times \dfrac{3}{20}$.

37. $\dfrac{2}{7} \times \dfrac{3}{10} \times \dfrac{14}{9}$.   **38.** $\dfrac{3}{11} \times \dfrac{13}{6} \times \dfrac{15}{39}$.   **39.** $\dfrac{5}{12} \times \dfrac{14}{15} \times \dfrac{27}{35}$.

40. $\dfrac{34}{26} \times \dfrac{65}{51} \times \dfrac{15}{35}$.   **41.** $\dfrac{2}{3} \div \dfrac{5}{7}$.   **42.** $\dfrac{4}{13} \div \dfrac{7}{17}$.

43. $\dfrac{26}{15} \div \dfrac{3}{2}$.   **44.** $\dfrac{7}{13} \div \dfrac{5}{17}$.   **45.** $\dfrac{3}{8} \div \dfrac{6}{4}$.

46. $\dfrac{10}{7} \div \dfrac{15}{14}$.   **47.** $\dfrac{21}{15} \div \dfrac{28}{30}$.   **48.** $\dfrac{34}{39} \div \dfrac{51}{26}$.

49. $\dfrac{6}{35} \times \dfrac{14}{12} \div \dfrac{21}{15}$.   **50.** $\dfrac{12}{15} \times \dfrac{20}{21} \div \dfrac{24}{7}$.

51. $\dfrac{2}{5} \times \dfrac{84}{91} \div \dfrac{6}{65}$.   **52.** $\dfrac{10}{33} \times \dfrac{39}{12} \div \dfrac{13}{66}$.

1.10. Addition and Subtraction of Fractions

The denominator of a fraction can be thought of as the number of equal parts into which the unit is divided. If the denominator is thought of in this way, we can see that in adding fractions we would be adding different kinds of things unless the denominators of the fractions to be added are the same. Consequently, in order to be able to add two or more fractions, they must have the same denominator. If they do not have the same denominator, it is a simple matter to replace them by equal fractions that do have the same denominator. Thus, if the fractions are a/b and c/d, we can multiply the first by $d/d = 1$ and the second by $b/b = 1$, and have

$$\frac{a}{b} + \frac{c}{d} = \frac{a\,d}{b\,d} + \frac{c\,b}{d\,b}$$

$$= \frac{ad + cb}{bd}.$$

This procedure does not give the smallest possible number as common denominator if b and d have a factor in common, but the smallest can be obtained by using only once any factor that is common to both denominators.

Thus, if the denominators are 6 and 15, they have the common factor 3 as seen by writing $6 = 3 \times 2$ and $15 = 3 \times 5$; hence, using the common factor only once, we see that $3 \times 2 \times 5$ is a common denominator. Now that we can express fractions with different denominators as fractions with the same denominator and equal to the given fractions, we are able to add them.

EXAMPLES

1. In order to add $\frac{2}{3}$ and $\frac{5}{7}$, we must find a number that contains each denominator as a factor, since such a number is a common denominator. Both are factors of their product; hence, 21 is a common denominator. Consequently, multiplying the first fraction by $\frac{7}{7}$ and the second by $\frac{3}{3}$, we have

$$\frac{2}{3} + \frac{5}{7} = \frac{2}{3} \times \frac{7}{7} + \frac{5}{7} \times \frac{3}{3}$$

$$= \frac{14}{21} + \frac{15}{21}$$

$$= \frac{29}{21} = 1\frac{8}{21}.$$

2. In order to evaluate

$$\frac{1}{3} + \frac{2}{5} - \frac{4}{15},$$

we multiply the first fraction by $\frac{5}{5}$, the second by $\frac{3}{3}$, and leave the last alone, since that gives all three fractions the same denominator. Thus,

$$\frac{1}{3} + \frac{2}{5} - \frac{4}{15} = \frac{1}{3} \times \frac{5}{5} + \frac{2}{5} \times \frac{3}{3} - \frac{4}{15}$$

$$= \frac{5 + 6 - 4}{15} = \frac{7}{15}$$

3. Find the sum indicated by $\frac{2}{15} + \frac{9}{35}$.

SOLUTION

$$\frac{2}{15} + \frac{9}{35} = \frac{2}{15}\frac{7}{7} + \frac{9}{35}\frac{3}{3} \text{ since the common denominator is } 5 \times 3 \times 7 = 105$$

$$= \frac{14 + 27}{105} = \frac{41}{105}$$

In the last article, we discussed multiplication and division of fractions, and so far in this one we have dealt with addition and subtraction. Quite often a problem requires the use of multiplication or division or both, along with addition or subtraction or both. We shall now make an agreement concerning the order of operations.

Multiplications and divisions of fractions shall be performed before additions and subtractions.

EXAMPLE

4. Perform the indicated operations in

$$\frac{2}{3} + \frac{5}{7} \div \frac{9}{14} - \frac{5}{6} \times \frac{1}{3}$$

SOLUTION

In accordance with our agreement, we shall perform the multiplication and division first. Thus,

$$\frac{5}{7} \div \frac{9}{14} = \frac{5}{7} \times \frac{14}{9}, \text{ inverting,}$$

$$= \frac{10}{9}$$

and

$$\frac{5}{6} \times \frac{1}{3} = \frac{5}{18};$$

consequently,

$$\frac{2}{3} + \frac{5}{7} \div \frac{9}{14} - \frac{5}{6} \times \frac{1}{3} = \frac{2}{3} + \frac{10}{9} - \frac{5}{18}$$

$$= \frac{2}{3} \times \frac{6}{6} + \frac{10}{9} \times \frac{2}{2} - \frac{5}{18}$$

$$= \frac{12 + 20 - 5}{18}$$

$$= \frac{27}{18} = 1\frac{1}{2}.$$

Exercise 1.3

Perform the indicated operations.

1. $\frac{2}{3} + \frac{5}{7}.$ **2.** $\frac{1}{4} + \frac{3}{5}.$ **3.** $\frac{3}{8} + \frac{1}{5}.$

4. $\dfrac{7}{11}+\dfrac{2}{5}.$

5. $\dfrac{6}{5}-\dfrac{2}{3}.$

6. $\dfrac{11}{6}-\dfrac{3}{5}.$

7. $\dfrac{4}{9}-\dfrac{3}{7}.$

8. $\dfrac{4}{7}-\dfrac{1}{6}.$

9. $\dfrac{2}{3}+\dfrac{5}{6}.$

10. $\dfrac{3}{4}-\dfrac{1}{2}.$

11. $\dfrac{4}{5}-\dfrac{7}{10}.$

12. $\dfrac{4}{9}+\dfrac{2}{3}.$

13. $\dfrac{3}{4}-\dfrac{5}{6}.$

14. $\dfrac{1}{6}+\dfrac{7}{9}.$

15. $\dfrac{3}{4}+\dfrac{3}{10}.$

16. $\dfrac{7}{10}-\dfrac{2}{15}.$

17. $\dfrac{1}{2}+\dfrac{1}{3}-\dfrac{1}{6}.$

18. $\dfrac{3}{4}-\dfrac{2}{5}+\dfrac{1}{20}.$

19. $\dfrac{1}{2}-\dfrac{2}{3}+\dfrac{5}{12}.$

20. $\dfrac{2}{3}-\dfrac{1}{5}-\dfrac{4}{15}.$

21. $\dfrac{1}{2}+\dfrac{1}{3}-\dfrac{2}{5}.$

22. $\dfrac{2}{5}-\dfrac{1}{3}+\dfrac{2}{7}.$

23. $\dfrac{5}{6}+\dfrac{1}{5}-\dfrac{1}{4}.$

24. $\dfrac{2}{9}-\dfrac{3}{7}-\dfrac{1}{5}.$

25. $\dfrac{2}{5}+\dfrac{3}{7}\times\dfrac{14}{5}.$

26. $\dfrac{3}{4}+\dfrac{1}{3}\times\dfrac{7}{2}.$

27. $\dfrac{5}{7}\times\dfrac{21}{8}+\dfrac{1}{4}.$

28. $\dfrac{3}{4}\times\dfrac{22}{5}+\dfrac{7}{10}.$

29. $\dfrac{7}{9}\div\dfrac{2}{3}+\dfrac{1}{2}.$

30. $\dfrac{4}{7}\div\dfrac{8}{21}-\dfrac{4}{5}.$

31. $\dfrac{5}{3}+\dfrac{2}{7}\div\dfrac{1}{21}.$

32. $\dfrac{4}{11}-\dfrac{3}{5}\div\dfrac{2}{9}.$

33. $\dfrac{2}{5}\div\dfrac{8}{9}+\dfrac{7}{10}\times\dfrac{25}{21}.$

34. $\dfrac{3}{8}\times\dfrac{16}{9}-\dfrac{2}{5}\div\dfrac{4}{15}.$

35. $\dfrac{2}{7}+\dfrac{3}{5}\times\dfrac{20}{9}-\dfrac{19}{21}.$

36. $\dfrac{3}{10}-\dfrac{2}{11}\div\dfrac{4}{33}+\dfrac{4}{5}.$

1.11. The Hindu-Arabic Number System and the Fundamental Operations

The number system which we use is based on 10 and powers thereof. The symbols 0, 1, 2, 3, 4, 5, 6, 7, 8, and 9 are called *digits* and are used in writing numbers, but the value of any one of the symbols, other than 0, depends in part on its position. In 3142, the symbol 3 represents 3000 since it indicates the number of thousands, whereas the 3 in 31 represents 30 since it signifies the number of tens. This idea of the number symbols having place value was well developed by the Hindus of India by A.D. 500, and the Arabs used it by A.D. 800. The same symbols are used to indicate tenths of a unit, hundredths, thousandths, etc. as indicated in Fig. 1-3. A dot or period is placed to the right of the units to separate the whole number and fraction and is called the *decimal point*. It and the numbers to the right of it are called a *decimal fraction*. The number in Fig. 1-3 is

$$7(10,000) + 4(1000) + 1(100) + 0(10) + 3(1) +$$
$$2(.1) + 8(.01) + 5(.001) + 9(.0001)$$

and is read seventy-four-thousand one-hundred three and two-thousand eight-hundred fifty-nine ten-thousandths.

As a matter of convenience in adding two or more numbers which contain decimal fractions, we place the decimal points under one another. This is done so that all the units, all the tens, all the tenths, all the hundredths, etc., will be in columns, thus making it easier to add. The same arrangement is used for subtraction.

ten thousands	thousands	hundreds	tens	units	decimal point	tenths	hundreds	thousands	ten thousands
7	4	1	0	3	•	2	8	5	9

FIGURE 1-3

EXAMPLE

1. Add 381.42, 27.314, and 10.38.

SOLUTION

If we arrange the numbers so that the decimal points are in a column, we have

$$\begin{array}{r} 381.42 \\ 27.314 \\ 10.38 \\ \hline 419.114 \end{array}$$

2. Subtract 23.74 from 39.26.

SOLUTION

In order to be able to do the subtraction in connection with the tenths, we must borrow a unit and then have 12 tenths from which to subtract the seven tenths. Thus,

$$\begin{array}{r} 39.26 \\ 23.74 \\ \hline 15.52 \end{array}$$

There are two methods used for determining the position of the decimal point in a multiplication. In one of them, we pay no attention to the position of the decimal point during the process of multiplication and then *place the decimal point in the result at the only place it can reasonably go.*

E X A M P L E

3. If we multiply 36.4 by 25.7, we get 93548 as the sequence of digits. In order to place the decimal point, we use the fact that we are multiplying about 40 by about 25 and should have a product near 1000; hence, we put the decimal point the only place it can reasonably go and have $(36.4)(25.7) = 935.48$

The other method of placing the decimal point is a mechanical one but, like many mechanical operations, can be justified. In order to help arrive at the rule, we shall consider

$$P = (3.4)(.72)$$
$$= (3 + .4)(.7 + .02)$$
$$= 3(.7) + 3(.02) + .4(.7) + .4(.01)$$

This includes some (4) tenths times some (2) hundredths; hence it involves some thousandths. This could have been arrived at by adding the number of places to the right of the decimal point in 3.4 and the number to the right of the decimal point in .72. This type of argument can be used in any case to arrive at the statement that *the number of decimal places in the product of two exact factors is equal to the number in one factor plus the number in the other.*

E X A M P L E

4. Find the product of 8.47 and 72.9.

S O L U T I O N

The sequence of digits is easily seen to be 617463 by multiplying 847 by 729. Since there are two decimal places in the first factor and one in the second there must be $2 + 1 = 3$ in the product; hence the product is 617.463.

Since division is the inverse of multiplication it follows that *the number of decimal places in the quotient of two numbers is equal to the number in the numerator minus the number in the denominator.* If there are fewer decimal places in the numerator than in the denominator, enough zeros must be added to make the two equal.

We can also use the system of putting the decimal point in the only reasonable place for it.

E X A M P L E

5. Find the quotient obtained by dividing 744.654 by 31.42.

SOLUTION

$$
\begin{array}{r}
23.7 \\
31.42\overline{)744.654} \\
628\ 4 \\
\hline
116\ 25 \\
94\ 26 \\
\hline
21\ 994 \\
21\ 994 \\
\end{array}
$$

The sequence of digits in the quotient is 237, and the decimal point belongs as placed, since there are three decimal places in the numerator and two in the denominator; hence $3 - 2 = 1$ in the quotient. If the method of putting the decimal point in the only place it can reasonably be put is used, we see that we are dividing about 750 by about 30; hence, we should have an answer that is approximately 25. Therefore, the decimal point is properly placed.

1.12. Changing to Decimal Form

We saw in Art. 1.9 and 1.10 how to perform the fundamental operations on common fractions and in Art. 1.11 how to perform them on numbers that include decimal fractions. It is often simpler to perform the fundamental operations if the numbers are in decimal form, and it is essential to have them in that form if a computer, desk or other, is to be used. Consequently, it is desirable to be able to change a common fraction to decimal form. To do this, we begin by dividing as in Example 5, but the quotient does not always come out exactly as in that problem. If the quotient does not come out exactly, it should be carried out to one more digit than we intend to keep. If this extra digit is 0, 1, 2, 3, or 4 we discard it and use the digit just before it as it is, but if the extra digit is 5, 6, 7, 8 or 9 we discard it and increase the digit just before it by one. This procedure is called *rounding off*.

EXAMPLES

1. If we want to divide 19 by 22 and keep the quotient to three digits, we divide to four digits and have .8261; hence, we write $\frac{19}{22} = .826$ by discarding the 1.

2. If we want to divide 21 by 32 and keep four digits, we divide to five digits and have .65625; hence, we use $\frac{21}{32} = .6563$ by discarding the 5 and increasing the digit 2 before it by 1.

3. Find the cost of 6.75 yards of material at $2.83 per yard.

SOLUTION

The cost of 6.75 yards is 6.75 times the cost of one yard; hence,
$$\text{cost} = 6.75(\$2.83) = \$19.10, \text{ to the nearest cent,}$$
since the sequence of digits is 191025 and the product is close to that of 7 and 3.

Exercise 1.4

Perform the indicated operations in Problems 1 through 20.

1. $13.26 + 2.49 + 11.417$.

2. $7.83 + 16.71 + 9.6$.

3. $.238 + 5.713 + 21.42$.

4. $86.21 + 2.739 + 14.7$.

5. $19.796 - 8.45$.

6. $78.39 - 13.416$.

7. $21.86 + 47.3 - 3.802$.

8. $59.73 + 15.721 - 72.9$.

9. $(31.76)(28.53)$.

10. $(284.7)(15.32)$.

11. $(3.415)(46.2)$.

12. $(729.6)(.302)$.

13. $66.066 \div 36.3$.

14. $580.65 \div 23.7$.

15. $1.24012 \div .172$.

16. $389.422 \div 57.1$.

17. $(3.21)(1.76) + 4.17$.

18. $(59.2)(3.14) + .5372$.

19. $(.0259)(6.03) - .305$.

20. $(6.21)(8.73) - 27.53$.

Perform the operations indicated in Problems 21 through 28 and carry each result to three digits.

21. $473.23 \div 13.2$.

22. $83.74 \div 17.43$.

23. $9.836 \div 47.21$.

24. $105.28 \div 38.65$.

25. $78.2 \div 27.7 + 1.7$.

26. $9.83 \div 27.6 + .782$.

27. $179 \div 3.84 - 57.3$.

28. $42.7 \div 5.77 - 3.4$.

29. If 2.3 yards of material at $7.36 per yard are needed to upholster a chair, what is the cost of the material?

30. If a train travels at 48 miles per hour for .42 hours, how far does it go?

31. If topsoil costs $3.75 per cubic yard, how much must one pay for a load that contains 2.8 cubic yards?

32. If lots in a subdivision cost 37.5 cents per square foot, what is the price of a lot that is 65.3 feet by 136.2 feet?

33. If 3.7 yards of material cost $8.14, what is the cost of one yard?

34. What is the cost of a dozen apples if a box of 64 costs $4.80?

35. Find the cost of one sheep if a herd of 236 costs $3211.96.

36. If 2 yards and 2 feet of material cost $8.80, what is the price per yard?

1.13. Significant Digits and Computations

The accuracy of a measurement depends on several factors, including the skill and care of the person who makes the measurement and the precision of the instrument used. Certainly a measurement with a steel tape should be more exact than one determined by stepping. The digits known to be correct in a measurement or other approximate value are called *significant digits*, and are not affected by the position of the decimal point. The position of the decimal point is determined by the unit used. Thus 8.26 meters and 826 centimeters both contain three significant digits and the position of the decimal point is determined by the unit used.

There is no way for a second person to tell whether a digit in a measurement or other approximation is significant; hence, we shall give the procedure that is generally used in deciding whether or not a digit is significant.

The digits 1, 2, 3, 4, 5, 6, 7, 8, and 9 are always significant if used in connection with a measurement and so are any zeros between any two of the nonzero digits. Zeros on the right may or may not be significant. Zeros between the decimal point and the first nonzero digit in positive numbers less than 1 are not significant.

EXAMPLES

1. Each digit in 3208 is significant since each is between 1 and 9, inclusive, or is a zero between two nonzero digits.

2. The zeros in 35,700 may or may not be significant.

3. The zero in .063 is not significant.

4. Read from left to right, the first zero in .045090 is not significant, the second is significant and the third may or may not be significant.

If the accuracy of a digit is in doubt, it is not used in computation. Instead of using the doubtful digit, the number is rounded off in keeping with the procedure given in Art. 1.12. Furthermore, in rounding off to n digits in a number with the decimal point to the right of the nth digit, we replace all digits between the nth and the decimal point by zeros and discard all digits to the right of the decimal point.

EXAMPLES

5. If 37.2681 is rounded off to four significant digits, we have 37.27 since the first digit omitted is 5 or greater.

6. If .027854 is rounded off to four significant digits, we obtain 0.2785 since the first digit left off is 4 or less.

7. If 25,376.3 is rounded off to four digits, we get 25,380 since the first digit omitted is 5 or greater and the decimal is to the right of the fourth digit.

We are now almost ready to use approximations in computation but first shall present three rules that are used in such cases. They are:

1. *If the number of significant digits in* **M** *is equal to the number in* **N** *or differs from it by* **1**, *round off the product* **MN** *and the quotient* **M/N** *to the number of significant digits in the one that has the lesser number of significant digits.*

2. *If the number of significant digits in* **M** *differs from the number in* **N** *by more than* **1**, *round off the one with the greater number of significant digits so that it contains one more significant digit than the other number and proceed as in rule* **1**.

We shall illustrate these two rules before giving the third one.

EXAMPLES

8. The number of significant digits in $M = 2.371$ differs from the number in $N = 4.73$ by one; hence, we use all digits in each in obtaining the product and get

$$MN = (2.371)(4.73)$$
$$= 11.21483 = 11.2 \text{ to three digits.}$$

9. The number of significant digits in $M = 720.386$ is two greater than the number in $N = 39.16$; hence, we round M off to $4 + 1 = 5$ digits and have $M = 720.39$. Therefore,

$$\frac{M}{N} = \frac{720.39}{39.16} = 18.394 = 18.39 \text{ to four digits.}$$

In obtaining a sum or difference we are not primarily interested in the number of significant digits in the data but rather in the precision of the measurement. That is to say we are interested in whether the data measure the number of units, tenths of a unit, hundredths of a unit or some other multiple or fractional part of a unit. Thus, we are interested in the number of significant digits in the decimal portions of the numbers to be added or subtracted since we add digits that have the same place value. Consequently, we shall use the following rule in adding and subtracting approximate numbers.

3. *Round off each number that is to be added to one more decimal place than there is in the number which has the least number of decimal places, add the resulting numbers, and round off the last digit in the sum.* A similar procedure is used in subtraction.

EXAMPLE

10. Find the sum of 13.6, 21.347, 8.56, and 38.7134.

SOLUTION

There is one decimals place in 13.6; hence, the other numbers are to be rounded off to two decimals if they contain more than that. If this is done, the addends and their sum are as shown below. Therefore, after rounding the sum off to one decimal place it is 82.2.

$$
\begin{array}{r}
13.6 \\
21.35 \\
8.56 \\
38.71 \\
\hline
82.22
\end{array}
$$

Approximate numbers are often written in a form that is called *scientific notation*. To express a number in scientific notation, we put down all the significant digits, place a decimal point after the first digit, and then multiply by the power of 10 that is needed to make the product the right size. In keeping with this, if no one of the zeros in 93,000,000 is significant, we write $9.3(10^7)$; furthermore, if only the first two are significant, we then write $9.300(10^7)$ and by 10^7 mean $10 \times 10 \times 10 \times 10 \times 10 \times 10 \times 10$. In general for n a positive integer, 10^n means that 10 is to be used n times as a factor. Consequently, for m and n positive integers, we have $10^m 10^n = 10^{m+n}$ and $10^m/10^n = 10^{m-n}$ for m larger than n.

Exercise 1.5

Perform the indicated operations in the problems below and round off to the proper number of significant digits or to the appropriate number of decimal places. Assume that all numbers are approximations, and use scientific notation if needed to show that final zeros are or are not significant.

1. (3.7)(7.2).   **2.** (28.3)(4.19).

3. (1.8)(3.42).   **4.** (23.7)(9.6).

5. (3.2)(4.037).   **6.** (5.6)(89.214).

7. (80.7)(76.546).   **8.** (63)(4.718).

9. $73 \div 22$.   **10.** $81.9 \div 32$.

11. $36 \div 9.05$.   **12.** $48.3 \div 5.6$.

13. $742.6 \div 8.1$.   **14.** $90 \div 58.73$.

15. $663.14 \div 285$.   **16.** $45.7 \div 63.927$.

17. $2.14 + 3.26 + 1.15$.   **18.** $58.2 + 7.34 + 14.73$.

19. $84.39 + 13.7 - 41.63$.   **20.** $3.027 + 5.61 - 7.411$.

21. $3.825 + 2.6 - 4.37$.   **22.** $7.309 - 2.8 + 1.83$.

23. $8.462 + 7.304 - 5.1$.   **24.** $9.883 + 3.2 + 7.124$.

25. (3.5)(4.2) + 5.73.   **26.** (7.4)(3.8) + 13.26.

27. $(9.8)(.23) - 4.3.$

28. $(6.8)(.29) - 1.74.$

29. $37.2 \div 23 + 1.6.$

30. $59 \div 37.6 + 2.43.$

31. $4963.2 \div 71.43 + 186.3.$

32. $38.427 \div 23.51 - .46.$

Perform the indicated operations and leave the result in scientific notation.

33. $(2.17)10^3(3.85)10^4.$

34. $(7.30)10^3(14.1)10^2.$

35. $(17.2)10^2(8.6)10^2.$

36. $(13.92)10^3(11.3)10^2.$

37. $\dfrac{(19.86)10^4}{(3.77)10^3}.$

38. $\dfrac{(473.9)10^4}{(17.26)10^3}.$

39. $\dfrac{(530.4)10^3}{(7.14)10^2}.$

40. $\dfrac{(74.92)10^4}{(27.1)10^3}.$

1.14. Percentage

We studied decimal fractions in Art. 1.12 and shall now continue that study with a change in language. It is common practice to use the term *per cent* when we mean hundredths. Thus, if we want to indicate .06, we often say 6 per cent and write 6%. Per cent and hundredths are used interchangeably. Since they are synonyms and since the decimal point is moved two places to the right to find the number of hundredths in a decimal fraction, it follows that *we move the decimal point two places to the right to change from a decimal fraction to per cent*; consequently, *we move it two places to the left to change from per cent to a decimal fraction*.

EXAMPLES

1. $.3258 = 32.58\%.$

2. $.017 = 1.7\%.$

3. $65.7\% = .657.$

4. $2.34\% = .0234.$

5. If the enrollment of a college was 8200 one year and increased 8% the next year, then the increase in enrollment was 8% of 8200 or $(.08)(8200) = 658.00$ and the increased student body was $8200 + 658 = 8858.$

Three quantities are involved in many situations in which per cent is used. There is a number called the *base* of which a per cent is to be taken. The per cent that is to be taken is called the rate and the number obtained by multiplying the base by the rate is called the *percentage*. Thus, we have the relation

$$\textbf{percentage} = \textbf{base} \times \textbf{rate}.$$

EXAMPLE

6. Find 32.3% of 784.

If we represent the percentage by P, we have

$$P = 32.3\% \times 784$$
$$= (.323)(784)$$
$$= 253.232.$$

1.15. Simple Interest

If we borrow money from a bank or other commercial lending agency, we are expected to pay for its use. This payment is called *interest* and is determined by the sum borrowed, the per cent of this sum that is charged for one year, and the number of years the sum is kept. The sum borrowed is called the *principal*, the per cent charged per year is the *annual rate* or merely the *rate*, and the number of years the principal is kept is called the *term* of the loan. The letters I, P, r, and t are used to designate the interest, principal, rate, and term, respectively. The interest for one year is the product Pr and the interest for a term of t years is given by

$$I = Prt.$$

The borrower should repay the principal and pay the interest at the end of the term. Their sum is called the *accumulated value*, is designated by S and given by

$$S = P + I.$$

In actual practice, the interest is often paid periodically so that the only interest due at the end of the term is that for the last period.

EXAMPLE

1. Find the interest and accumulated value if $460 is borrowed for 15 months at 5% per year.

SOLUTION

In this problem, $P = \$460$, $r = 5\% = .05$, and $t = 15$ months $= 1\frac{1}{4}$ years $= 1.25$ years; hence,

$$I = Prt$$

becomes

$$I = \$460(.05)(1.25) = \$28.75$$

and

$$S = P + I = \$460 + \$28.75 = \$488.75.$$

It is common practice among lending institutions to count the exact number of days in the term of a loan that is for less than a year and to divide this by 360, not 365, to determine the fractional part of a year in the term.

2. Mr. Stafford borrowed $500 on May 13 and repaid it on August 16 of the same year. How much interest did he pay if the rate was 6%?

SOLUTION

In counting the number of days, we include the day on which the loan was made or the day on which it was repaid but not both. We shall include the date of repayment; hence, the time is 18 days in May, plus 30 in June, plus 31 in July, plus 16 in August for a total of 95 days. Therefore,

$$I = Prt$$
$$= \$500(.06)\frac{95}{360} = \$7.92.$$

It is not unusual for a lending agency to deduct the interest at the time a loan is made. They refer to this as interest in advance but the term *discount* is also used, since the payment is based on the accumulated value instead of on the principal. The effect is to get a higher interest rate than the stated one.

EXAMPLE

3. If one borrows $300 for six months and pays interest in advance at the rate of 7%, what is the equivalent interest rate based on the sum actually received?

SOLUTION

The interest paid is

$$300(.07)\left(\frac{1}{2}\right) = \$10.50;$$

consequently the borrower receives $300 − $10.50 = $289.50 and pays $10.50 for its use for six months. Therefore, substituting in the interest formula, we have

$$10.50 = 289.50r\left(\frac{1}{2}\right) = 144.75r.$$

Finally, dividing each member of this equation by 144.75, we have

$$r = \frac{10.50}{144.75} = 7.25\% \text{ to three digits.}$$

1.16. Property Tax and Income Tax

There is a tax on most property, and it is determined by the assessed value of the property and the tax rate. The *assessed value* is a fixed per cent of a reasonable sale price. This fixed per cent may vary between two political subdivisions but is or should be the same for all property in one. The unit of tax rate is called the *mil* and it is one-thousandth of the assessed value.

EXAMPLE

1. Find the tax on a $16,000 house if it is assessed at 30% of its value and the rate is 39.4 mils.

SOLUTION

The assessed value is (.30)($16,000) = $4800 and the tax is .0394 of that value since the rate is 39.4 mils. Consequently the tax is

$$\$4800(.0394) = \$189.12.$$

The federal government and some states require that a tax be paid on all personal income beyond a specified amount. This tax is called *income tax*. The amount paid depends on several factors, including the number of dependents the tax payer has and the amount of his taxable income. In many states, the state income tax is paid on the taxable income that remains after the federal tax is subtracted.

EXAMPLE

2. In a recent year, the federal income tax on a taxable income between $2000 and $4000 was $400 plus 22% of the excess of taxable income over $2000 for a single person. Find the federal tax that had to be paid by a single person with a taxable income of $3759.23. If the state income tax was 3% of the amount left after subtracting the federal tax from the taxable income, find the state tax.

SOLUTION

The excess of taxable income over $2000 is $3759.23 − $2000 = $1759.23 and the rate thereon is 22%; hence, the total federal tax is

$$\$400 + .22(1759.23) = \$400 + \$387.03$$
$$= \$787.03.$$

In order to find the sum on which a state tax must be paid, we subtract the federal tax from the taxable income and have

$$\$3759.23 - \$787.03 = \$2972.20.$$

Since the state rate is 3%, the state tax is

$$(.03)(\$2972.20) = \$89.17$$

Exercise 1.6

Change each of the following to a decimal.

1. 3%, 22%.

2. .8%, 91%.

3. 36.7%, 82.9%.

4. 17.31%, .872%.

Change each of the following to a per cent.

5. .07, .63.   **6.** .15, .71.

7. .086, .629.   **8.** .902, .0026.

Find the interest and accumulated value if the principal in each of Problems 9 through 16 is invested at the given rate for the given time or between the given dates. Both dates are in the same year.

9. $700, 6%, $1\frac{1}{2}$ years.   **10.** $800, 5%, 9 months.

11. $600, 4.5%, 5 months.   **12.** $320, 6%, 8 months.

13. $840, 5%, June 13, August 24.

14. $1730, 5.5%, January 3, April 3, 1965.

15. $3640, 5.25%, March 14, May 1.

16. $350, 4.5%, September 3, December 17.

Find the rate if the sum in each of Problems 17 through 20 is invested for the given time and produces the specified amount of interest.

17. $400 produces $15 in 9 months.

18. $800 produces $12 in 3 months.

19. $760 produces $39.90 in 10.5 months.

20. $450 produces $4.95 in 72 days.

If the interest in each of Problems 21 through 24 is paid in advance, what is the equivalent rate, to the nearest hundredth of a per cent, based on the sum actually received?

21. $800, 6%, 3 months.   **22.** $1200, 5%, 2 years.

23. $360, 4%, 105 days.   **24.** $1920, 6%, 216 days.

Find the state, county, and city tax in each of Problems 25 through 28 if the assessed value is as given and the rates are in the order named above.

25. $15,000, 7 mils, 16.3 mils, 4.6 mils.

26. $7,300, 6.4 mils, 3.8 mils, 14.7 mils.

27. $4,350, 5.5 mils, 11.3 mils, 4.8 mils.

28. $2,960, 7.2 mils, 12.4 mils, 11.6 mils.

The following is taken from a tax table printed by the United States government for use in determining income tax for a recent year. Make use of it to determine the tax in each of Problems 29 through 36; furthermore, find the state tax if the rate is as given and if the federal tax is a deduction.

TAXABLE INCOME		TAX COMPUTATION
Over	*Not Over*	*For a Single Person*
$ 2,000	$ 4,000	$ 400 plus 22% of the excess over $ 2,000
$ 4,000	$ 6,000	$ 840 plus 26% of the excess over $ 4,000
$ 6,000	$ 8,000	$1,360 plus 30% of the excess over $ 6,000
$ 8,000	$10,000	$1,960 plus 34% of the excess over $ 8,000
$10,000	$12,000	$2,640 plus 38% of the excess over $10,000

29. $2782.19.   **30.** $5632.47.

31. $7623.36.   **32.** $6253.72.

33. $3471.28, state tax of 3%.

34. $6729.84, state tax of 3.5%.

35. $9775.51, state tax of 4%.

36. $11,253.79, state tax of 4.5%.

37. A refinery employee used to receive $300 per month but has twice received a 10% raise. The second raise was based on his salary after the first raise. Find the current salary.

38. A suit of clothes was originally marked $70 but, due to a sale, was reduced by 20%. It still was not sold and was further reduced by 15% of the price after the 20% reduction. How much did the suit cost after the second reduction?

39. A store bought a lawn mower for $70 and marked it to sell for 130% of that amount. It was not sold and was reduced by 20% of the intended selling price and then sold. What price was received?

40. A store advertised that a stove could be bought for $300 on the first day of a sale or for 10% less than on the preceding day each day thereafter. What was the price on the fourth day of the sale?

1.17. Roman Numerals

We know from archaeology and history that the Romans, Babylonians, Incas, and others had systems of writing numbers. The Roman system of numerals is the only one of these that is used to any appreciable extent today. However, it has certain inherent disadvantages that caused it to lose out for general use to the Hindu-Arabic system: the lack of a symbol for zero kept it from being able to use place value as we now use it; a one-letter symbol always has the same value regardless of its position. The symbols and their values in terms of our numbers are:

I   V   X   L   C   D   M

1   5   10   50   100   500   1000

There is an agreement that if two symbols are written next to one another they are to be added unless the smaller is on the left and then they are to be subtracted.

EXAMPLES

1. $III = I + I + I = 3.$ **2.** $IV = -I + V = 4.$

3. $VII = V + I + I = 7.$ **4.** $XL = -X + L = 40.$

5. $MCDLXXVI = M - C + D + L + X + X + V + I.$
$$= 1000 - 100 + 500 + 50 + 20 + 5 + 1 = 1476.$$

There is also an agreement that a number is to be multiplied by 1000 if a bar is above it. Thus $\overline{M}$ represents 1,000,000 and $\overline{X}$ represents 10,000.

1.18. The Abacus

The abacus, still used extensively in Asia and eastern Europe, is one of the oldest calculating machines. It consists of a rectangular framework, several parallel metal bars, beads on the bars, and a reservoir for the beads for each bar. Each bead on the bar on the right represents one, each bead on the next bar represents ten, each bead on the next bar represents one hundred, and each bead on any bar further to the left represents ten times as much as a bead on the bar immediately to the right. The number represented on the abacus in Fig. 1-4 is 2413, since there are 3 beads on the bar on the right, one on the next, 4 on the hundreds' bar, and 2 on the thousands' bar. In order to add 3259 to 2413, we begin by bringing down 9 more beads from the units' reservoir. This makes 12 beads on the units' bar; hence, we return 10 of them

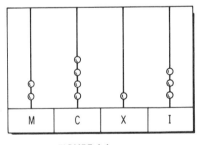

FIGURE 1-4

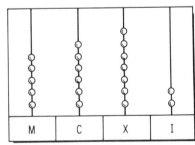

FIGURE 1-5

to the reservoir and bring down one additional bead on the tens' bar. This makes 2 beads on that bar, and we bring down 5 more corresponding to the 5 in 3259; thus, we have 7 beads on the tens' bar. Now we bring down 2 more beads on the hundreds' bar and 3 more on the thousands' bar. Consequently, the sum is 5672, as shown in Fig. 1-5.

Subtraction is performed by returning the proper number of beads from each bar to the reservoir. If at any time there are fewer beads on a bar than are due to be returned to the reservoir, we bring down ten before returning any and offset this by returning one from the bar to the left to the reservoir.

Multiplication can be considered as repeated addition and performed accordingly. Finally, division can be performed as repeated subtraction.

1.19. Numbers to Various Bases

If a number is expressed as the sum of multiples of powers of a number b, we say that the number is expressed in terms of the base b. Thus, $N = 5 + 7b + 3b^2 + 0b^3 + 8b^4$ is expressed to the base b. We are accustomed to the use of the base 10 since we express numbers as the sum of multiples of powers of 10. For example, 3426 is $6 + 2(10) + 4(10^2) + 3(10^3)$, and the number N given above would be 80,375. We shall use the practice of writing the base as a subscript or understanding that the base is 10 if none is given. In keeping with this, $2103_4 = 3 + 0(4) + 1(4^2) + 2(4^3) = 3 + 16 + 2(64) = 147$.

In order to write a number to a given base, we must find the largest integral power of the base that is less than the number, the number of times this power of the base goes into the number, and the remainder when this multiple of the base is subtracted from the number; we then treat this remainder just as the number was treated and continue the process until we have a remainder smaller than the base.

EXAMPLES

1. Write 147 in terms of base 4.

SOLUTION

We can see readily that $4^3 = 64$ is the largest power of the base that is less than the number 147 since $4^4 = 256$ is larger than 147; furthermore, $2(4^3)$ is less than 147. If we subtract $2(4^3) = 128$ from 147, we find the remainder to be 19. We now find that 4^2, the next lower power of 4, goes into 19 one time with a remainder of 3. We now notice that 4^1, the next lower power of 4, does not go into 3; hence, there are no 4's and a remainder of 3; i.e., 3 units. Consequently, in 147, 4^3 goes 2 times, then 4^2 goes 1 time into the remainder 19, then 4^1 goes zero times into the remainder 3. Then $147 = 2103_4$.

We shall now see how to get the result in another manner that may be simpler for some to follow. The procedure consists of dividing 147 by 4, dividing the quotient by 4, dividing the next quotient by 4, and continuing until a quotient of zero is obtained. We must keep a record of the remainders at all steps since they are the digits in the number to the new base. If this procedure is followed, we have

4	147	*remainder*	*use*	*number*
4	36	3	3	
4	9	0	0×4^1	
4	2	1	1×4^2	
	0	2	2×4^3	2103_4

2. By use of the division scheme, change 1939 to base 9.

SOLUTION

9	1939	*remainder*	*use*	*number*
9	215	4	4	$2584_9 = 1939$
9	23	8	8×9	
9	2	5	5×9^2	
	0	2	2×9^3	

These two examples illustrate how to change a number from base 10 to any other base. If that new base is greater than 10, we need further number symbols besides the usual ones. If the base is 12, we need a symbol for 10 and one for 11. We shall use *t* for ten and *e* for eleven. The method used in examples 1 and 2 can also be used if the base is greater than 10 but then we must use the unfamiliar symbols.

EXAMPLE

3. Change 2147 to base 12.

SOLUTION

We shall use the division procedure.

12	2147	*remainder*	*use*	*number*
12	178	*e*	*e*	
12	14	*t*	$t \times 12$	$12te_{12} = 2147$
12	1	2	2×12^2	
	0	1	1×12^3	

It is a relatively simple matter to change a number from any other base to base 10, and the procedure will be illustrated.

EXAMPLE

4.
$$452_7 = 4 \times 7^2 + 5 \times 7 + 2$$
$$= (196 + 35 + 2) = 233$$

If we have a number to any base other than 10 and want to express it in terms of any other base besides 10, we need only change it from the original form to base 10 and then to the desired form.

1.20. Arithmetic in Bases Other Than 10

The following figures are tables for use in addition and multiplication with a system of numbers to the base 6. The entry in the addition table, Fig. 1-6, across from 4 and under 5, is the sum of those two numbers to base 6; hence it is 13 to indicate $1 \times 6 + 3$. Similarly, the entry across from 4 and under 5 in the multiplication table is the product of 4 and 5 to base 6; consequently it is written as 32 to indicate $3 \times 6 + 2$. All numbers in each table are to base 6.

0	1	2	3	4	5
1	2	3	4	5	10
2	3	4	5	10	11
3	4	5	10	11	12
4	5	10	11	12	13
5	10	11	12	13	14

FIGURE 1-6

Basic 6 addition.

0	1	2	3	4	5
1	1	2	3	4	5
2	2	4	10	12	14
3	3	10	13	20	23
4	4	12	20	24	32
5	5	14	23	32	41

FIGURE 1-7

Basic 6 multiplication.

We now have before us in Fig. 1-6 the number combinations that are needed in base 6 addition, and in Fig. 1-7 those needed in base 6 multiplication.

EXAMPLES

1. The sum of the numbers given at the right is as shown below them.   221
The steps in finding the sum are as follows. The sum of 1, 5, and 2 is 12,   105
since it is 2 more than 1 times the base; hence we put down the 2 and   422
carry the 1. Next we add this 1 to 2 and 2 and get 5; hence we write   1152
5 as a part of the sum without having anything to carry; finally, we add
the numbers in the column on the left and write 11 since their sum is 1 more than
1 times the base.

2. Find the product of 24_6 and 135_6.

SOLUTION

In determining the product, we shall make use of the tables in Figs. 1-6 and 1-7 just as we make use of number combinations and multiplication tables in multiplication to base 10.

To begin the multiplication, we see from Fig. 1-7 that $4 \times 5 = 32$; hence we record the 2 and carry 3. Then $4 \times 3 = 20$ and the 3 to carry makes 23; therefore we record the 3 and carry the 2. Next, $4 \times 1 = 4$ and the 2 to carry makes 10. Now multiplying by 2 gives 314 as recorded. Finally, by adding we see that the desired product is 4212 and this means it is $4 \times 6^3 + 2 \times 6^2 + 1 \times 6 + 2$.

$$
\begin{array}{r}
135 \\
24 \\
\hline
1032 \\
314 \\
\hline
4212
\end{array}
$$

Exercise 1.7

Express each of the following in the one of Hindu-Arabic and Roman numerals that is not given.

1. V, 7.   **2.** C, 101.

3. LX, 53.   **4.** MCXII, 63.

5. IV, 92.   **6.** XCI, 742.

7. DCXII, 919.   **8.** CDXLIV, 1492.

Sketch an abacus that shows the first number in each of Problems 9 through 16 and another that shows the sum of the two numbers in the problems.

9. 143, 254.   **10.** 302, 685.

11. 415, 372.   **12.** 836, 153.

13. 268, 613.   **14.** 574, 248.

15. 377, 745.   **16.** 291, 839.

Express each of the following in terms of base 10.

17. 325_8.   **18.** 436_7.

19. 147_{12}.   **20.** 1243_5.

Change each of the following to the indicated base.

21. 734 to base 7.   **22.** 562 to base 3.

23. 2371 to base 12.   **24.** 186 to base 2.

25. 342_7 to base 5.   **26.** 5146_7 to base 9.

27. 475_8 to base 11.   **28.** 703_6 to base 4.

Each number in Problems 29 through 36 is to the base 6. Perform the indicated operations by means of the tables given in Figs. 1-6 and 1-7.

29. $243 + 101 + 211$.   **30.** $121 + 213 + 211$.

31. $153 + 214 + 425$.   **32.** $235 + 523 + 352$.

33. (43)(245).   **34.** (25)(341).

35. (124)(234).   **36.** (513)(542).

2

FORMULAS, CHARTS, FUNCTIONS, GRAPHS

2.1. Formulas

There are many situations in which we want to find the value of a quantity by making use of the value of another or even the values of several others. This was the case in Arts. 1.5 and 1.6. There we found the simple interest by using values of principal, rate, and term and found the amount of property tax in terms of assessed value and tax rate. We shall now consider some more formulas as worked-out examples and as problems to be worked.

EXAMPLES

1. The volume of a right circular cylinder with h as height and r as radius of base is given by

$$V = \pi r^2 h, \qquad \pi = 3.1416 \text{ approximately.}$$

Find the volume of a right circular cylinder with $h = 17$ inches and $r = 11$ inches to the nearest cubic inch.

SOLUTION

If we substitute the given values in the formula for volume, we have

$$V = \pi r^2 h$$
$$= (3.1416)(11^2)17$$
$$= 6462.2712$$
$$= 6462 \text{ to the nearest cubic inch.}$$

2. If a person with a net income between \$4000 and \$8000 pays a tax

$$T = \$800 + .22E,$$

where E is the net income minus \$4000, find the tax paid if one has a net income of \$6827.32.

SOLUTION

In this problem,
$$E = \$6827.32 - \$4000 = \$2827.32;$$
consequently,
$$T = \$800 + (.22)(\$2827.32)$$
$$= \$800 + \$622.0104 = \$1422.01.$$

2.2. Charts

In the last article we studied formulas and found how to determine the value of one quantity when one or more other associated quantities were given. In this article, we shall be given a table of corresponding values of two quantities and see how to represent this situation graphically. It can be done in several ways; we shall present two of them. The following table shows the number of million pupils enrolled in public schools in the United States at five year intervals from 1900 through 1960.

Year	Enrollment	Year	Enrollment
1900	15.5	1930	25.7
1905	16.5	1935	25.8
1910	17.8	1940	25.4
1915	19.7	1945	23.2
1920	21.6	1950	25.1
1925	24.7	1955	27.3
		1960	36.1

In order to represent this data graphically, we shall begin by drawing a horizontal line segment and marking the dates at equal intervals along it, beginning at the left end with 1900. We shall now draw a vertical line segment above each point that designates a date. The first segment can be of any length but that length represents 15.5 million and determines our vertical unit. It should be so chosen that segments to represent the other enrollments fit into the conveniently available space. The segment which represents the enrollment of 24.7 million in 1925 should be drawn above the point which marks 1925 and should be $24.7/15.5 = 1.59$ times as long as the segment which indicates an enrollment of 15.5. Instead of deciding how long to draw a segment as above, we can decide on the length that is to represent 1 million and then draw each segment accordingly. Thus if 1 centimeter represents 5 million, then 15.5 million would be indicated by 15.5/5, or 3.1 centimeters.

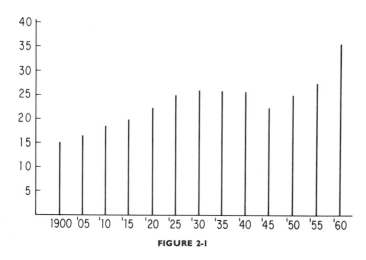

FIGURE 2-1

Instead of representing the data given in the table as we did in Fig. 2-1, it could have been done by marking the dates along a horizontal segment as before, next marking the enrollments along a vertical segment erected at the left end of the horizontal, then putting a dot or cross above each date and across from the point which represents the enrollment for that year, and, finally, drawing a smooth curve through the points. It may be easier to locate the points for the dots or crosses by drawing a vertical segment through each date and horizontal segments at regular intervals as was done in Fig. 2-2.

Exercise 2.1

1. The number of bushels that a rectangular container will hold is given by $b = lwh/1848$, where l, w, and h represent the length, width, and height and are measured in inches. How many bushels will a box hold if it is 33 inches long, 28 inches wide and 18 inches high?

2. The area of the walls of a rectangular room is given by $A = 2h(l + w)$, where h, l, and w represent the height, length, and width, respectively. Find the number of square feet of wall space in a room that is 9 feet high, 20 feet long, and 14 feet wide.

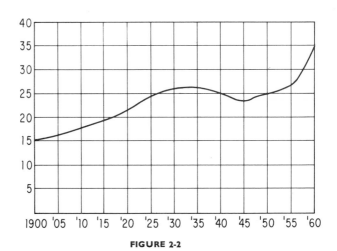

FIGURE 2-2

3. The time in seconds required to make a photographic enlargement of area A from a particular plate is $t = .32A$. Find the time required to make a 5 by 7 enlargement; a 7 by 11 enlargement.

4. The lateral surface of a cylinder is $L = 2\pi rh$, where r is the radius and h is the height. Find the lateral surface of a cylinder that is 17 inches in radius and 35 inches high.

5. The mass M of a block of metal is given by $M = lwhd$, where l is the length, w the width, h the height, and d the density. Find the mass of a block that is 14 by 15 by 20 inches if its density is 2.6.

6. If the tax paid by a person with a taxable income between \$4000 and \$6000 is given by $T = \$840 + .26E$, where E is the excess of income over \$4000, find the tax due by a person with a taxable income of \$5138.74.

7. If $l = a + nd - d$, find l for $a = 13$, $n = 9$, and $d = 4$.

8. If the normal blood pressure of a person is given by $P = .5A + 110$, where A is the age, find the normal pressure for a person whose age is 26, 40, 64.

9. The distance that a freely falling body will fall in t seconds is given by $s = 16t^2$. Find the distance that such a body will fall in 5 seconds, 7 seconds, 10 seconds.

10. The volume V of a pyramid of height h and square base of length s is given by $V = \frac{1}{3}s^2h$. Find the volume of such a pyramid if $s = 17$ inches and $h = 21$ inches; if $s = 197$ feet and $h = 138$ feet.

11. The volume V of a sphere of radius r is given by $V = 4\pi r^3/3$. Find the volume if the radius is 2 inches; 6 inches; 4000 miles. Use $\pi = 3.1416$.

12. The surface area S of a sphere of radius r is given by $S = 4\pi r^2$. Find the surface area if the radius is 3 inches, 4 feet, 4000 miles. Use $\pi = 3.1416$.

13. The principal P invested at simple interest that will accumulate to A dollars in t years at rate r per year is given by $P = A/(1 + rt)$. Find the principal that would accumulate to \$1200 in 4 years at 6%. By how much would the principal be decreased if the time were extended to 7 years?

14. If $s = a(1 - r^n)/(1 - r)$, find s for $a = 1$, $r = 2$, $n = 5$; for $a = 3$, $r = 3$, $n = 4$.

15. If $s = (a - rl)/(1 - r)$, find s for $a = 2$, $r = 3$, $l = 54$; for $a = 1$, $r = 2$, $l = 32$.

16. The area A of a circle of radius r is given by $A = \pi r^2$. Find the area if $r = 2$, 5, 2.7. Use $\pi = 3.14$.

17. The area A of a triangle of base b and altitude h is given by $A = .5bh$. Find, to as many digits as justified, the area of a triangle if its base is measured to be 13.6 centimeters and its altitude 11.5 centimeters.

18. The volume V of a cone of altitude h and base radius r is given by $V = \frac{1}{3}\pi r^2 h$. If the base radius is measured to be 23.7 centimeters and the altitude to be 16.42 centimeters, find the volume to the justifiable degree of accuracy with $\pi = 3.142$.

19. The area of a trapezoid with bases b and B is given by $A = .5h(b + B)$, where h is the distance between the parallel sides. The bases and altitude were measured to be 9.23 feet, 12.4 feet, and 5.38 feet, respectively. Find the area to the justifiable degree of accuracy.

20. The volume V of a prolate spheroid (football) is given by $V = 4\pi ab^2/3$, where $2a$ is the length and b is the radius midway between the ends. Find the volume if, by measurement, $2a = 12.66$ inches and $b = 3.12$ inches.

Show the data in each of Problems 21 through 32 graphically. Use a bar graph, a continuous curve, or both as directed by your instructor.

21. The minimum number of hundreds of calories required by boys of various ages is as follows.

Age	Calories	Age	Calories
8	17	13	25
9	19	14	26
10	21	15	27
11	21	16	27
12	23	17	28

22. The distance in feet that a car traveling at a specified number of miles per hour will move after the brakes are applied is as follows.

Speed	Distance	Speed	Distance
10	29	60	230
20	39	70	300
30	70	80	380
40	110	90	470
50	160	100	560

23. This table shows the normal weight for a man who is 30 years of age and is the indicated height.

Height, in inches	Weight, in pounds	Height, in inches	Weight, in pounds
60	126	66	144
61	128	68	152
62	130	70	161
63	133	72	172
64	136	74	184

24. This table shows the normal weight for a woman who is 20 years of age and is the indicated height.

Height, in inches	Weight, in pounds	Height, in inches	Weight, in pounds
60	106	65	125
61	110	66	128
62	114	67	131
63	117	68	135
64	121	69	139

25. The following table shows the U.S. population in millions at 10 year intervals beginning in 1870.

Year	Population	Year	Population
1870	39	1920	106
1880	50	1930	123
1890	63	1940	132
1900	76	1950	151
1910	92	1960	179

26. The following table shows the per cent of single men who marry within a year after reaching the specified age.

Age	Per cent	Age	Per cent
21	9.4	26	17.3
22	12.5	27	17.3
23	15.3	28	17.1
24	15.9	29	16.8
25	17.0	30	15.9

27. The following table shows the per cent of single men who eventually get married after having reached a specified age.

Age	Per cent	Age	Per cent
21	92.3	26	85.9
22	91.8	27	83.4
23	90.9	28	80.3
24	89.6	29	76.6
25	88.0	30	72.3

28. The following table shows the number of divorces in the U.S. per 1000 population at various dates.

Year	Divorces	Year	Divorces
1915	1.0	1940	2.0
1920	1.6	1945	3.5
1925	1.5	1950	2.6
1930	1.6	1955	2.3
1935	1.7	1960	2.2

29. The per capita public debt to the nearest dollar in the U.S. for various years is shown.

Year	Debt	Year	Debt
1935	226	1950	1697
1938	286	1953	1667
1941	367	1956	1621
1944	1452	1959	1607
1947	1792	1962	1600

30. The retail food purchasing power of the dollar for several years is given below with the 1947-1949 average as 100.

Year	Food	Year	Food
1943	146	1953	89
1945	145	1955	90
1947	104	1957	87
1949	100	1959	85
1951	89	1961	83

31. The total passenger car production in millions in the U.S. is shown for various years here.

Year	Cars	Year	Cars
1900	.004	1930	2.787
1905	.024	1935	3.274
1910	.181	1940	3.717
1915	.896	1945	.069
1920	1.906	1950	6.666
1925	3.735	1955	7.920
		1960	6.675

32. The following table shows the U.S. tobacco production for several years in billions of pounds.

Year	Tobacco	Year	Tobacco
1952	2.3	1957	1.7
1953	2.1	1958	1.7
1954	2.2	1959	1.8
1955	2.2	1960	1.9
1956	2.2	1961	2.1

2.3. Constants and Variables

We worked with formulas in Art. 2.1 and found the value of one quantity when one or more others were given.

As in Example 1 of Art. 2.1, we shall use the formula for the volume of a right circular cylinder. It is

(1) $$V = \pi r^2 h.$$

We shall consider a beaker in this case that is 8 centimeters in radius; hence $V = 64\pi h$. We shall assume that a liquid is being poured into the beaker; hence the height of the liquid is rising. We now have formula (1) in which π always has the same value, r has the value 8 cm in this discussion, and h may take on any value from zero to the height of the beaker.

The things considered in this discussion illustrate the types of quantities described in the following definitions.

DEFINITION. *A **constant** is a symbol which does not change in value during a discussion or situation and may never change.*

DEFINITION. *A **variable** is a symbol which may have any value within a given range.*

EXAMPLES

1. The values of π and 3 never change; hence, they are called absolute constants.
2. The value of the radius of a circle does not change; hence, it is a constant so long as we are talking about a particular circle.
3. The time of day and the amount of water in a leaky radiator are variables.

2.4. Functions and Functional Notation

We can see from the equation $V = 64\pi h$ that a value of V is determined if a value is assigned to h. Thus, if $h = 3$, we see that $V = 64\pi 3 = 192\pi$. We describe this type of relation in mathematical language by saying that the one variable is a function of the other. The situation is often described more formally by the statement given in the following definition.

DEFINITION. *If **D** is a set of numbers and if for each number **x** in **D** there exists a relation or rule which specifies exactly one corresponding number **y** in a set **R**, and if each number in **R** is taken on, then **y** is a **function of x**; furthermore, **D** is the **domain** of **x** and **R** is the **range** of **y**.*

Thus, in the equation $V = 64\pi h$ for the volume of liquid in a cylinder of radius 8, the domain for h is the set of numbers from zero to the height of the cylinder while the range on V is the set of numbers from zero to 64π times the height.

The variable to which values are assigned is called the *independent* variable and the one whose value is thereby determined is called the *dependent* variable.

The independent variable is not always expressed explicitly. For example, it is not so expressed in $V = 64\pi h$. If we want to show that the function y depends on the independent variable x, we write $y(x)$ and read "the y function of x" or merely "y of x". If we want to show not only that y depends on x but the way in which it depends we put $y(x)$ equal to the specific function in terms of x. Thus we write $y(x) = x^2 - x + 3$ to show that the function y depends on the variable x and is equal to $x^2 - x + 3$. If we want to indicate that the function y has been or is to be evaluated for a specific value of x we replace x by that number.

EXAMPLE

1. If $y(x) = x^2 - 4x + 2$, find $y(3)$ and the value of the function for $x = -2$.

SOLUTION

The value of $y(3)$ is obtained by replacing x by 3 in the expression for $y(x)$ and is

$$y(3) = 3^2 - 4(3) + 2$$
$$= 9 - 12 + 2 = -1.$$

We indicate that y is to be evaluated for $x = -2$ by writing $y(-2)$. Thus,

$$y(-2) = (-2)^2 - 4(-2) + 2$$
$$= 4 + 8 + 2 = 14.$$

2.5. Functions of More Than One Variable

It is not unusual for one quantity to depend on several others. For example, the simple interest I produced by an investment of P for n years at rate r per year is $I = Prn$. This illustrates the definition given below.

DEFINITION. *If a variable W is so related to several others that a value for W is determined if a value is assigned to each of the others, then W is said to be a function of the others. The set of values that can be assigned to each of the others is called its* **domain** *and the set of values taken on by W is called its* **range**.

We indicate that W is a function of several independent variables by following W by the other variables enclosed in parentheses and separated by commas. In keeping with this, $W(x, y, r, s)$ indicates that W is a function and depends on x, y, r, and s for its value.

EXAMPLES

1. If $$W(x, y, r, s) = x^2 - 6xyr - y^2s,$$

then

$$W(2, 3, -1, 5) = 2^2 - 6(2)(3)(-1) - 3^2(5)$$
$$= 4 + 36 - 45$$
$$= -5$$

indicates that we intend to evaluate W for $x = 2$, $y = 3$, $r = -1$, and $s = 5$ and have done so.

2. If

$$V(s, h) = \frac{1}{3} s^2 h,$$

then

$$V(5, 6) = \frac{1}{3}(5^2)6$$

$$= 50.$$

3. If

$$s(x) = \frac{3x - 1}{2x + 3}$$

then

$$s(\tfrac{2}{5}) = \frac{3(\tfrac{2}{5}) - 1}{2(\tfrac{2}{5}) + 3} = \frac{\tfrac{6}{5} - 1}{\tfrac{4}{5} + 3}$$

$$= \frac{\tfrac{1}{5}}{\tfrac{19}{5}} = \frac{1}{5}\frac{5}{19}, \text{ inverting,}$$

$$= \frac{1}{19}, \text{ multiplying.}$$

Exercise 2.2

1. If $f(x) = 3x - 2$, find $f(1)$, $f(3)$, $f(-5)$.

2. If $h(t) = 1 - 5t$, find h for $t = 3, -2, -5$.

3. If $m(s) = 6s - 7$, find m for $s = 2.5, 0, -3.25$.

4. If $a(r) = 4r + 3$, find $a(4)$, $a(0)$, $a(-3)$.

5. If $g(m) = 4 - 3m$, find g for $m = \frac{2}{3}, \frac{3}{5}, -\frac{1}{6}$.

6. If $d(b) = 5 - 2b$, find d for $b = \frac{3}{2}, 0, -\frac{5}{4}$.

7. If $s(a) = 7a + 3$, find $s(\frac{2}{7})$, $s(0)$, $s(-\frac{3}{7})$.

8. If $t(u) = 6u - 1$, find t for $u = \frac{2}{3}, \frac{1}{6}, -\frac{1}{2}$.

9. If $s(t) = \dfrac{2t-1}{3t+2}$, find $t(\frac{1}{2})$, $t(0)$, $t(-\frac{2}{3})$.

10. If $h(m) = \dfrac{5m+2}{2m+1}$, find $m(-\frac{2}{5})$, $m(\frac{1}{5})$, $m(-\frac{1}{2})$.

11. If $g(h) = \dfrac{6h-5}{5h+1}$, find $g(\frac{5}{6})$, $g(\frac{1}{3})$, $g(-\frac{1}{5})$.

12. If $L(a) = \dfrac{12a+7}{7a-12}$, find L for $a = \frac{2}{5}, \frac{3}{4}, -\frac{7}{12}$.

13. If $q(x) = x^2 - 3x + 1$, find q for $x = 5, 0, -2$.

14. If $r(y) = 2y^2 + y - 3$, find r for $y = 3, -1, -5$.

15. If $s(z) = -z^2 - z + 4$, find $s(-3)$, $s(0)$, $s(2)$.

16. If $t(a) = -3a^2 + 2a - 1$, find $t(-5)$, $t(-2)$, $t(3)$.

17. If $A(f) = f^2 + f - 1$, find A for $f = \frac{2}{3}, \frac{1}{2}, -\frac{2}{5}$.

18. If $B(g) = 2g^2 - g - 3$, find B for $g = \frac{2}{5}, -\frac{1}{3}, -\frac{3}{4}$.

19. If $C(h) = 2 - 7h + 6h^2$, find C for $h = \frac{2}{3}, \frac{1}{2}, -\frac{1}{3}$.

20. If $D(i) = 10i^2 + i - 3$, find D for $i = \frac{2}{5}, -\frac{3}{5}, -\frac{2}{3}$.

21. If $L(m) = \dfrac{2m-3}{m^2-m+1}$, find L for $m = \frac{3}{2}, \frac{2}{3}, -\frac{1}{2}$.

22. If $M(n) = \dfrac{3n-1}{2n^2-3n-2}$, find M for $n = \frac{1}{3}, -\frac{1}{2}, -\frac{2}{5}$.

23. If $N(a) = \dfrac{6a^2-13a-5}{3a-4}$, find N for $a = \frac{5}{2}, \frac{4}{3}, -\frac{2}{3}$.

24. If $Q(p) = \dfrac{6p^2-5p-6}{2p+5}$, find Q for $p = -\frac{2}{3}, -\frac{5}{2}, \frac{3}{5}$.

25. If $H(C) = C^2 - C + 2$, find $H(\frac{2}{3}) \div H(\frac{1}{2})$.

26. If $D(S) = S^2 - 2S + 1$, find $D(\frac{2}{5}) \div D(\frac{3}{4})$.

27. If $L(W) = W^2 - 5W + 3$, find $L(\frac{2}{7}) \div L(\frac{1}{5})$.

28. If $N(R) = 2R^2 - 3R + 5$, find $N(\frac{1}{2}) \div N(\frac{5}{6})$.

29. If $H(e, a, t) = 2ea - 3at + et$, find H for $e = 2, a = 3, t = -2$ and for $e = \frac{1}{2}$, $a = \frac{1}{3}, t = 2$.

30. If $W(x, y, z) = 3x^2y - xyz - 2yz^2$, find $W(2, 1, 3)$ and $W(\frac{1}{3}, \frac{1}{2}, 6)$.

31. If $K(n, i, t) = n^2i - i^2t + t^2n$, find $K(2, 3, -1)$ and $K(\frac{2}{3}, \frac{3}{4}, \frac{4}{5})$.

32. If $E(a, s, y) = 2a^2 - as + sy^2$, find $E(1, 0, -2)$ and $E(\frac{1}{2}, \frac{2}{3}, \frac{3}{4})$.

2.6. The Rectangular Coordinate System

In order to be able to sketch the graph of a function, we shall make use of the rectangular Cartesian coordinate system. It was named for the French mathematician René Descartes (1596–1650) and consists of a pair of perpendicular lines with a positive direction chosen on each as indicated by the arrows. The lines are called the X and Y axes, and their intersection is the *origin*. A point is located by giving its distance and direction from each axis. Distances measured to the right from the Y axis are considered positive, as are those up from the X axis. Distances measured to the left from the Y axis and those down from the X axis are negative. The distance and direction from the Y axis to a point is called the *abscissa* of the point. The distance and direction from the X axis to a point is known as the *ordinate* of the point. The abscissa and ordinate are called the *coordinates* of the point. They are enclosed in parentheses, separated by a comma, and the abscissa is written first. The four portions into which the axes divide the plane are called *quadrants* and are numbered as in Fig. 2-3.

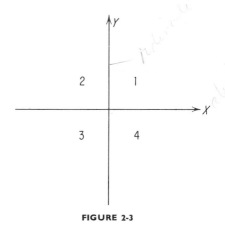

FIGURE 2-3

EXAMPLES

1. The point that is 3 units to the right of the Y axis and 2 units below the X axis is written $(3, -2)$.
2. The point $(-1,4)$ is 1 unit to the left of the Y axis and 4 units above the X axis.
3. The point $(5,0)$ is 5 units to the right of the origin.

2.7. Graphs and Zeros of Functions

Since we can find the value of the dependent variable when a value of the independent in its domain is given and can locate a point if its coordinates are given, we are now in a position to sketch the graph of a function. If $y = F(x)$ is the function whose graph we wish to sketch, we assign several values to x, determine each corresponding value of y, locate the points with coordinates x and y, and draw a smooth curve through them.

EXAMPLE

4. Sketch the graph of
$$y(x) = 3x - 4.$$

SOLUTION

We shall assign three values to x and record them and each corresponding value of y in a table like the one shown

here. If we assign the value -1 to x, then $y(x)$ becomes $y(-1)$ and is $3(-1) - 4$ $= -7$. Similarly, $y(1) = 3(1) - 4 = -1$ and $y(4) = 3(4) - 4 = 8$. Now, entering these pairs of values in the table, it becomes

x	-1	1	4
y	-7	-1	8

Finally, locating each point (x, y) indicated in the table and drawing a smooth curve through them, we have the graph shown in Fig. 2-4.

A value of the independent variable for which the function is zero is called a *zero of the function*. Geometrically this is a value of x for which the graph crosses the X axis, since the ordinate of any such point is zero. The zero or zeros of a function can be estimated after the graph is sketched. The graph in Fig. 2-4 crosses the X axis slightly nearer 1 than 2 and we shall estimate that $x = 1.3$ is the zero of the function.

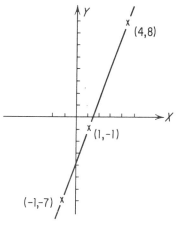

FIGURE 2-4

EXAMPLE

5. Sketch the graph and estimate the zeros of
$$y(x) = 2x^2 - 6x + 3.$$

SOLUTION

We shall assign several values to x beginning with -2, then compute each corresponding value of y and enter all pairs in the table before locating the points and drawing the curve through them. If $x = -2$, then $y(x) = y(-2) = 2(-2)^2$ $- 6(-2) + 3 = 23$. The other pairs of values in the table can be obtained in a similar

x	-2	-1	0	1	2	3	4	5
y	23	11	3	-1	-1	3	11	23

manner. We shall use 4 times as much space to represent a unit on the X axis as on the Y axis since the range of values on y is nearly 4 times that on x. Now, locating the points indicated in the table and drawing a smooth curve through them, we have the graph shown in Fig. 2-5. The graph appears to cross the X axis at the points for which $x = .7$, and $x = 2.3$; hence we say these are the zeros of the function.

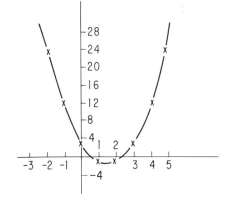

FIGURE 2-5

Exercise 2.3

1. Locate the following points: $(3, 2)$, $(3, -1)$, $(-2, 4)$, $(-5, -2)$, $(4, 0)$.

2. Locate the following points: $(2, 5)$, $(-1, 4)$, $(-2, -3)$, $(0, 3)$, $(6, -2)$.

3. In which quadrant is each of the following points if k is positive? (k, k), $(k^2, -k)$, $(k, -k^2)$, $(-k, -k)$, $(k + 1, -k - 2)$.

4. In which quadrant is each of the following points if k is negative? $(-k, k)$, (k^2, k), $(-k + 1, k^3)$, $(k, -k + 3)$, $(-k^2, k)$.

Sketch the part of the graph of each function given below for values of x in the specified interval and estimate the zeros to the nearest tenth of a unit.

5. $y(x) = 2x + 1$ from $x = -4$ to $x = 3$.

6. $y(x) = 3x - 5$ from $x = -2$ to $x = 4$.

7. $y(x) = 2 - x$ from $x = -3$ to $x = 5$.

8. $y(x) = 4 - 5x$ from $x = -2$ to $x = 3$.

9. $y(x) = -\frac{2}{3}x + 5$ from $x = -3$ to $x = 6$.

10. $y(x) = -\frac{1}{2}x - 3$ from $x = -8$ to $x = 4$.

11. $y(x) = \frac{3}{5}x + 2$ from $x = -5$ to $x = 5$.

12. $y(x) = \frac{3}{4}x - 5$ from $x = -2$ to $x = 6$.

13. $y(x) = x^2 + x - 3$ from $x = -4$ to $x = 3$.

14. $y(x) = x^2 - 3x + 1$ from $x = -1$ to $x = 5$.

15. $y(x) = -x^2 + 5x + 2$ from $x = -2$ to $x = 7$.

16. $y(x) = -x^2 + 2x + 1$ from $x = -3$ to $x = 5$.

17. $y(x) = 2x^2 - 3x - 4$ from $x = -2$ to $x = 4$.

18. $y(x) = 3x^2 + 2x - 1$ from $x = -3$ to $x = 2$.

19. $y(x) = 2x^2 + 2x - 3$ from $x = -4$ to $x = 3$.

20. $y(x) = 4x^2 - x - 5$ from $x = -2$ to $x = 3$.

21. $y(x) = -2x^2 - 5x + 2$ from $x = -5$ to $x = 2$.

22. $y(x) = -3x^2 + 4x + 9$ from $x = -4$ to $x = 4$.

23. $y(x) = -5x^2 + 3x + 3$ from $x = -2$ to $x = 3$.

24. $y(x) = -4x^2 + 7x + 9$ from $x = -2$ to $x = 4$.

3

INTRODUCTION TO SETS

3.1. Some Definitions

New concepts are introduced into mathematics from time to time. Some are relatively unimportant, but others give mathematics life, vigor, a potential for growth, and a new look at the older concepts. The idea of a set is one of the recent concepts that is now widely used in developing and interpreting mathematics.

Any collection of objects is called a *set*; the objects are called *elements* or *members* of the set. Thus, California, Texas, and Utah are elements of the set of states that are west of the Mississippi River.

As further examples of sets, we present

EXAMPLES

1. The set of all nations in the world.
2. The set of all men over six feet tall.
3. The set of all past and present U.S. Senators.

4. The set of all positive integers: 1, 2, 3, ...

5. The set of all numbers between zero and one; inclusive.

6. The set of all numbers that can be expressed as the quotient of two positive integers. These are called *positive rational numbers*.

7. The set of the six sets listed above.

We indicate that a is an element of set A by writing $a \in A$.

There is no end to the number of sets. Some of them have only a few elements, some have a considerable number, and some have an infinite number. The set in 7 has only six elements; there are a considerable number of men over six feet tall; and the number of positive integers is infinite. There are sets that do not have any elements. Such a set is called a *null set*. Thus, the set of all living men over a thousand years of age is a null set.

We shall be primarily interested in sets whose elements are numbers as in 4, 5, and 6 above. Each of the integers can be expressed as the quotient of two integers by using 1 as the denominator; consequently, we see that the set of positive integers is part of the set of positive rational numbers. If each element of one set is also an element of a second set, then the first set is called a *subset* of the second. Thus if each element of set A is also an element of set B, then A is a subset of B and we write $A \subseteq B$ to indicate this. If we want to indicate that A is not a subset of B, we write $A \not\subseteq B$. If, in addition to each element of A being an element of B, there is an element of B that is not an element of A, then A is called a *proper subset* of B and we write $A \subset B$.

Quite often a subset of a set A is defined as those elements of A that have a common property. This common property may be determined in a variety of ways. If the set A, often called the *universal set* or *universe*, is all of the people in your home town, then all of them who are over 73 years of age is a subset a. We indicate this symbolically by writing

$$\{a|a \text{ is over 73 years of age}\}$$

and reading: The set a such that a is over 73 years of age. There are many subsets of A and some of them are defined below.

$$\{a|a \text{ has exactly 3 sisters}\}$$
$$\{a|a \text{ is a millionaire}\}$$
$$\{a|a \text{ is a farmer}\}$$

EXAMPLE

8. If the universe for set A is 1, 2, 3, 4, 5, 6, state the common property of a subset B and list the members thereof.

SOLUTION

We shall take divisibility by 2 as the common property of the subset. Its members are 2, 4, 6.

The symbol $\emptyset$ is used to designate a null set and $\{a, b, c, d, e\}$ is used to indicate the set whose elements are a, b, c, d, and e.

EXAMPLE

9. The set A indicated by $\{1, 2, 5, 9\}$ is the four numbers 1, 2, 5, and 9 and it is a proper subset of $B = \{1, 2, 5, 7, 9\}$ since each element of A is an element of B and B contains an element that is not an element of A.

Two sets are said to be *identical* if and only if each is a subset of the other. If A and B are identical, we write $A = B$.

EXAMPLE

10. The set A represented by $\{a, b, c, d\}$ and the set B designated by $\{a, d, b, c\}$ are *identical* since each element of each is also an element of the other.

Another relation between two sets is described by saying that they are in 1 to 1 correspondence. Two sets A and B are said to be in 1 *to* 1 *correspondence* if it is possible to associate the elements of the two sets in such a way that each element of each set is associated with exactly one element of the other.

EXAMPLES

11. Establish a 1 to 1 correspondence between the letters of the Roman alphabet and the numbers from 7 through 32. A 1 to 1 correspondence is established by associating 7 with a, 8 with b, 9 with c, and so on, as indicated by the next two lines.

7	8	9	10	11	. . .	30	31	32
a	b	c	d	e		x	y	z

There are other ways to establish a 1 to 1 correspondence. One more is indicated below

10	11	12	13	14	...	32	7	8	9
a	b	c	d	e		w	x	y	z

12. Set up a 1 to 1 correspondence between $\{1, 2, 3, \dots\}$ and $\{3, 6, 9, \dots\}$.

SOLUTION

If we represent an element of the first set by n and pair it with the element $3n$ of the second set, we have the 1 to 1 correspondence given below.

1	2	3	4	...	n	...
3	6	9	12	...	$3n$	...

Exercise 3.1

Find a proper subset of the set given in each of Problems 1 through 4.

1. $\{a, b, c, d, e\}$.

2. {Herbert, Franklin, Harry, Ike, John, Lyndon}.

3. $\{x | x$ is a rational number$\}$.

4. $\{x | x$ is between 1 and 3, inclusive$\}$.

Determine and state whether the set A in each of Problems 5 through 8 is a subset of the set B and give a reason for your answer.

5. $A = \{5, 7, 9, 11, 12\}$, $B = \{5, 6, 7, 8, 9, 11, 12\}$.

6. $A = $ {John Garner, Henry Wallace, Harry Truman, Lyndon Johnson, John Jones}.

$B = \{x | x$ is or was a Vice-President of the U.S.$\}$.

7. $A = \{x | x$ is a positive rational number$\}$, $B = \{x | x$ is a positive integer$\}$.

8. $A = $ {Tom, Bob, Jack, Hulen}, $B = $ {Bob, Hulen, Tom, Jack}.

In each of Problems 9 through 12, give a set that has A as a subset.

9. $A = \{x | x$ is a resident of Alaska$\}$.

10. $A = \{1, 3, 5, 7, 9\}$.

11. The elements of A are a, c, e, g, and i.

12. $A = \{x | x$ is between 4 and 7, inclusive$\}$.

State the common property of some subset of the set given in each of Problems 13 through 16 and list the members of the subset in Problems 13 and 14. An answer is given, but there are others.

13. {George Washington, Zachary Taylor, Ulysses Grant, Dwight D. Eisenhower, John Kennedy, Herbert Hoover, Woodrow Wilson}.

14. $\{2, 6, 8, 9, 12\}$.

15. $\{x | x$ is between 0 and 1, inclusive$\}$.

16. $\{x | x$ is a positive rational number$\}$.

Establish a 1 to 1 correspondence between the two sets given in each of Problems 17 through 20 or tell why it cannot be done.

17. $\{1, 3, 5, 7, 9\}$, $\{2, 5, 8, 11, 23\}$.

18. $\{a, b, c, d, e\}$, $\{2, 7, 11, 12\}$.

19. $\{2, 4, 6, \quad \ldots \quad, 2n, \quad \ldots \}$, $\{3, 6, 9, \quad \ldots \quad, 3n, \ldots \}$.

20. {5, 11, 16, 20, 23, 25}, {5, 7, 10, 14, 19, 25}.

21. Is the null set a subset of all sets?

22. Show that {3, 8, −1, 2, 7} and {8, 2, 3, 7, −1} are identical.

23. Show that {Wilson, Harding, Coolidge, Hoover, Roosevelt, Truman, Eisenhower, Kennedy, Johnson } and {$p|p$ was President of the U.S. between 1915 and 1963} are identical.

24. Show $A = \{x|x$ is a positive integer} is a proper subset of $B = \{x|x$ is a rational number}.

3.2. Union, Intersection, Cartesian Product

We pointed out in the last article that a subset of the universe under consideration can be determined by listing its elements or by stating a rule that determines the elements. Thus, {1, 2, 3, 4} and {$x|x$ is a positive integer less than 5} are two ways of indicating the same subset of the positive integers.

In this article, we shall be interested in (1) the set that is a subset of A or of B or of both and (2) the set that is a subset of both A and B. Instead of describing these as often as we want to refer to them, we shall use the following definitions and symbols.

The set of elements that are members of A or of B or of both is called the *union* of A and B and is symbolized by $A \cup B$. The union of A and B is shown in Fig. 3.1. Part (a) of the figure shows A and B disjoint; i.e., with no elements in common, part (b) shows A and B with elements in common and elements not in common, and part (c) shows the situation with $B \subset A$.

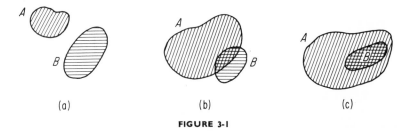

(a)   (b)   (c)

FIGURE 3-1

EXAMPLE

1. If $A = \{x|x$ is an integer between 2 and 8, inclusive} and if $B = \{x|x$ is an integer between 5 and 9, inclusive}, then $A \cup B$ is {$x|x$ is an integer between 2 and 9, inclusive } since that is the set of elements that are in A or in B or in both.

The set of elements that are members of both A and B is called the *intersection* of A and B and is symbolized by $A \cap B$. The sets A and B and their intersection are shown in Fig. 3.2. The intersection is the shaded area. As in Fig. 3.1, part (a) of the figure shows the situation with A and B disjoint, part (b) shows A and B with elements in common and elements not in common and part (c) shows the situation with $B \subset A$.

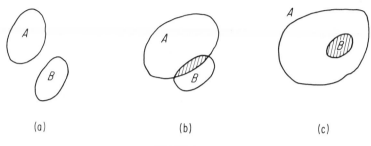

(a)   (b)   (c)

FIGURE 3-2

EXAMPLE

2. If A and B are the sets given in Example 1, then $A \cap B$ is $\{x \mid x$ is an integer between 5 and 8, inclusive$\}$ since 5, 6, 7, 8 are the elements that are common to $\{2, 3, 4, 5, 6, 7, 8\}$ and $\{5, 6, 7, 8, 9\}$.

If we have three sets A, B, and C, we can form other sets from them by use of the operations of union and intersection.

EXAMPLE

3. If $A = \{1, 2, 5\}$, $B = \{2, 3, 4\}$ and $C = \{3, 4, 5\}$, find $(C \cap B) \cup (C \cap A)$.

SOLUTION

As a first step, we find that $C \cap B$ is $\{3, 4\}$ since 3 and 4 are the elements in both B and C; furthermore, $C \cap A$ is $\{5\}$ since that is the only element common to C and A. Finally,

$$(C \cap B) \cup (C \cap A) = \{3, 4\} \cup \{5\} = \{3, 4, 5\}$$

A pair of numbers in which x is the first and y is the second is called an *ordered pair* and is written (x, y). Since the order in which the numbers occur is considered as well as the numbers, it follows that (y, x) and (x, y) are different ordered pairs unless $y = x$.

If A and B are sets, then $A \times B = \{(x, y) \mid x \in A$ and $y \in B\}$ is called the *Cartesian Product* of A and B.

EXAMPLE

4. If $A = \{2, 4, 5\}$ and $B = \{1, 3, 6\}$, then $A \times B = \{(2, 1), (2, 3), (2, 6), (4, 1), (4, 3), (4, 6), (5, 1), (5, 3), (5, 6)\}$.

Exercise 3.2

If $A = \{3, 6, 9, 12\}$, $B = \{2, 4, 6, 8, 10, 12\}$, and $C = \{2, 3, 4, 5\}$, calculate the set called for in each of Problems 1 through 12.

1. $A \cup B$. **2.** $A \cup C$. **3.** $B \cup C$. **4.** $C \cup A$.

5. $A \cap B$. **6.** $A \cap C$. **7.** $B \cap C$. **8.** $C \cap A$.

9. $(A \cup B) \cap B$. **10.** $(B \cup C) \cup C$.

11. $(A \cap B) \cup B$. **12.** $(B \cap C) \cap C$.

If $A = \{-2, 0, 5\}$, $B = \{4, 5, 6\}$, and $C = \{-2, 4, 6\}$, find the set called for in each of Problems 13 through 36.

13. $A \times B$. **14.** $B \times C$. **15.** $A \times C$. **16.** $C \times A$.

17. $A \times (B \cup C)$. **18.** $(A \cup B) \times C$.

19. $A \times (B \cap C)$. **20.** $(A \cap B) \times C$.

21. $(A \cup B) \cup C$. **22.** $(A \cup B) \cap C$.

23. $A \cap (B \cup C)$. **24.** $A \cap (B \cap C)$.

25. $A \cup (B \cup C)$. **26.** $A \cup (B \cap C)$.

27. $(A \cap B) \cap C$. **28.** $(A \cap B) \cup C$.

29. $(A \cup B) \cap (B \cup C)$. **30.** $(A \cup B) \cap (B \cap C)$.

31. $(A \cap B) \times (A \cup C)$. **32.** $(A \cap B) \times (A \cap C)$.

33. $(A \times B) \cap (B \times C)$. **34.** $(A \times B) \cap (A \times C)$.

35. $(A \cup C) \cap (B \cup C)$. **36.** $(A \cap B) \cup (B \cap C)$.

4

LINEAR EQUATIONS AND SYSTEMS
OF LINEAR EQUATIONS

4.1. Introduction

Each formula and function that was studied in Chapter 2 is a statement that two expressions are equal. The two expressions are called *members* and the statement that they are equal is called an *equation*. If an equation is a true statement for all values of the unknown in its domain, it is called an *identity*. We, however, shall be interested primarily in equations that are true statements for some values of the unknown and not for others. Equations of this type are called *conditional equations*. We shall be interested just now in a special type of conditional equation that is described in the following definition. An equation of the form $ax + b = 0$, $a \neq 0$, is called a *linear equation in one unknown*.

Any number that makes an equation a true statement if substituted for the unknown is called a *root* or *solution* of the equation.

EXAMPLE

1. We see that $x = 3$ is a root of $4x + 2 = 5x - 1$ since $4(3) + 2 = 14$ and $5(3) - 1 = 14$ are equal. Furthermore, $x = 5$ is not a root of the equation, since $4(5) + 2 = 22$ and $5(5) - 1 = 24$ are not equal.

4.2. Solution of Linear Equations

Now that we know what a root is and how to tell whether a given number is one or not, our next job is to see how to find each root of an equation. If two equations have the same solutions or roots, they are called *equivalent* equations. We shall use the following statements in deciding whether two equations are equivalent.

AXIOM I. *If the same quantity is added to or subtracted from each member (or side) of an equation, the new equation is equivalent to the original.*

AXIOM II. *If each member of an equation is multiplied or divided by the same nonzero constant, the new equation is equivalent to the original.*

EXAMPLE

1. If we add $c - bx$ to each member of

(1) $$ax - c = bx + d,$$

we see that it and

(2) $$ax - bx = c + d$$

are equivalent.

If we examine these equivalent equations, we see that (2) could be obtained from (1) by moving $-c$ and bx from one member to the other and changing the sign of each term that is moved to the other member. This procedure is justified by use of Axiom I and is known as *transposing*.

In solving a linear equation, we should first get all the unknowns in one member and the constants in the other by use of Axiom I above or by transposing. We then have an equation of the form $Ax = B$ and can solve for x by dividing both sides by A as justified by Axiom II. The number A is called the *coefficient* of x.

EXAMPLES

2. Solve

$$5x - 3 = 2x + 9.$$

SOLUTION

The first step in solving is to add $3 - 2x$ to each member and obtain

$$5x - 2x = 9 + 3$$
$$3x = 12, \text{ collecting terms,}$$
$$x = 4, \text{ dividing by 3.}$$

In order to check to determine whether 4 is a solution, we shall evaluate the left and right members of the given equation and see if they are equal:

$$\text{left member} = 5x - 3, \qquad \text{right member} = 2x + 9$$
$$= 5(4) - 3 \qquad\qquad\qquad = 2(4) + 9$$
$$= 17 \qquad\qquad\qquad\qquad = 17.$$

Consequently $x = 4$ is a solution, since the two members are equal for $x = 4$.

3. Solve

$$\frac{1}{2}x - 2 = 3 - \frac{1}{3}x$$

and check.

SOLUTIONS

1. Our first step is to add $2 + \frac{1}{3}x$ to each member of the equation so as to have the unknowns in one member and the knowns in the other. Thus we get

$$\frac{1}{2}x + \frac{1}{3}x = 3 + 2$$

$$\frac{5}{6}x = 5, \text{ performing the additions.}$$

We now have a fractional coefficient for x and must recall that, in dividing by a fraction, we invert the fraction and multiply; hence

$$x = 5\left(\frac{6}{5}\right) = 6.$$

As a check, the value of the left member is $(\frac{1}{2})(6) - 2 = 1$ and the right member is $3 - (\frac{1}{3})(6) = 1$; consequently $x = 6$ is the solution since the two members are equal for that value.

2. We can begin by multiplying through by a common denominator. If 6 is used, we get

$$3x - 12 = 18 - 2x$$

$$5x = 30 \qquad \text{adding } 12 + 2x \text{ to each member}$$

$$x = 6 \qquad \text{dividing by 5}$$

Exercise 4.1

Solve the following equations and check the answers.

1. $3x = 12.$

2. $-5x = 30.$

3. $2x = -14.$

4. $-7x = -21.$

5. $5x = 2x + 6.$

6. $-4x = 2x - 18.$

7. $-3 = 4x + 5.$

8. $-7 = -9x + 2.$

9. $3x - 2 = 5x - 12.$

10. $9 - x = 3x - 7.$

11. $6 + 2x = 5x + 27.$

12. $6x + 19 = 3x + 10.$

13. $5x + 2 = 2x + 4.$

14. $16x + 11 = 2 - 2x.$

15. $7x + 2 = 2x - 1.$

16. $4x - 1 = 1 - 3x.$

17. $3x - \dfrac{2}{9} = \dfrac{1}{9}.$

18. $5x - \dfrac{2}{7} = \dfrac{3}{7}.$

19. $7x + \dfrac{3}{11} = \dfrac{10}{11}.$

20. $4x - \dfrac{3}{2} = \dfrac{1}{2}.$

21. $\dfrac{2}{5}x - \dfrac{2}{5} = \dfrac{4}{5}.$

22. $\dfrac{3}{7}x - \dfrac{3}{7} = \dfrac{6}{7}.$

23. $\dfrac{3}{4}x - \dfrac{7}{4} = 2.$

24. $\dfrac{5}{9}x - \dfrac{2}{9} = \dfrac{8}{9}.$

25. $2x - \dfrac{3}{5} = \dfrac{7}{10}.$

26. $3x + \dfrac{3}{7} = \dfrac{5}{14}.$

27. $5x - \dfrac{6}{11} = \dfrac{3}{22}.$

28. $7x - \dfrac{2}{5} = \dfrac{8}{15}.$

29. $\dfrac{2}{5}x - \dfrac{7}{15} = 3.$

30. $\dfrac{4}{9}x + \dfrac{2}{3} = 4.$

31. $\dfrac{2}{7}x + \dfrac{3}{14}x = \dfrac{5}{7}.$

32. $\dfrac{4}{11}x - \dfrac{7}{33}x = \dfrac{15}{11}.$

33. $\dfrac{3}{5}x + \dfrac{1}{3}x = 14.$

34. $\dfrac{5}{7}x - \dfrac{1}{2} - = 3.$

35. $\dfrac{2}{3}x - \dfrac{1}{7}x = 11.$

36. $\dfrac{4}{5}x + \dfrac{1}{2}x = 13.$

37. $\dfrac{1}{3}x + \dfrac{3}{5}x = 13 + \dfrac{1}{2}x.$

38. $\dfrac{5}{7}x - \dfrac{1}{3}x = \dfrac{2}{7}x + 4.$

39. $\dfrac{2}{11}x + \dfrac{3}{5}x = \dfrac{1}{2}x + 31.$

40. $\dfrac{1}{2}x + \dfrac{2}{7}x = \dfrac{2}{3}x + 5.$

41. $\dfrac{1}{2}x + \dfrac{2}{3}x = \dfrac{5}{6}x + \dfrac{1}{4}.$

42. $\dfrac{2}{5}x + \dfrac{1}{4}x = \dfrac{1}{3}x + \dfrac{19}{24}.$

43. $\dfrac{1}{3}x + \dfrac{2}{7}x = \dfrac{1}{12}x + \dfrac{5}{7}.$

44. $\dfrac{1}{5}x - \dfrac{2}{3}x = \dfrac{17}{72} - \dfrac{3}{4}x.$

4.3. Introduction to Systems of Equations

The formulas of Art. 2.1 and the functions of Art. 2.4 are examples of equations in two or more unknowns although that name was not used. In Art. 2.4, we would have referred to $y = \frac{5}{2}x - 2$ by saying that y was a function of x but we shall now say that it is an equation in x and y, since it is a statement that two expressions in x and y are equal. If $f(x, y) = 0$ is a true statement for $x = a$ and $y = b$, we say that the number pair (a, b) is a *solution*; furthermore, two equations are *equivalent* if every solution of each equation is a solution of the other.

We can change the form of an equation and obtain an equivalent one by applying either or both of Axioms I and II of Art. 4.2. Now, multiplying $y = \frac{5}{2}x - 2$ by 2, we have $2y = 5x - 4$, and adding $-5x$ to each member gives $-5x + 2y = -4$ as an equation that is equivalent to the original one.

Any equation that is equivalent to one of the form $ax + by = c$, where a, b, and c are constants, is called a *linear equation in two unknowns*. We shall now study systems of two such equations and follow that by a study of systems of three equations of the form $ax + by + cz = d$.

4.4. Graphical Solution of Two Linear Equations

We saw how to sketch the graph of a function and how to estimate its zeros in Art. 2.7 and shall need the same concepts and skills here as were used there. All that is necessary in order to obtain the solution of two linear equations graphically is to draw the two graphs about the same pair of axes and then estimate the coordinates of their point of intersection since the solution is a point that is on both graphs. A number pair that satisfies both of two linear equations is often referred to as their *intersection*.

EXAMPLE

1. Solve the system

(1) $\qquad\qquad\qquad 3x + 2y = 15$

(2) $\qquad\qquad\qquad 4x - 3y = 7$

graphically.

SOLUTION

We want the point of intersection of the lines that are represented by the given equations; hence, we shall draw the lines. As an intermediate step we shall solve each equation for y in terms of x, since it is then easier to find the value of y corresponding to each arbitrarily assigned value of x. In order to solve (1) for y, we add $-3x$ to each member, then divide through by 2 and have

(3)
$$y = -\frac{3}{2}x + \frac{15}{2}.$$

Similarly, from (2), we obtain

(4)
$$y = \frac{4}{3}x - \frac{7}{3}.$$

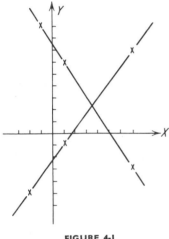

FIGURE 4-1

If we assign the values to x that are given in the tables, we find each corresponding value of y to be as shown. The tables are

x	-1	1	7
y	9	6	-3

from (3)

and

x	7	1	-2
y	7	-1	-5

from (4)

We now locate these points and draw a line through each set and obtain the graphs shown in Fig. 4-1. These graphs appear to intersect for $x = 3.5$ and $y = 2.3$; hence we say that $(3.5, 2.3)$ is their point of intersection. We can check by substitution in the given equations. Thus, from (1),

$$3(3.5) + 2(2.3) = 10.5 + 4.6 = 15.1;$$

hence $(3.5, 2.3)$ is not a solution of (1) but it is very nearly one and is probably correct to one decimal place. Similarly, from (2),

$$4(3.5) - 3(2.3) = 14 - 6.9 = 7.1$$

and the point is not on (2) but is near it.

It may happen that a pair of linear equations represent parallel lines; consequently, they have no solution. Such a pair of equations is called *inconsistent*. This term may have been chosen since the left member of one of a pair of inconsistent equations is a constant times the left member of the other but the right member of the second is not that constant times the right member of the first. In symbols this statement becomes, if the two equations are $ax + by = c$ and $kax + kby = d$, $d \neq kc$; then the equations are inconsistent and represent a pair of parallel lines. Furthermore, the equations are called *dependent* and represent coincident lines if one of them can be obtained by multiplying the other by a constant.

EXAMPLES

2. The pair of equations

$$2x + 5y = 6 \quad \text{and} \quad 4x + 10y = 7$$

are inconsistent since $4x + 10y = 2(2x) + 2(5y)$ but $7 \neq 2(6)$.

3. The equations

$$3x - 2y = 4 \quad \text{and} \quad 9x - 6y = 12$$

are a dependent pair since the second can be obtained from the first by multiplying by the constant 3.

Exercise 4.2

Solve the following pairs of equations graphically to one decimal place or identify them as inconsistent or dependent.

1. $3x + y = 6$
$x - 2y = 1.$

2. $2x - y = 6$
$2x + 3y = 8.$

3. $3x - 5y = 2$
$6x - 10y = 3.$

4. $x + 3y = 4$
$3x + 9y = 12.$

5. $2x + y = 1$
$4x + 2y = 2.$

6. $5x - 2y = -4$
$10x - 4y = 8.$

7. $5x + 7y = -2$
$2x - y = -4.$

8. $4x + 5y = 10$
$2x + 3y = 6.$

9. $3x + 2y = 0$
$2x + 3y = -2.$

10. $5x + 2y = 2$
$3x + y = 2.$

11. $3x - y = -1$
$-6x + 2y = 2.$

12. $-x + 2y = 2$
$2x - 4y = 5.$

13. $2x - 3y = 1$
$4x - 6y = -2.$

14. $2x - 3y = 1$
$4x - 6y = 2.$

15. $3x + y = 1$
$7x + 2y = 2.$

16. $5x + 3y = 4$
$3x - 5y = -4.$

17. $x - 5y = 3$
$-2x + 10y = -6.$

18. $x - 5y = 3$
$-3x + 15y = 9.$

19. $3x + 4y = -1$
$5x + 2y = 13.$

20. $4x + 2y = 1$
$3x - y = 15.$

21. $7x + 6y = 6$
$3x - 2y = 2.$

22. $2x + 5y = 0$
$3x + y = 5.$

23. $8x + 4y = -20$
$2x + y = -5.$

24. $5x + 7y = 1$
$10x + 14y = 3.$

25. $x - 4y = -3$
$3x - 12y = -8.$

26. $7x - 2y = 5$
$-14x + 4y = -10.$

27. $2x + 3y = 1$
$3x - y = 10.$

28. $5x + 2y = -1$
$2x - y = -8.$

29. $2x - 3y = -1$

$3x - 2y = 4.$

30. $5x - 3y = 2$

$7x + y = 0.$

31. $5x - 15y = -10$
$-x + 3y = -2.$

32. $6x - 3y = -21$
$-2x + y = 7.$

4.5. Use of Parentheses

We often want to indicate that a sum or difference of several terms is to be considered as a single term. This is done by enclosing them in parentheses ( ). If the quantity to be enclosed is $ax + cy$, we write $(ax + cy)$; furthermore, if the quantity enclosed is to be multiplied by k, we write $k(ax + cy)$ and mean that k is to be multiplied by both ax and cy and then the products added.

EXAMPLE

1. Remove parentheses from

$$3(2x - 4y) - 2(4x - 3y)$$

and collect terms.

SOLUTION

We must multiply both $2x$ and $-4y$ by 3 and both $4x$ and $-3y$ by -2 and then add the products. Thus,

$$3(2x - 4y) - 2(4x - 3y) = 6x - 12y - 8x + 6y$$

$$= -2x - 6y.$$

4.6. Solution of Two Linear Equations by Substitution

The graphical method of Art. 4.4 has the advantage that it can be used when some others cannot but also has the disadvantage that it gives only an approximation to the solution. We shall now discuss a method which cannot always be used but is easily applied in the case of linear equations. The procedure consists of

(1) *solving one of the equations for x or y in terms of the other*;

(2) *substituting the expression found in step* (1) *in the equation not used in it*;

(3) *solving the resulting equation for the only unknown in it*;

(4) *finding the value of the other unknown by substituting the value found in step* (3) *in either of the original equations*.

EXAMPLE

1. Solve the pair

(1) $$2x + 5y = -4$$
(2) $$3x + 2y = 5$$

simultaneously by substitution.

SOLUTION

We shall solve (2) for y and have

$$y = \frac{5 - 3x}{2}.$$

Now, putting this expression for y in (1), we have

$$2x + 5\left(\frac{5 - 3x}{2}\right) = -4$$

$4x + 5(5 - 3x) = -8$, multiplying by 2,
$4x + 25 - 15x = -8$, removing parentheses,
$-11x = -33$, adding -25 to each member and collecting,
$x = 3$, dividing through by -11.

In order to find the value of y, we shall substitute 3 for x in (1) and have

$2(3) + 5y = -4$
$5y = -10$, adding -6 to each member and collecting,
$y = -2$, dividing through by 5.

Hence the solution is $x = 3$, $y = -2$ and it can be checked by substituting in either original equation. Using (2) gives

$$3(3) + 2(-2) = 9 - 4 = 5$$

and the solution checks. We sometimes say that the lines intersect at (3, 2) or that the number pair (3, 2) satisfies the pair of equations.

4.7. *Solution of Two Linear Equations by Addition and Subtraction*

In solving a pair of equations simultaneously by addition or subtraction we:

(1) *decide which unknown to eliminate;*

(2) *multiply each equation through by the coefficient of* **y** *in the other if* **y** *is being eliminated and by the coefficient of* **x** *in the other if* **x** *is being eliminated; this gives a pair of equations that is equivalent to the given pair;*

(3) *subtract corresponding members of the equations obtained in* (2);

(4) *solve the equation obtained in* (3);

(5) *substitute the value obtained in* (4) *in either of the original equations and solve the resulting equation.*

EXAMPLE

1. Solve

(1) $$3x + 4y = 3$$
(2) $$2x + 7y = -11$$

simultaneously by addition and subtraction.

SOLUTION

We shall eliminate x, since to eliminate y would require the use of relatively large numbers. To do this, we multiply (1) by 2 and (2) by 3 and have

(3) $$6x + 8y = 6,$$
(4) $$6x + 21y = -33.$$

This pair of equations is equivalent to the given pair and the coefficients of x are equal; hence, we can eliminate x by subtracting each member of (4) from the corresponding member of (3). Thus

$$-13y = 39$$
$$y = -3, \text{ dividing through by } -13.$$

Now, substituting this in (1) gives

$$3x + 4(-3) = 3$$
$$3x - 12 = 3, \text{ removing parentheses,}$$
$$3x = 15, \text{ adding 12 to each member and collecting,}$$
$$x = 5, \text{ dividing by 3.}$$

Since (1) was used in obtaining this value of x, we shall check by using (2). Substituting $x = 5$ and $y = -3$ in (2) gives

$$2(5) + 7(-3) = 10 - 21 = -11;$$

hence, the solution checks, and the lines intersect at $(5, -3)$.

Exercise 4.3

Remove the parentheses and collect terms in Problems 1 through 8.

1. $2(2x + 3y)$.

2. $5(3x - 4y)$.

3. $-7(-2x + 5y)$.

4. $-3(-6x - 5y)$.

5. $2(3x - 4y) + 3(-4x + 3y)$.

6. $5(-2x + 7y) - 3(4x - 11y)$.

7. $-4(2x + 5y) + 5(x - 2y)$.

8. $-2(3x - 4y) - 5(-2x + y)$.

Solve the pair of equations in each of Problems 9 through 20 simultaneously by substitution. Check each solution.

9. $2x + y = 5$
$3x + 2y = 8.$

10. $x - 2y = 5$
$2x + 5y = 1.$

11. $5x + 4y = -2$
$x + 2y = -4.$

12. $3x + 7y = 5$
$2x - y = 9.$

13. $3x + 2y = 0$
$5x + 3y = -1.$

14. $5x + 7y = -3$
$-3x + 2y = -23.$

15. $7x - 6y = -2$
$2x + 3y = -10.$

16. $5x + 2y = -3$
$3x - 2y = 11.$

17. $4x + 3y = 3$
$2x + 9y = 4.$

18. $3x + 4y = 0$
$9x - 2y = 7.$

19. $5x + 2y = 3$
$7x - 6y = 2.$

20. $4x - y = 1$
$2x + 2y = 1.$

Solve the pair of equations in each of Problems 21 through 32 simultaneously by addition or subtraction. Check each solution.

21. $2x - 5y = -4$
$x + 2y = 7.$

22. $4x + y = 1$
$3x - 2y = 9.$

23. $3x + 2y = 2$
$5x + 4y = 2.$

24. $5x + 2y = 1$
$3x - 4y = -15.$

25. $4x + 3y = 0$
$3x + 2y = 1.$

26. $5x + 4y = 10$
$3x - 5y = 6.$

27. $7x + 6y = -6$
$3x - 5y = 5.$

28. $5x + 2y = 1$
$4x + 3y = -2.$

29. $2x + 5y = 2$
$4x + y = 1.$

30. $3x + 2y = 3$
$4x + 4y = 5.$

31. $5x + 3y = 1$
$3x + 7y = -2.$

32. $2x + 5y = 1$
$7x - 2y = -3.$

Solve the following pairs of equations by any algebraic method and check.

33. $3x + y = 3$
$5x + 3y = 1.$

34. $5x - 3y = 1$
$x - 5y = 9.$

35. $4x + 3y = 0$
$2x + y = 2.$

36. $x + 4y = 3$
$3x + 7y = 9.$

37. $3x + 5y = 1$
$5x + 3y = 7.$

38. $4x + 3y = 1$
$3x + 2y = 0.$

39. $3x - 8y = 5$
$5x - 6y = 12.$

40. $6x + 5y = -3$
$9x + 2y = 1.$

41. $2x + 3y = 0$
$4x + 9y = -1.$

42. $3x + 8y = 1$
$9x + 4y = -2.$

43. $9x + 10y = 12$
$6x - 15y = -5.$

44. $2x + 8y = -1$
$4x + 4y = 1.$

4.8. Systems of Three Linear Equations

There are several* methods which can be used for solving a system of three linear equations in three unknowns, but we shall discuss only that of addition and subtraction. In using this method we eliminate one of the unknowns between two of the equations just as we did with two equations and two unknowns. Next, we eliminate that same unknown between the third equation and either of the two previously used. We then have a system of two equations in two unknowns which can be solved by substitution or by addition and subtraction. The solution of this pair of equations can be put into either of the original equations and the resulting equation solved for the only unknown in it. Finally, the solution should be checked by substituting in one of the original equations. We shall now illustrate the procedure.

In order to see the situation geometrically, we shall make use of a fact that is proved in analytic geometry: a linear equation in three variables represents a plane in three dimensional space. The solution, if any, of three linear equations in three unknowns is a point. It is represented by a value for each of x, y, and z and is written as (x, y, z).

EXAMPLE

1. Solve the system

(1) $2x + 3y + 2z = 14$
(2) $3x + 4y + 5z = 26$
(3) $5x - 2y + 3z = 10$

* See Art. 6.5 and 6.9 of Rees and Sparks *College Algebra*, 4th ed. (New York: McGraw-Hill Book Company, 1961).

SOLUTION

We shall reduce this system to two equations in two unknowns by eliminating x between equations (1) and (2) and then eliminating it between equations (2) and (3). Thus,

$$
\begin{array}{lll}
(4) & 6x + 9y + 6z = 42 & \text{eq. (1) multiplied by 3} \\
(5) & 6x + 8y + 10z = 52 & \text{eq. (2) multiplied by 2} \\
\hline
(6) & y - 4z = -10 & \text{eq. (4) } - \text{ eq. (5)}
\end{array}
$$

If we now multiply (2) by 5 and subtract the result from 3 times (3), we get

$$-26y - 16z = -100$$

Now, dividing through by -2, we have

$$
(7) \qquad\qquad 13y + 8z = 50
$$

This equation and (6) form a pair of equations in 2 unknowns. To eliminate y between them, we shall solve (6) for y in terms of z and substitute in (7). Thus, we get

$$
\begin{aligned}
13(4z - 10) + 8z &= 50 \\
52z - 130 + 8z &= 50, \text{ removing parenthesis,} \\
60z &= 180, \text{ adding 130 and collecting,} \\
z &= 3
\end{aligned}
$$

We can now find that $y = 2$ by substituting $z = 3$ in (6) and solving for y. Finally, putting these values for y and z in either of the given equations, we find that $x = 1$; hence, the point of intersection of the planes represented by the given equations is $(1, 2, 3)$.

Exercise 4.4

Take each of the following systems of equations simultaneously.

1. $2x + y - z = 2$
$3x - y + 2z = 11$
$x + y - z = 0.$

2. $3x + 2y - z = 8$
$2x - 3y + z = -5$
$5x + y - z = 8.$

3. $2x + 3y - 4z = 3$
$3x - 3y + 5z = -14$
$x + 3y + z = 0.$

4. $3x + 5y - z = -6$
$3x - y + 2z = 12$
$-3x + 4y - 3z = -21.$

5. $2x + 3y + 4z = 9$
$2x - y + 3z = 5$
$x + 2y - 3z = 5.$

6. $3x - 2y - 5z = 5$
$x - 2y + 7z = 13$
$2x + 4y - 3z = -7.$

7. $3x + 3y + 2z = -1$
$2x + y - 2z = -8$
$2x + 4y + z = 7.$

8. $5x + 4y - z = 1$
$-5x - y + 8z = -2$
$2x + 3y - z = -5.$

9. $2x + 5y + z = -1$
$3x + 2y + 3z = 7$
$4x - 3y - 2z = 3.$

10. $6x + 7y + z = 4$
$3x - 5y + 2z = -1$
$2x - 3y + 5z = -8.$

11. $2x + 3y + 6z = 2$
$x - 2y + z = -15$
$3x - 4y - 2z = 1.$

12. $4x + 6y + z = 2$
$3x + 5y + 2z = 7$
$-2x + 2y + 3z = 0.$

13. $2x + 3y - 2z = -3$
$6x + 4y + 3z = -3$
$4x - 2y + 5z = 9.$

14. $3x + 4y + 5z = -4$
$5x - 2y - 2z = 7$
$4x + 6y + 7z = -5.$

15. $3x + 4y + 6z = -5$
$2x + 5y + 4z = 6$
$5x - 9y - 2z = 1.$

16. $3x + 2y + z = -6$
$6x + 3y - 2z = 4$
$9x + 5y - 3z = 8.$

17. $2x + 3y - 4z = 10$
$3x + 4y + 3z = -3$
$5x + 7y - 5z = 15.$

18. $5x + 3y + 2z = 15$
$6x + 7y + 3z = 12$
$3x + 8y + 4z = 10.$

19. $2x + 3y + 5z = 0$
$3x + 5y + 2z = 30$
$5x + 2y + 3z = 20.$

20. $3x + 6y + 5z = 1$
$4x + 5y + 2z = -1$
$2x + 7y + 8z = 3.$

21. $2x + 3y + 3z = 1$
$3x + 7y - 7z = 5$
$7x + 5y - 9z = 7.$

22. $2x + 4y + 3z = 2$
$4x + 2y + 6z = 3$
$3x - 3y + 12z = 4.$

23. $3x + 6y + z = 1$
$2x + 10y - 3z = 2$
$5x + 4y + 2z = 1.$

24. $4x + 2y + 4z = -1$
$6x + 3y + 2z = -2$
$3x + 4y + 4z = 0.$

4.9. Problems Solved by Use of Linear Equations

In order to solve a problem by means of an equation or a system of equations, we must in general have as many equations as we use unknown letters. We represent each desired quantity by a letter or a function of a letter and equate two expressions for the same thing in order to obtain each of the necessary equations. The mistake most often made in attempting to solve a problem by means of equations is in not being sure of the facts given in the statement of the problem. Read the problem, study it, and know what it says before trying to work it.

EXAMPLES

1. The sum of two numbers is 94. Find them if one is 5 less than twice the other.

SOLUTIONS

1. (*a*) If we decide to represent one of the members by x, then the other is $2x - 5$, since this is 5 less than twice the first. Their sum is $x + 2x - 5$ but 94 is given as the sum; hence, equating these two expressions for the same thing, we have

$$x + 2x - 5 = 94$$
$$3x = 99, \text{ adding 5 to each member and collecting,}$$
$$x = 33, \text{ dividing by 3.}$$

Consequently,

$$2x - 5 = 2(33) - 5 = 61.$$

Therefore, the numbers are 33 and 61.

1. (*b*) If x represents one of the numbers and y represents the other, then

(1) $$x + y = 94,$$

since their sum is 94; furthermore,

(2) $$y = 2x - 5,$$

since one is 5 less than twice the other. If we solve (1) and (2) simultaneously, we get $x = 33$ and $y = 61$.

EXAMPLE

2. A collection plate contained 41 coins. Only nickels, dimes, and quarters were included and the total value was $5.20. How many of each denomination were there if the number of nickels was 7 less than the total of dimes and quarters?

SOLUTION

In order for the meaning of each symbol to be easily remembered, we shall

let $n =$ the number of nickels,
$d =$ the number of dimes, and
$q =$ the number of quarters;

hence,

(1) $$n + d + q = 41$$

since there were 41 coins. Furthermore,

(2) $$d + q = n + 7,$$

since the number of nickels was 7 less than the sum of the number of dimes and quarters.

We must have a third equation, since we are using three unknowns. In order to get it, we shall make use of the value of the coins. Each nickel is worth 5 cents, each dime 10 cents, and each quarter 25 cents; hence, the value of the money in cents was

$5n + 10d + 25q$ in terms of n, d, and q. Therefore,

$$(3) \qquad\qquad 5n + 10d + 25q = 520,$$

since the money was worth $\$5.20 = 520$ cents.

The solution of the system can be started by subtracting each member of (2) from the corresponding member of (1). Thus,

$$
\begin{array}{ll}
(1) & n + d + q = 41 \\
(2) & \underline{\quad d + q = n + 7} \\
& \quad\; n \quad\;\; = 41 - n - 7, \quad \text{Eq. (1)} - \text{Eq. (2),} \\
& \;\; 2n \quad\; = 34, \\
& \quad\; n \quad\;\; = 17.
\end{array}
$$

To complete the solution, we shall use (2) and (3). If we put $n = 17$ in each of these, we have

$$
\begin{array}{ll}
(4) & d + q = 24 \quad \text{from (2) and} \\
(5) & 85 + 10d + 25q = 520 \quad \text{from (3).}
\end{array}
$$

The form of (5) can be simplified by adding -85 to each member and dividing through by 5. Thus,

$$
\begin{array}{ll}
2d + 5q = 87 & \\
\underline{2d + 2q = 48} & \text{Eq. (4) times 2,} \\
\quad\;\; 3q = 39, & \text{subtracting,} \\
\quad\;\;\; q = 13, & \text{dividing by 3.}
\end{array}
$$

Now that we know the values of n and q, we can find d from either of the original equations. Using (1) gives

$$
\begin{array}{l}
17 + d + 13 = 41, \\
\qquad\quad\; d = 11.
\end{array}
$$

We can check the solution by finding the value in cents of 17 nickels, 11 dimes, and 13 quarters. It is

$$
\begin{array}{l}
17(5) + 11(10) + 13(25) = 85 + 110 + 325 \\
\qquad\qquad\qquad\qquad\qquad\;\; = 520;
\end{array}
$$

consequently, the solution checks.

3. A grocer mixed 12 pounds of one grade of coffee with 8 pounds of another grade and obtained a blend worth a total of $\$13.20$. He made a second blend to sell at $\$.63$ per pound by mixing 9 pounds of the first grade and 21 pounds of the second. Find the price per pound of each grade.

SOLUTION

Since there are two numbers to determine, we must form two equations. In order to do that, we shall let $x =$ the price per pound in dollars of the first grade and $y =$ the price per pound in dollars of the second grade. Consequently, $12x + 8y$ is the value in dollars of the first blend since it consists of 12 pounds at x dollars per pound and 8 pounds at y dollars per pound. Therefore,

(1) $$12x + 8y = 13.20;$$

furthermore,

(2) $$9x + 21y = 30(.63) = 18.90$$

since each is the value of the second blend.

The pair of equations (1) and (2) can be solved by either of the methods we have studied. If we multiply each member of (1) by 3 and each member of (2) by 4, we get

(3) $$36x + 24y = 39.60$$
(4) $$36x + 84y = 75.60$$
$$60y = 36.00 \quad \text{Eq. (4)} - \text{Eq. (3)}$$
$$y = .60$$

To find the value of x, we shall substitute this value of y in (1) and have

$$12x + 8(.60) = 13.20$$
$$12x = 8.40, \text{ adding } -4.80 \text{ to each member,}$$
$$x = .70$$

Therefore, the first grade sold for $.70 per pound and the second for $.60 per pound.

Exercise 4.5

1. The sum of two numbers is 43. Find each if they differ by 13.

2. The difference of two numbers is 17 and their sum is 45; find them.

3. A farmer bought a pig and a cow for $106. Find the cost of each if the cow costs $10 more than twice as much as the pig.

4. Mr. Timpkin bought two lots with a total area of 17,700 square feet. Find the area of each if one contained 3300 square feet more than the other.

5. Mr. Fescue earned $16,700 in two consecutive years. How much did he earn each year if his income was $2100 more the second year than the first?

6. A student made 7 points more than a passing mark on one test and 3 points under a passing mark on the other. Find the grade on each if the total was 150.

7. A football team scored a total of 122 points on touchdowns and one point conversions. How many points were scored by each method if the number of conversions was 4 less than the number of touchdowns.

8. A fisherman drove to a lake by one route and returned home by another that was 17 miles longer. Find the length of each route if their total was 179 miles.

9. James is twice as old as John. In 4 years he will be 1.5 times as old. Find the age of each now.

10. Thomas and Thompkins together owned 760 sheep. After the latter bought 70 from the former, each had the same number. How many did each own originally?

11. Cam is three times as old as Sam. In 6 years, Cam will be only twice as old as Sam. Find the age of each now.

12. A contractor has 27 men on his payroll and a daily payroll of $300. How many receive each wage if some get $12 per day and the others $10 per day?

13. John earned $100 less than Fred from his summer job. John saved 20% of his earnings and Fred 30% of his. How much did each earn if together they saved $280?

14. Mrs. Salter had two rooms for rent. During one year the cheaper was vacant 2 months and the other vacant one month and she received $685 in rent. Find the monthly rent on each if the total was $65.

15. Farmer Zelinski sold his calves at $110 each and his pigs at $45 per head and received $5635 for 79 animals. How many of each did he sell?

16. Two grades of tobacco that sell for $1.60 per pound and $2.20 per pound were mixed so as to sell for $1.84 per pound. How many pounds of each should be in a mixture of 25 pounds?

17. A dealer mixed two grades of coffee that sold for $.55 and $.80 per pound in order to obtain a mixture to sell at $.65 per pound. If he used 33 pounds of the cheaper grade, how much of the other was used?

18. Baker Zweibach mixed two grades of cookies to get a combination to sell for 30 cents per dozen. If the mixture contained 9 dozen at 36 cents per dozen, how many dozen at 21 cents per dozen should he have used?

19. Mr. Teller travels for a gas company and is allowed 9 cents per mile for use of his car and $11 per day for food and lodging. On a recent trip, his food and lodging bill was $27 more than his mileage allowance. Find the length of the trip in days and in miles if the total for mileage, food, and lodging was $171.

20. Mr. Jiminez worked 89 days one summer and made $651. Part of the time he got $7 per day and the remainder he got $8. How many days did he work at each rate?

21. Mr. Petri invested part of $2900 at 5% and the remainder at 6%. How much was invested at each rate if the total interest was $162 per year?

22. Mr. Sauer invested $4000 and received $4178 back at the end of the year. If part of the money was invested at 4% and the remainder at 5%, how much was at each rate?

23. Mr. Bourbaki invested some money at 5% and twice that amount at 6%. How much was invested at each rate if his income from the two sources was $238 per year?

24. Mr. Real received $133 per year from two investments. Part of the investment was at 6% and $200 less at 5%. Find the amount invested at each rate.

25. Mr. Relgis made investments at 4%, 5%, and 6% and received $127 in interest annually. Find the amount invested at each rate if the total was $2400 and the sum at 6% was equal to the total at 4% and 5%.

26. Mr. Chatura invested a certain sum at 3%, three times that much at 4% and some more at 5%. The total annual interest was $155 and the total investment was $3700. How much was invested at each rate?

27. Mr. Scholz made three investments that totaled $4100. Some was at 5%, $800 less than that at 4% and the remainder at 6%. Find the amount invested at each rate if the annual interest was $223..

28. Mr. Decell received $190 from three investments. Some was at 3%, $300 less than twice that amount was at 5% and the remainder of the $3700 was at 6%. How much was at each rate?

29. The value of the 144 half dollars, quarters and dimes in a collection plate was $44. How many coins of each type were in the plate if the number of quarters was 16 less than the sum of the number of half dollars and dimes?

30. A tray contained a total of 130 quarters, dimes, and nickels that were worth an average of 10 cents each. Find the number of each if there were 10 more dimes than quarters.

31. The average age of the 30 boys on a high school football squad was 17.1 years. Find the number of each age if all were 16, 17, or 18 and the number of 17-year-olds was 4 less than the sum of those of the other ages.

32. A boy took three tests and made an average grade of 79 on them. Find the grade on each if he made 6 more on the second than on the first and 9 more on the third than on the second.

4.10. Use of Determinants in Solving Two Linear Equations in Two Unknowns

We have found how to solve a system of two linear equations in two unknowns by algebraic and geometric methods. The amount of physical labor involved in solving a system of equations can be materially reduced if we make use of an invention* that is described and illustrated in the remainder of this article

* The invention was first made by Seki Kowa, a Japanese mathematician, who lived in then isolated Japan from 1647 to 1708. It was made some ten years later by the German mathematician Leibnitz, who is generally credited with it by the Western World.

and in the next. We shall develop the method by solving the system

(1) $$ax + by = f$$

(2) $$cx + dy = s$$

for x and y by addition and subtraction since we can use the solution obtained as a formula for solving any pair of consistent equations in two unknowns. If we multiply (1) by d and (2) by b, we have

(3) $$adx + bdy = fd$$

(4) $$bcx + bdy = bs$$

$(ad - bc)x = fd - bs$, subtracting (4) from (3). Now dividing by $ad - bc$, we have

(5) $$x = \frac{fd - bs}{ad - bc}, \qquad ad - bc \neq 0.$$

Similarly,

(6) $$y = \frac{as - fc}{ad - bc}, \qquad ad - bc \neq 0.$$

If $ad - bc = 0$, the equations are dependent or inconsistent as seen from Art. 4.4.

Equations (5) and (6) can be used as formulas for solving any consistent pair of linear equations in two unknowns. They can be more readily remembered if we use

(7) $$\begin{vmatrix} a & b \\ c & d \end{vmatrix}$$

as a symbol for $ad - bc$. The reader should notice that the coefficients a, b, c, and d enter in the same relative positions in this symbol as in the given equations. The square array given in (7) is called a *determinant of order two* and its value, $ad - bc$, can be readily obtained by making use of

$$\begin{vmatrix} a & b \\ c & d \end{vmatrix} = ad - bc.$$

Each arrow indicates that the product of the two numbers connected by it is to be taken and prefixed by the indicated sign.

EXAMPLES

1.
$$\begin{vmatrix} 2 & 5 \\ 3 & 9 \end{vmatrix} = (2)(9) - (5)(3) = 18 - 15 = 3$$

2.
$$\begin{vmatrix} 2 & -7 \\ 3 & 4 \end{vmatrix} = (2)(4) - (-7)(3) = 8 + 21 = 29$$
$$ - +$$

If we make use of (5), (6), and (7), we see that the solutions of (1) and (2) can be put in the form

(8)
$$x = \frac{\begin{vmatrix} f & b \\ s & d \end{vmatrix}}{\begin{vmatrix} a & b \\ c & d \end{vmatrix}} \quad \text{and} \quad y = \frac{\begin{vmatrix} a & f \\ c & s \end{vmatrix}}{\begin{vmatrix} a & b \\ c & d \end{vmatrix}}$$

The determinant (7) is called the determinant of the coefficients and the determinant in the numerator of each of equations (8) can be obtained from it by replacing the coefficients of the unknown for which we are solving by the constant terms as they appear in the right hand members of the given equations.

EXAMPLE

3. Use determinants to solve the system

$$3x + 4y = -6$$
$$2x - 3y = 13$$

SOLUTION

Since these equations are in the form of (1) and (2), we can use (8) to solve them. Thus,

$$x = \frac{\begin{vmatrix} -6 & 4 \\ 13 & -3 \end{vmatrix}}{\begin{vmatrix} 3 & 4 \\ 2 & -3 \end{vmatrix}} = \frac{(-6)(-3) - (4)(13)}{(3)(-3) - (4)(2)} = \frac{18 - 52}{-9 - 8} = \frac{-34}{-17} = 2$$

and

$$y = \frac{\begin{vmatrix} 3 & -6 \\ 2 & 13 \end{vmatrix}}{-17} = \frac{(3)(13) - (-6)(2)}{-17} = \frac{39 + 12}{-17} = \frac{51}{-17} = -3$$

The solution is $x = 2$, $y = -3$ and is sometimes written in the form $(2, -3)$.

4.11. Use of Determinants in Solving Three Linear Equations in Three Unknowns

If we make use of determinants of order three, we can develop a formula for use in solving a system of three linear equations in three unknowns.

The square array

$$\begin{vmatrix} a_1 & b_1 & c_1 \\ a_2 & b_2 & c_2 \\ a_3 & b_3 & c_3 \end{vmatrix}$$

is called a *determinant of order* 3 since it has 3 columns and 3 rows. Its expansion or value is $a_1b_2c_3 + a_2b_3c_1 + a_3b_1c_2 - a_3b_2c_1 - a_2b_1c_3 - a_1b_3c_2$. There is one number from each row and one from each column in each of the six products in the expansion; furthermore, the product is preceded by a plus or minus sign according to whether a larger subscript is preceded by a smaller one an even or an odd number of times when the letters are in the order a, b, c. Thus in the term $a_3b_2c_1$, the 3 precedes both 2 and 1 and the 2 precedes the 1; consequently, a larger subscript precedes a smaller one three times and we write $-a_3b_2c_1$ in the expansion.

A relatively easy way to remember or obtain the expansion of a third order determinant consists of writing the determinant, recopying the first two columns, forming each product of terms connected by the arrows, and prefixing each product by a plus or minus sign as indicated. Thus,

$$\begin{vmatrix} a_1 & b_1 & c_1 \\ a_2 & b_2 & c_2 \\ a_3 & b_3 & c_3 \end{vmatrix} = \begin{bmatrix} a_1 & b_1 & c_1 & a_1 & b_1 \\ a_2 & b_2 & c_2 & a_2 & b_2 \\ a_3 & b_3 & c_3 & a_3 & b_3 \end{bmatrix}$$
$$- \quad - \quad - \quad + \quad + \quad +$$

$$= a_1b_2c_3 + a_3b_1c_2 + a_2b_3c_1 - a_3b_2c_1 - a_1b_3c_2 - a_2b_1c_3$$

The reader may prefer not to draw the arrows. If we solve the system

$$a_1x + b_1y + c_1z = d_1$$
$$a_2x + b_2y + c_2z = d_2$$
$$a_3x + b_3y + c_3z = d_3$$

for x by addition and subtraction, we find that

$$x = \frac{d_1b_2c_3 + d_2b_3c_1 + d_3b_1c_2 - d_3b_2c_1 - d_2b_1c_3 - d_1b_3c_2}{a_1b_2c_3 + a_2b_3c_1 + a_3b_1c_2 - a_3b_2c_1 - a_2b_1c_3 - a_1b_3c_2}$$

or, in terms of determinants,

$$x = \frac{\begin{vmatrix} d_1 & b_1 & c_1 \\ d_2 & b_2 & c_2 \\ d_3 & b_3 & c_3 \end{vmatrix}}{\begin{vmatrix} a_1 & b_1 & c_1 \\ a_2 & b_2 & c_2 \\ a_3 & b_3 & c_3 \end{vmatrix}} = \frac{N_x}{D}, \qquad D \neq 0.$$

If we continue the solution, we find that

$$y = \frac{\begin{vmatrix} a_1 & d_1 & c_1 \\ a_2 & d_2 & c_2 \\ a_3 & d_3 & c_3 \end{vmatrix}}{D} \quad \text{and} \quad z = \frac{\begin{vmatrix} a_1 & b_1 & d_1 \\ a_2 & b_2 & d_2 \\ a_3 & b_3 & d_3 \end{vmatrix}}{D}, \quad D \neq 0.$$

We should notice here, as in the solution of two linear equations in two unknowns, that each numerator can be obtained from the denominator D by replacing the coefficients of the unknowns for which we are solving by the constant terms as they appear in the right members of the equations.

EXAMPLE

1. Solve the following system by use of determinants

$$3x + 2y + 5z = -2$$
$$x - 3y + 2z = -5$$
$$4x + 2y + 5z = 0$$

SOLUTION

The determinant of the coefficients is

$$D = \begin{vmatrix} 3 & 2 & 5 \\ 1 & -3 & 2 \\ 4 & 2 & 5 \end{vmatrix} \begin{matrix} 3 & 2 \\ 1 & -3 \\ 4 & 2 \end{matrix}$$

$$ \quad - \quad - \quad - \quad + \quad + \quad +$$

$$= (3)(-3)(5) + (2)(2)(4) + (5)(1)(2) - (5)(-3)(4) - (3)(2)(2) - (2)(1)(5)$$

$$= -45 + 16 + 10 + 60 - 12 - 10$$

$$= 19$$

We now replace the appropriate column in D by the constant terms in order to find the numerator of the desired unknown. If we replace the coefficients of x, we get

$$N_x = \begin{vmatrix} -2 & 2 & 5 \\ -5 & -3 & 2 \\ 0 & 2 & 5 \end{vmatrix} \begin{matrix} -2 & 2 \\ -5 & -3 \\ 0 & 2 \end{matrix}$$

$$= -2(-3)(5) + (2)(2)(0) + (5)(-5)(2) - (5)(-3)(0) - (-2)(2)(2) - (2)(-5)(5)$$

$$= 30 + 0 - 50 + 0 + 8 + 50$$

$$= 38$$

Consequently, $x = \dfrac{N_x}{D} = \dfrac{38}{19} = 2$

Similarly,

$$N_y = \begin{vmatrix} 3 & -2 & 5 \\ 1 & -5 & 2 \\ 4 & 0 & 5 \end{vmatrix} \begin{matrix} 3 & -2 \\ 1 & -5 \\ 4 & 0 \end{matrix}$$

$$= (3)(-5)(5) + (-2)(2)(4) + (5)(1)(0) - (5)(-5)(4) - (3)(2)(0) - (-2)(1)(5)$$

$$= -75 - 16 + 0 + 100 - 0 + 10$$

$$= 19$$

Therefore, $y = \dfrac{N_y}{D} = \dfrac{19}{19} = 1$

We could obtain the value of z by evaluating N_z and then dividing it by D but instead shall find it by substituting the determined values of x and y in one of the given equations and then solving for z. Thus, if we substitute in the first equation, we have

$$(3)(2) + (2)(1) + 5z = -2$$

$$6 + 2 + 5z = -2$$

$$z = -2$$

The solution is $x = 2$, $y = 1$, $z = -2$ and is often written as $(2, 1, -2)$.

Exercise 4.6

Find the value of each determinant in Problems 1 through 16.

1. $\begin{vmatrix} 2 & 3 \\ 1 & 4 \end{vmatrix}$ **2.** $\begin{vmatrix} 3 & 1 \\ 2 & -1 \end{vmatrix}$ **3.** $\begin{vmatrix} 2 & 3 \\ 4 & 9 \end{vmatrix}$

4. $\begin{vmatrix} -2 & 0 \\ 3 & 1 \end{vmatrix}$ **5.** $\begin{vmatrix} -3 & 1 \\ -2 & 0 \end{vmatrix}$ **6.** $\begin{vmatrix} 0 & 2 \\ 5 & 7 \end{vmatrix}$

7. $\begin{vmatrix} 1 & -2 \\ 0 & 4 \end{vmatrix}$ **8.** $\begin{vmatrix} -3 & 5 \\ 1 & -2 \end{vmatrix}$

9. $\begin{vmatrix} 2 & 3 & 1 \\ 1 & 2 & 3 \\ 3 & 1 & 2 \end{vmatrix}$ **10.** $\begin{vmatrix} 1 & 4 & 2 \\ 3 & 1 & 2 \\ 1 & 2 & 5 \end{vmatrix}$ **11.** $\begin{vmatrix} 0 & 2 & 3 \\ 4 & 1 & 2 \\ 0 & 2 & 5 \end{vmatrix}$

12. $\begin{vmatrix} 2 & 0 & 3 \\ 0 & 2 & 1 \\ 1 & 0 & 5 \end{vmatrix}$ **13.** $\begin{vmatrix} -1 & 4 & 0 \\ 3 & 0 & -2 \\ -2 & 1 & 3 \end{vmatrix}$ **14.** $\begin{vmatrix} 5 & 0 & -2 \\ -3 & 2 & 1 \\ 1 & 0 & 3 \end{vmatrix}$

15. $\begin{vmatrix} 3 & -2 & 4 \\ -6 & 4 & -8 \\ 1 & 1 & 1 \end{vmatrix}$ **16.** $\begin{vmatrix} 1 & 0 & -3 \\ 3 & -2 & 1 \\ 2 & 0 & -6 \end{vmatrix}$

Solve the following systems of equations by use of determinants.

17. $3x - 2y = 4$
$2x + y = 5.$

18. $2x + 3y = 3$
$5x + 6y = 9.$

19. $5x - 4y = 7$
$3x + 2y = 13.$

20. $6x + 7y = 2$
$3x - y = -8.$

21. $4x + 3y = -1$
$6x + y = 2.$

22. $2x + 3y = 1$
$5x - 6y = 7.$

23. $5x - 2y = 2$
$6x + 8y = 5.$

24. $2x + 4y = 0$
$5x - 2y = -2.$

25. $2x + 3y - 5z = 3$
$3x - y + 2z = 3$
$-x + 2y = 3.$

26. $2x + 3y + z = -2$
$2x - y = 5$
$5x + 2y + 3z = -1.$

27. $x + 2y + 3z = -1$
$3y - 2z = 12$
$3x + 5z = -3.$

28. $2x - 3z = 4$
$3x + y = 0$
$2y + z = 4.$

29. $x - 2y - z = 0$
$3x + 2y + z = 2$
$7x - 6y = 5.$

30. $x + y + 5z = 1$
$2x + y + z = 1$
$3x + y - 3z = 1.$

31. $3x + 2y + 6z = 3$
$3x - 2y = 1$
$x - 4y + 3z = 0.$

32. $5x + y + 5z = 1$
$3x + 6z = 1$
$y - 10z = -1.$

5

PRODUCTS AND FACTORS

5.1. Multiplication and Division of Monomials

We have seen in our previous work that $a^2 = a \times a$ and $a^3 = a \times a \times a$. A generalization of this is given in the following definition. *If n is a positive whole number, then*

$$a^n = a \times a \times a \cdots \text{ on to } n \text{ factors.}$$

The number n is called an *exponent* and a is called the *base*.

Using the definition of a positive whole number exponent, we have

$$a^m a^n = (a \times a \cdots \text{ to } m \text{ factors})(a \times a \cdots \text{ to } n \text{ factors})$$

$$= a \times a \cdots \text{ to } (m + n) \text{ factors}$$

$$= a^{m+n}.$$

Hence, we have shown that

$$a^m a^n = a^{m+n}.$$

This may be put in words as: *In multiplying, we add exponents of like bases.*

In getting the product of ba^m and ca^n, we use the axiom which states that multiplication is commutative and have

$$ba^m \times ca^n = bca^{m+n}.$$

EXAMPLES

1. $2^3 \times 2^4 = 2^{3+4} = 2^7.$ **2.** $b^5 \times b^6 = b^{5+6} = b^{11}.$

3. $2a^4 \times 6a^9 = 12a^{13}.$

In order to see what to do with exponents of the same base in division, we shall consider the fraction a^m/a^n with m greater than n. By use of the meaning of a positive integral exponent, we get

$$\frac{a^m}{a^n} = \frac{a \times a \cdots \text{to } m \text{ factors}}{a \times a \cdots \text{to } n \text{ factors}} \qquad m > n$$

$$= a \times a \cdots \text{to } (m - n) \text{ factors, removing}$$

$$= a^{m-n}. \qquad\qquad \text{common factors,}$$

Now we have shown that

$$\frac{a^m}{a^n} = a^{m-n}, \quad m > n, a \neq 0.$$

In words, it is: *The quotient of two powers of the same nonzero base is that base with an exponent equal to the exponent of the dividend minus the exponent of the divisor.*

If we apply this rule to a^n/a^n, we get $a^{n-n} = a^0$, $a \neq 0$. Since we may need to use its value we shall give the usual definition by saying that $a^0 = 1$, $a \neq 0$.

EXAMPLES

4. $\dfrac{3^6}{3^4} = 3^{6-4} = 3^2.$ **5.** $\dfrac{b^8}{b^5} = b^{8-5} = b^3.$

6. $\dfrac{10a^7}{2a^3} = \dfrac{10}{2}\, a^{7-3} = 5a^4.$ **7.** $\dfrac{c^4}{c^4} = c^{4-4} = c^0 = 1.$

Example 6 illustrates the fact that

$$\frac{ba^m}{ca^n} = \frac{b}{c}\, a^{m-n}.$$

5.2. Multiplication of Polynomials

By the product of two polynomials, we mean the sum of all products obtained by multiplying each term of the one factor by each term of the other. In order to rearrange terms and shorten the labor involved, we use the commutative laws of addition and of multiplication and the distributive law of multiplication with respect to addition.

EXAMPLE

Find the product of

$$2x^3 - 5xy^2 - 3x^2y + 2y^3 \text{ and } 3y^2 + 4x^2 - 7xy.$$

SOLUTION

We shall first rearrange terms in each factor so that the exponents of x occur in descending order and then perform the multiplication. In the multiplication, we multiply $4x^2$ by each term in the first factor, then $-7xy$, and last $3y^2$ by each term in the first factor. In performing the multiplication by $-7xy$ and by $3y^2$, we place each partial product below similar terms that have already been obtained. For example, we place $(-7xy)(-3x^2y) = 21x^3y^2$ below the $-20x^3y^2$ that was obtained by multiplying $4x^2$ by $-5xy^2$. Finally we add coefficients of like terms. If this procedure is followed, the work of the problem appears as shown below.

$$
\begin{array}{r}
2x^3 - 3x^2y - 5xy^2 + 2y^3 \\
4x^2 - 7xy + 3y^2 \\
\hline
8x^5 - 12x^4y - 20x^3y^2 + 8x^2y^3 \\
-\,14x^4y + 21x^3y^2 + 35x^2y^3 - 14xy^4 \\
+6x^3y^2 - 9x^2y^3 - 15xy^4 + 6y^5 \\
\hline
8x^5 - 26x^4y + 7x^3y^2 + 34x^2y^3 - 29xy^4 + 6y^5
\end{array}
$$

We pointed out in Art. 3.5 that an expression is sometimes enclosed in parentheses to indicate that it is to be considered a single term. Furthermore, each of two expressions is often written in parentheses and the two pairs of parentheses written next to one another to indicate that the two expressions are to be multiplied together. Thus,

$$(2x^3 - 5xy^2 - 3x^2y + 2y^3)(3y^2 + 4x^2 - 7xy)$$

indicates that the two expressions are to be multiplied together. The product would be found exactly as in the example given earlier in this article.

5.3. Division of Polynomials

We shall consider a dividend $b - c + d$ and a divisor a and define the quotient as the number x such that

$$ax = b - c + d.$$

If we multiply each member of this equation by $1/a$ and make use of the fact that multiplication is distributive with respect to addition, we get

$$x = \frac{1}{a}(b - c + d) = \frac{b}{a} - \frac{c}{a} + \frac{d}{a}.$$

Therefore *the quotient obtained by dividing a polynomial by a monomial is the algebraic sum of the quotients obtained by dividing each term of the dividend by the divisor.*

EXAMPLE

1.
$$\frac{5x^5 - 20x^4y + 15x^3y^2 + 10x^2y^3}{5x^2} = \frac{5x^5}{5x^2} - \frac{20x^4y}{5x^2} + \frac{15x^3y^2}{5x^2} + \frac{10x^2y^3}{5x^2}$$

$$= x^3 - 4x^2y + 3xy^2 + 2y^3.$$

We shall now illustrate and explain a method to be followed in dividing one polynomial by another.

EXAMPLE

2. Divide $8x^5 - 16x^4 - 23x^2 - 16x^3 + 5 - 4x$ by $4x^3 + 3x - 1 + 2x^2$.

SOLUTION

We shall arrange both dividend and divisor in descending powers of x as a first step. Then divide the first term $4x^3$ of the divisor into the first term

$$
\begin{array}{r}
2x^2 - 5x - 3 = \text{quotient} \\
\text{divisor} = \underline{4x^3 + 2x^2 + 3x - 1}\,\big|\,8x^5 - 16x^4 - 16x^3 - 23x^2 - 4x + 5 = \text{dividend} \\
(1) \qquad 8x^5 + 4x^4 + 6x^3 - 2x^2 \\
(2) \qquad -20x^4 - 22x^3 - 21x^2 - 4x + 5 \\
(3) \qquad -20x^4 - 10x^3 - 15x^2 + 5x \\
(4) \qquad -12x^3 - 6x^2 - 9x + 5 \\
(5) \qquad \; -12x^3 - 6x^2 - 9x + 3 \\
2 = \text{remainder}
\end{array}
$$

$8x^5$ of the dividend. This gives the first term $2x^2$ of the quotient. This $2x^2$ is then multiplied by each term of the divisor and line (1) is obtained. Line (2) is obtained by subtracting line (1) from the dividend. Divide $4x^3$ into the first term of line (2) and obtain $-5x$, which is put in the quotient and multiplied by the divisor. Thus we obtain line (3), which is subtracted from line (2), producing line (4). Divide $4x^3$ into the first term of line (4) and obtain -3, which is put in the quotient and multiplied by the divisor. Thus we obtain line (5), which is subtracted from line (4), producing the remainder 2.

Exercise 5.1

Perform the indicated operations.

1. x^3x^4.

2. x^6x^2.

3. aa^3.

4. a^0a^4.

5. $2y3y^2$.

6. $5y^37x^0$.

7. $3y^22y^3$.

8. $4y^55y^4$.

9. $\dfrac{x^5}{x^2}$.

10. $\dfrac{x^8}{x^3}$.

11. $\dfrac{x^5}{x^0}$.

12. $\dfrac{x^3}{x}$.

13. $\dfrac{6x^6}{2x^0}$.

14. $\dfrac{9x^5}{3x^2}$.

15. $\dfrac{8x^7}{2x^3}$.

16. $\dfrac{12x^5}{6x^3}$.

17. $5a^3(a^2 - 3a + 4)$.

18. $-3a^2(2a^2 - 7a + 4)$.

19. $2a^5(3a^2 - 4a + 7)$.

20. $7a^4(a^2 - 5a + 3)$.

21. $(3x - y)(2x^2 - xy + y^2)$.

22. $(2x - 3y)(3x^2 - 4xy - 2y^2)$.

23. $(5a + 3b)(a^2 - 3ab - 2b^2)$.

24. $(3a - 4b)(2a^2 + 3ab - 3b^2)$.

25. $(x^2 - 3x + 5)(2x^2 - 5x + 2)$.

26. $(2x^2 - 7x + 4)(3x^2 - 2x - 3)$.

27. $(5x^2 + 2x - 3)(2x^2 + 5x - 4)$.

28. $(3x^2 + 7x - 5)(x^2 - 4x - 3)$.

29. $\dfrac{15x^5 - 20x^4 + 10x^3}{5x^2}$.

30. $\dfrac{18x^4 + 12x^3 - 30x^2}{6x^2}$.

31. $\dfrac{12x^6 - 8x^5 + 4x^4}{2x^3}$.

32. $\dfrac{9x^7 - 12x^6 - 6x^5}{3x^4}$.

33. $(6x^2 + 7x - 3) \div (2x + 3)$.

34. $(6x^2 + x - 2) \div (3x + 2)$.

35. $(15x^2 - 4x - 3) \div (5x - 3)$.

36. $(10x^2 + 21x - 10) \div (2x + 5)$.

37. $(2x^4 - x^3 - 10x^2 + 8x + 3) \div (x^2 + x - 3)$.

38. $(6x^4 + x^3 + 5x^2 + 7x - 3) \div (2x^2 - x + 3)$.

39. $(10x^4 + 13x^3 - 4x^2 + 7x - 2) \div (5x^2 - x + 2)$.

40. $(12x^4 + 5x^3 - 9x^2 + 10x - 12) \div (3x^2 + 2x - 4)$.

5.4. The Product of Two Binomials

The product of expressions of the form $ax + by$ and $cx + dy$ arises sufficiently often in mathematics that it is desirable to know or be able to get it without having to put one factor below the other and multiply. We shall first obtain the product in the usual manner and then examine a diagram which should help us to remember the product. If we multiply, we have

$$ax + by$$
$$cx + dy$$
$$\overline{acx^2 + bcxy}, \quad \text{multiplying } ax + by \text{ by } cx,$$
$$\quad\quad adxy + bdy^2, \quad \text{multiplying } ax + by \text{ by } dy,$$
$$\overline{acx^2 + (ad + bc)xy + bdy^2}, \quad \text{adding.}$$

Hence, we see that

(1) $(ax + by)(cx + dy) = acx^2 + (ad + bc)xy + bdy^2.$

This can be put in words: *There are three terms in the product of two binomials:* (1) *the product of the two first terms of the binomials,* (2) *the algebraic sum of the products obtained by multiplying the first term in each binomial by the second term in the other*, and* (3) *the product of the two second terms of the binomials.*

The product can be recalled or obtained readily if we study the following diagram. The arrows connect the terms that are multiplied together.

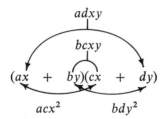

EXAMPLE

1. Find the product of $2x + 3y$ and $4x - 5y$.

SOLTUION

If we perform the steps given in italics above and indicated in the diagram, we have $(2x)(4x) = 8x^2$ as the product of the first terms of the binomials, $(2x)(-5y) + (3y)(4x) = 2xy$ as the sum of the cross products, and $(3y)(-5y) = -15y^2$ as the product of the second terms; hence, we know that

$$(2x + 3y)(4x - 5y) = 8x^2 + 2xy - 15y^2.$$

* These products are often called the *cross products*.

There are two special cases of the product of two binomials which occur sufficiently often to deserve separate consideration. One of them is that in which $c = a$ and $d = b$. If these replacements are made in (1), and if we make use of the facts that $(ax)(ax) = aaxx = a^2x^2$ and $(by)(by) = b^2y^2$, we get

$$(2) \qquad (ax + by)(ax + by) = a^2x^2 + 2abxy + b^2y^2.$$

We can show by multiplication that

$$(2') \qquad (ax - by)(ax - by) = a^2x^2 - 2abxy + b^2y^2.$$

We call (2) or (2') the *square of a binomial*. We can put them in words: *The square of a binomial is the square of the first term plus or minus twice the product of the two terms plus the square of the second.*

EXAMPLES

2. $(2x + 3y)^2 = (2x)^2 + 2(2x)(3y) + (3y)^2$
$\qquad\qquad = 4x^2 + 12xy + 9y^2.$

3. $(3x - 5y)^2 = (3x)^2 + 2(3x)(-5y) + (-5y)^2$
$\qquad\qquad = 9x^2 - 30xy + 25y^2.$

The other special case of (1), which deserves attention is that in which $a = b = c = -d = 1$. If these replacements are made in (1), it becomes

$$(3) \qquad (x + y)(x - y) = x^2 - y^2$$

and is the product of the sum and the difference of two numbers. In words, it is: *The product of the sum and the difference of two numbers is the square of the first minus the square of the second.*

EXAMPLES

4. $(s + t)(s - t) = s^2 - t^2.$

5. $(3x + 4y)(3x - 4y) = (3x)^2 - (4y)^2$
$\qquad\qquad = 9x^2 - 16y^2.$

Exercise 5.2

Find the indicated products.

1. $(x + 3)(x - 3).$ **2.** $(x - 5)(x + 5).$

3. $(w - 2)(w + 2).$ **4.** $(t + 6)(t - 6).$

5. $(2x + y)(2x - y).$ **6.** $(3a - b)(3a + b).$

7. $(4x - y)(4x + y).$ **8.** $(5x + w)(5x - w).$

9. $(3x + 2y)(3x - 2y)$.   **10.** $(2x + 5y)(2x - 5y)$.

11. $(5x - 3y)(5x + 3y)$.   **12.** $(4x - 3y)(4x + 3y)$.

13. $(x - 3)^2$,   **14.** $(x - 4)^2$.   **15.** $(x + 5)^2$.

16. $(x + 2)^2$.   **17.** $(3a - 1)^2$.   **18.** $(2a + 1)^2$.

19. $(5w + 1)^2$.   **20.** $(4w + 1)^2$.   **21.** $(3x - 2y)^2$.

22. $(2a - 5b)^2$.   **23.** $(5s - 3t)^2$.   **24.** $(4x + 3w)^2$.

25. $(x + 3)(x - 2)$.   **26.** $(a - 4)(a - 3)$.

27. $(b + 5)(b + 3)$.   **28.** $(y + 4)(y - 2)$.

29. $(2x + 1)(3x - 1)$.   **30.** $(3a - 1)(4a - 1)$.

31. $(5b + 1)(2b + 1)$.   **32.** $(2y - 1)(4y + 1)$.

33. $(3a - 2)(2a + 3)$.   **34.** $(5x - 3)(2x - 5)$.

35. $(4b + 3)(2b + 5)$.   **36.** $(2x - 7)(3x + 5)$.

37. $(2a + 3b)(3a - 2b)$.   **38.** $(4x - 3b)(5x + 4b)$.

39. $(4x - 5y)(3x - 4y)$.   **40.** $(7x + 4y)(2x + 5y)$,

5.5. Factors of $Ax^2 + Bxy + Cy^2$ *Know*

In the last article, we were given the factors and were asked to find the product. We shall now consider the problem of finding the factors provided the product is given.

If we interchange right and left members of equation (3) of the last article it becomes

$$(1) \qquad\qquad x^2 - y^2 = (x + y)(x - y)$$

in symbols and can be put in words: *The factors of the difference of the squares of two numbers are the sum of the two numbers and the difference of them.*

EXAMPLE

1.
$$36a^2 - 25b^2 = (6a)^2 - (5b)^2$$
$$= (6a + 5b)(6a - 5b).$$

If we interchange right and left members of both (2) and (2′) of Art. 5.4, we get

$$(2) \qquad\qquad (ax)^2 \pm 2abxy + (by)^2 = (ax \pm by)^2$$

since

$$(ax + by)(ax + by) = (ax + by)^2$$

and
$$(ax - by)(ax - by) = (ax - by)^2.$$

Hence *if a trinomial is made up of the square of one number plus the square of another plus or minus twice the product of the two numbers, then it is a perfect square and is the square of the sum or the difference of the two numbers.*

EXAMPLES

2. $w^2 + 6w + 9 = w^2 + 2(w)3 + 3^2$
$$= (w + 3)^2.$$

3. $9s^2 - 30st + 25t^2 = (3s)^2 - 2(3s)(5t) + (5t)^2$
$$= (3s - 5t)^2.$$

If we interchange the members of (1) of Art. 5.4, we get

(3) $$acx^2 + (ad + bc)xy + bdy^2 = (ax + by)(cx + dy).$$

A study of this equation shows that *in order to factor a quadratic trinomial, we must find two linear functions* $ax + by$ *and* $cx + dy$ *such that*

    (i) *the product of the coefficients of* x *is the coefficient of* x^2,

    (ii) *the product of the coefficients of* y *is the coefficient of* y^2, *and*

    (iii) *the sum of the two products obtained by multiplying the coefficient of* x *in one trial factor by the coefficient of* y *in the other is the coefficient of* xy.

It is a relatively easy job to find two numbers as trial coefficients of x such that their product is the coefficient of x^2 and also to find trial coefficients of y such that their product is the coefficient of y^2, but it may require a systematic attack on the problem to determine these four coefficients in such a way that the sum of the cross products is the coefficient of xy. There may not be integers which satisfy the condition, but if there are, we can be certain of finding the right combination of integral coefficients of x and y by using each possible pair of factors for x with each possible pair for y.

EXAMPLE

4. Find the factors of $6x^2 + xy - 15y^2$.

SOLUTION

The coefficients of x must have the same sign, since the coefficient of x^2 is positive; furthermore, the coefficients of y must be of opposite sign since the coefficient of y^2 is negative. If we use positive numbers for the coefficients of x, the possibilities are 3 and 2 or 6 and 1. The possible coefficients for y are 15 and -1, -15 and 1, 3 and -5 and -3 and 5. If we use 3 and 2 as trial coefficients of x along with 15 and -1 for y, the sum of the cross products is $(3)(-1) + 2(15) = 27 \neq 1$ if 3 and 15 are used in the same factor and $3(15) + 2(-1) = 43 \neq 1$ if 3 and -1 are in the same factor; hence, these combinations are not what we want. We could try 3 and 2 along with -15 and 1 or along with 3 and -5 but would find that neither gives the correct sum for the cross products. If we use 3 and 2 as coefficients for x along with -3 and 5 for y, the sum of the cross products is $3(5) + 2(-3) = 9 \neq 1$ if 3 and -3 are in the same factor and $3(-3) + 2(5) = 1$ if 3 and 5 are in the same factor. Since this is the desired coefficient of xy, we know that

$$6x^2 + xy - 15y^2 = (3x + 5y)(2x - 3y).$$

It would be comforting to know before beginning work that the given expression has factors of the form $ax + by$ and $cx + dy$, with integral coefficients. There is a test which can be applied to an expression of the form $Ax^2 + Bxy + Cy^2$ to determine whether it can be factored into factors of the type mentioned above and we shall now give and then apply it even though we will not be in a position to justify it until after we have studied equations of the form $Ax^2 + Bx + C = 0$.

Test. *The expression $Ax^2 + Bxy + Cy^2$, where A, B, C are integers, can be factored into two expressions of the form $ax + by$ and $cx + dy$ with integral coefficients if and only if $B^2 - 4AC$ is a perfect square.*

Know
Rule

This test is applicable to $Ax^2 + Bx + C$ since $Ax^2 + Bxy + Cy^2$ reduces to that if $y = 1$.

EXAMPLES

5. The expression $3x^2 + 2xy - 5y^2$ can be factored, since $B^2 - 4AC = 2^2 - 4(3)(-5) = 64 = 8^2$ is a perfect square. The factors are $3x + 5y$ and $x - y$.
6. In $3x^2 + 6x - 5$, $B^2 - 4AC = 6^2 - 4(3)(-5) = 96$ is not a perfect square; hence, we cannot factor the given expression into linear factors with integral coefficients.

Exercise 5.3

Factor the expression in each of Problems 1 through 36.

1. $x^2 - 4$.   **2.** $x^2 - 9$.   **3.** $x^2 - 16$.

4. $x^2 - 25$.   **5.** $4a^2 - b^2$.   **6.** $9a^2 - b^2$.

7. $36b^2 - c^2$.   **8.** $49x^2 - y^2$.   **9.** $4x^2 - 9y^2$.

10. $25x^2 - 16y^2$.   **11.** $9x^2 - 49y^2$.   **12.** $16x^2 - 81y^2$.

13. $x^2 - 4x + 4$.   **14.** $x^2 - 6x + 9$.   **15.** $x^2 + 10x + 25$.

16. $x^2 + 14x + 49$.   **17.** $9t^2 + 6t + 1$.   **18.** $25y^2 + 10y + 1$.

19. $16a^2 - 8a + 1$.   **20.** $49a^2 - 14a + 1$.   **21.** $9x^2 - 12xy + 4y^2$.

22. $4x^2 + 20xy + 25y^2$.

23. $25s^2 + 30st + 9t^2$.

24. $16x^2 - 24xy + 9y^2$.

25. $x^2 + x - 6$.

26. $a^2 - 7a + 12$.

27. $x^2 + 8x + 15$.

28. $x^2 + 2x - 8$.

29. $6x^2 + 5x + 1$.

30. $12a^2 - 7a + 1$.

31. $10b^2 + 7b + 1$.

32. $8y^2 - 2y - 1$.

33. $6a^2 + 5a - 6$.

34. $10a^2 - 31a + 15$.

35. $8a^2 + 26a + 15$.

36. $6a^2 - 11a - 35$.

Test each of the following to see if it is factorable into two linear factors with integral coefficients. If it can be factored, find the factors. If it cannot be so factored, make a statement to that effect.

37. $x^2 - 5x + 3$.

38. $x^2 + 8x + 11$.

39. $x^2 + 8x + 12$.

40. $x^2 - 6x + 5$.

41. $4x^2 - 12x + 9$.

42. $9x^2 + 30x + 25$.

43. $16x^2 - 10x - 1$.

44. $9x^2 + 7x + 1$,

45. $4x^2 - 20xy + 27y^2$.

46. $16x^2 - 24xy + 11y^2$.

47. $6x^2 - 11xy + 4y^2$.

48. $15x^2 + 4xy - 4y^2$.

5.6. A Power of a Power and of a Product

We discussed multiplying two powers of the same base and dividing one power of a base by another power of the same base in Art. 5.1 and shall now continue our study of exponents by raising a power to a power. In order to see how to do this, we must recall that a positive integral exponent indicates the number of times the base is to be used as a factor. Hence the exponent q in $(a^p)^q$ indicates that a^p is to be used q times as a factor. Thus

$$(a^p)^q = a^p a^p \cdots \text{ to } q \text{ factors}$$
$$= a^{p+p+} \cdots \text{ to } q \text{ terms}$$
$$= a^{pq}.$$

Now we have shown for p and q positive integers that

$$(a^p)^q = a^{pq}$$

and can put it in words: *The exponent of a power of a power of a base is the product of the exponents in the two powers.*

EXAMPLES

1. $(a^3)^2 = a^{(3)(2)} = a^6$.   **2.** $(b^4)^5 = b^{(4)(5)} = b^{20}$.

There are times when a power of a power is given and we want to break it down. This may be possible in more than one way.

EXAMPLES

3. $a^{10} = (a^5)^2 = (a^2)^5$, since $(5)(2) = (2)(5) = 10$.

4. $a^{12} = (a^3)^4 = (a^4)^3 = (a^2)^6 = (a^6)^2$, since each of the indicated products is 12. The one of these forms to be used depends on the conditions under which we are working.

5. We can think of $x^6 - y^8$ as the difference of two squares if we use $x^6 = (x^3)^2$ and $y^8 = (y^4)^2$; then

$$x^6 - y^8 = (x^3)^2 - (y^4)^2$$
$$= (x^3 + y^4)(x^3 - y^4).$$

It is readily seen that

$$(ab)^n = a^n b^n,$$

since

$$(ab)^n = ab \times ab \cdots \text{ to } n \text{ factors}$$
$$= (aa \cdots \text{ to } n \text{ factors})(bb \cdots \text{ to } n \text{ factors})$$
$$= a^n b^n.$$

EXAMPLE

6. $(2b^3)^2 = 2^2 b^6$.

5.7. Factors of $x^3 \pm y^3$

We shall divide $x^3 - y^3$ by $x - y$ to see if the latter is a factor of the former and to find the other factor if $x - y$ is one.

$$\begin{array}{r}
x^2 + xy + y^2 \\
x - y \overline{\smash{\big)}\ x^3 - y^3} \\
\underline{x^3 - x^2 y } \\
x^2 y - y^3 \\
\underline{x^2 y - xy^2 } \\
xy^2 - y^3 \\
\underline{xy^2 - y^3}
\end{array}$$

We have shown that

$$x^3 - y^3 = (x - y)(x^2 + xy + y^2).$$

In words, this states that *the difference of the cubes of two numbers has the difference of the numbers as one factor and the square of the first plus the product of the two numbers plus the square of the second as the other.*

EXAMPLE

1. Since $x^3 - 27$ can be thought of as $x^3 - 3^3$, its factors are $x - 3$ and $x^2 + 3x + 3^2$.

We could find by dividing $x^3 + y^3$ by $x + y$ that

$$x^3 + y^3 = (x + y)(x^2 - xy + y^2)$$

and could put this in words by saying that *the sum of the cubes of two numbers has the sum of the numbers as one factor and the square of the first minus the product of the two numbers plus the square of the second as the other.*

EXAMPLES

2. We can think of $8x^3 + 1$ as $(2x)^3 + 1^3$ since the cube of the product of two factors is the product of their cubes; hence,

$$\begin{aligned}
8x^3 + 1 &= (2x)^3 + 1^3 \\
&= (2x + 1)[(2x)^2 - (2x)(1) + 1^2]* \\
&= (2x + 1)(4x^2 - 2x + 1).
\end{aligned}$$

3. $\begin{aligned}[t]
64a^3 - (b - c)^3 &= (4a)^3 - (b - c)^3 \\
&= [4a - (b - c)][(4a)^2 + (4a)(b - c) + (b - c)^2] \\
&= [4a - b + c][16a^2 + 4ab - 4ac + b^2 - 2bc + c^2].
\end{aligned}$

4. $\begin{aligned}[t]
m^9 - n^6 &= (m^3)^3 - (n^2)^3 \\
&= (m^3 - n^2)[(m^3)^2 + m^3 n^2 + (n^2)^2] \\
&= (m^3 - n^2)[m^6 + m^3 n^2 + n^4].
\end{aligned}$

5.8. Common Factors, Factors by Grouping

Each term of an expression that is to be factored may contain a factor in common with every other term, or it may be possible to group the terms so that this is the situation. If either of these conditions exist, the common factor should be removed but not discarded and then any other possible factoring done. If a common factor does not exist and the expression to be factored contains more than four terms, factoring if possible can usually be accomplished by grouping the terms.

* The symbols [ ] are called *brackets* and are used and removed just as parentheses are.

EXAMPLES

1. The expression $a^3 + 6a^2b + 8ab^2$ contains the common factor a. If this is removed, we have

$$a^3 + 6a^2b + 8ab^2 = a(a^2 + 6ab + 8b^2)$$
$$= a(a + 2b)(a + 4b).$$

2. If, in $x^3 + x^2 - y^3 - y^2$, we group x^3 and $-y^3$ and also group x^2 and $-y^2$, we get

$$x^3 + x^2 - y^3 - y^2 = (x^3 - y^3) + (x^2 - y^2), \text{ grouping}$$
$$= (x - y)(x^2 + xy + y^2) + (x - y)(x + y),$$

factoring each group,

$$= (x - y)(x^2 + xy + y^2 + x + y),$$

removing the common factor.

3. If we break $p^2 - q^2 + 10q - 25$ into two groups consisting of the first term and of the other three terms, we have

$$p^2 - q^2 + 10q - 25 = p^2 - (q^2 - 10q + 25), \text{ grouping,}$$
$$= p^2 - (q - 5)^2, \text{ factoring one group,}$$
$$= (p + q - 5)(p - q + 5).$$

Exercise 5.4

Raise to the indicated power in each of Problems 1 through 12.

1. $(x^3)^2$.   **2.** $(x^4)^3$.   **3.** $(x^5)^4$.   **4.** $(x^3)^3$.

5. $(3a)^2$.   **6.** $(2b)^3$.   **7.** $(6a)^2$.   **8.** $(2c)^5$.

9. $(2^3x^2)^3$.   **10.** $(3^2x^4)^2$.   **11.** $(a^2b^3)^3$.   **12.** $(a^4b^5)^2$.

Express each power in Problems 13 through 20 as an integral power of an integral power in as many ways as possible without using one as an exponent.

13. a^6.   **14.** a^{15}.   **15.** a^9.   **16.** a^{35}.

17. a^{12}.   **18.** a^{30}.   **19.** a^{27}.   **20.** a^{45}.

Factor each of the following.

21. $a^3 - b^3$.   **22.** $x^3 - y^3$.   **23.** $c^3 + d^3$.

24. $m^3 + n^3$.   **25.** $8c^3 + 27d^3$.   **26.** $27a^3 + 64d^3$.

27. $125a^3 - 8b^3$.   **28.** $64x^3 - 125y^3$.

29. $(a - b)^3 - c^3$ **30.** $(2a - b)^3 + 8c^3$.

31. $27a^3 + (2b + c)^3$.   **32.** $64a^3 - (3b + 2c)^3$.

33. $ax^2 - 5ax + 6a$.

34. $ay^2 - 3ay + 2a$.

35. $ax^2 + 2a^2x + a^3$.

36. $ax^2 + 3a^2x + 2a^3$.

37. $x^2 + 6x + 9 - 4y^2$.

38. $a^2 - 4a + 4 - 9b^2$.

39. $4x^2 - 4y^2 + 12y - 9$.

40. $9b^2 - 9a^2 + 12a - 4$.

41. $6a - 3ba + 2b^2 - 4b$.

42. $2a^2 + 4ab + 3a + 6b$.

43. $6a^2 - 4a - 9ab + 6b$.

44. $8a^2 + 6ab - 15b - 20a$.

45. $a^2 - 2ab + ac - bc + b^2$.

46. $x^2 - 4xy + zx - 2zy + 4y^2$.

47. $a^3 - a^2 - 2ab - b^2 + b^3$.

48. $8a^3 + 4a^2 - 4ab + b^2 - b^3$.

49. $2x^2 + 3x + xy - 2 + 2y$.

50. $3x^2 - x - 3yx - 2 - 2y$.

51. $4x^2 - 16x + 2xy + 15 - 3y$.

52. $6x^2 + 5x - 2xy - 3y - 6$.

6

ALGEBRAIC FRACTIONS; FRACTIONAL EQUATIONS

6.1. Multiplication of Fractions

We discussed arithmetic fractions in Art. 1.9 and pointed out that the value of a fraction is not changed if numerator and denominator are both multiplied by the same nonzero number since that, in effect, is multiplying the fraction by one. We shall use the same principle in dealing with algebraic fractions; in fact, we shall merely replace nonzero number by nonzero algebraic expression in the statement. Here, as in working with numbers, the product of any number of fractions is a fraction whose numerator is the product of the numerators of the separate fractions and whose denominator is the product of their denominators. We should remove any factor that occurs in both a numerator and a denominator before finding the product; consequently, we should factor each numerator and each denominator if possible.

EXAMPLES

1.
$$\frac{x^2 - 5x + 6}{x^2 + 2x - 3} \left\lvert \frac{x^2 + 5x + 6}{x^2 - 4} \right. = \frac{(x-2)(x-3)}{(x-1)(x+3)} \frac{(x+2)(x+3)}{(x+2)(x-2)},$$

factoring,

$$= \frac{x-3}{x-1}, \text{ removing common factors,}$$

provided that $x + 3$, $x + 2$, $x - 2$ are all nonzero.

2. If we factor each numerator and each denominator of

$$\frac{3x^2 - 7x - 6}{9x^2 - 4} \frac{4x^2 - 1}{2x^2 - 7x + 3} \frac{x^2 + 7x + 10}{4x^2 + 8x + 3}$$

we get

$$\frac{(3x+2)(x-3)}{(3x+2)(3x-2)} \frac{(2x+1)(2x-1)}{(x-3)(2x-1)} \frac{(x+2)(x+5)}{(2x+3)(2x+1)}.$$

In order to make it easier to see what remains after taking out each term that is common to a numerator and a denominator, we have marked through (canceled out) each such term. The remaining factors in the numerator and denominator are

$$\frac{(x+2)(x+5)}{(3x-2)(2x+3)};$$

hence, that is the product of the given fractions, provided that $3x + 2$, $x - 3$, $2x - 1$, $2x + 1$ are all nonzero.

6.2. Division of Fractions

Just after Example 5 in Art. 1.9, we derived a procedure for dividing one fraction by another and shall repeat it here. If we multiply both numerator and denominator by d/c, we have

$$\frac{a}{b} \div \frac{c}{d} = \frac{ad}{bc} \div \frac{cd}{dc}$$

$$= \frac{ad}{bc} \div 1, \quad \frac{cd}{dc} = 1,$$

$$= \frac{ad}{bc}.$$

Consequently, *in order to divide one fraction by another, we invert the one we are dividing by and multiply.*

EXAMPLES

1. $$\frac{2a^4}{5b^5} \div \frac{6a^2}{7b^6} = \frac{2a^4}{5b^5} \cdot \frac{7b^6}{6a^2}, \text{ inverting,}$$

$$= \frac{\cancel{2}\ 7}{\cancel{5}(2)(3)} a^{4-2} b^{6-5} = \frac{7}{15} a^2 b.$$

2. $$\frac{6x^2 - 13x + 5}{3x^2 + x - 10} \div \frac{4x^2 - 1}{3x^2 + 11x + 10} = \frac{6x^2 - 13x + 5}{3x^2 + x - 10} \cdot \frac{3x^2 + 11x + 10}{4x^2 - 1},$$

inverting,

$$= \frac{(3x-5)(2x-1)}{(3x-5)(x+2)} \cdot \frac{(3x + 5)(x+2)}{(2x-1)(2x + 1)},$$

factoring,

$$= \frac{3x + 5}{2x + 1},$$

provided that no factor removed is zero.

Exercise 6.1

Perform the indicated operations.

1. $\dfrac{a^5}{b^2} \times \dfrac{a^3}{b^4}.$
    2. $\dfrac{a^7}{b^0} \times \dfrac{a^4}{b^2}.$

3. $\dfrac{a^4}{b^3} \times \dfrac{b^2}{a}.$
    4. $\dfrac{a^8}{b^5} \times \dfrac{b^7}{a^3}.$

5. $\dfrac{3c^3}{2d^4} \times \dfrac{6c^2}{3d^5}.$
    6. $\dfrac{5c^5}{12d^2} \times \dfrac{6c^2}{15d^3}.$

7. $\dfrac{7c^4}{3d^3} \times \dfrac{12d^5}{14c^0}.$
    8. $\dfrac{13c^7}{7d^5} \times \dfrac{14d^2}{39c^6}.$

9. $\dfrac{5a^4}{3x^2} \div \dfrac{15a^4}{6x^3}.$
    10. $\dfrac{12b^7}{35y^4} \div \dfrac{18b^3}{7y^6}.$

11. $\dfrac{11c^7}{7w^6} \div \dfrac{33c^5}{14w^3}.$
    12. $\dfrac{6d^6}{8z^7} \div \dfrac{18d}{12z^5}.$

13. $\dfrac{4x - 6y}{9x - 6y} \times \dfrac{15x - 10y}{8x - 12y}.$
    14. $\dfrac{4x + 8y}{4x - 2y} \times \dfrac{2x - y}{3x + 6y}.$

15. $\dfrac{12x + 8y}{x - y} \times \dfrac{3x - 3y}{9x + 6y}.$
    16. $\dfrac{6x - 3y}{20x - 8y} \times \dfrac{25x - 10y}{12x - 6y}.$

17. $\dfrac{3x + 6y}{3x + 3y} \div \dfrac{4x + 8y}{6x + 6y}.$

18. $\dfrac{9x + 15y}{18x + 30y} \div \dfrac{4x - 6y}{6x - 9y}.$

19. $\dfrac{10x + 35y}{7y + 2x} \div \dfrac{8x + 28y}{28y + 8x}.$

20. $\dfrac{30x + 6y}{6x - 8y} \div \dfrac{10x + 2y}{9x - 12y}.$

21. $\dfrac{2x - 5y}{9x^2 - y^2} \times \dfrac{3x + y}{4x^2 - 25y^2}.$

22. $\dfrac{9a^2 - 16b^2}{4a + 3b} \times \dfrac{16a^2 - 9b^2}{3a - 4b}.$

23. $\dfrac{5x - a}{3x + 2a} \times \dfrac{9x^2 - 4a^2}{25x^2 - a^2}.$

24. $\dfrac{3a - 5y}{49a^2 - 4y^2} \times \dfrac{7a + 2y}{9a^2 - 25y^2}.$

25. $\dfrac{x(x^2 - y^2)}{(x - y)^3} \div \dfrac{(x + y)^2}{(x - y)^2 y}.$

26. $\dfrac{2x + 6y}{x^2 - 9y^2} \div \dfrac{x^2 + 3xy}{3x - 9y}.$

27. $\dfrac{x^2 + 2x}{2x^2 - x} \div \dfrac{3x + 6}{4x - 2}.$

28. $\dfrac{2x - 4y}{6x - 3y} \div \dfrac{x - 2y}{4x - 2y}.$

29. $\dfrac{x^2 + x - 6}{x^2 - x - 2} \times \dfrac{x^2 + 4x + 3}{x^2 + 7x + 12}.$

30. $\dfrac{x^2 - x - 6}{x^2 + 3x + 2} \times \dfrac{x^2 + 5x + 4}{x^2 + 2x - 15}.$

31. $\dfrac{10x^2 - 21x - 10}{6x^2 - 17x + 5} \times \dfrac{3x^2 + 5x - 2}{10x^2 + 9x + 2}.$

32. $\dfrac{12x^2 - 17x + 6}{12x^2 - x - 6} \times \dfrac{3x^2 + 7x - 6}{9x^2 + 9x + 2}.$

33. $\dfrac{x^2 + x - 2}{x^2 + 4x + 3} \div \dfrac{x^2 - x - 6}{x^2 - x - 1}.$

34. $\dfrac{x^2 + 2x + 3}{x^2 + 2x - 3} \div \dfrac{x^2 + x + 5}{x^2 + x - 6}.$

35. $\dfrac{x^2 + x - 6}{3x^2 - 5x - 2} \div \dfrac{2x^2 + 5x - 3}{3x^2 - 5x - 2}.$

36. $\dfrac{6x^2 - x - 2}{2x^2 + x - 6} \div \dfrac{6x^2 + x - 1}{4x^2 - 8x + 3}.$

37. $\dfrac{a^2 - b^2}{a^3 - b^3} \times \dfrac{a^3 + a^2 b + ab^2}{a^2 + 3ab + 2b^2}.$

38. $\dfrac{a^3 + b^3}{a^2 - b^2} \times \dfrac{a^2 + 2ab - 3b^2}{a^2 b - ab^2 + b^3}.$

39. $\dfrac{a^3 - 8b^3}{8a^3 + b^3} \times \dfrac{2a^2 + 3ab + b^2}{a^2 - ab - 2b^2}.$

40. $\dfrac{a^3 + 27}{a^3 - 64} \times \dfrac{2a^2 - 9a + 4}{2a^2 + 5a - 3}.$

41. $\dfrac{3x^2 + 2x - 1}{3x^2 + 8x - 3} \div \dfrac{x^3 + 1}{x^3 + 27}.$

42. $\dfrac{2x^2 - 5x + 2}{2x^2 + 9x - 5} \div \dfrac{x^3 - 8}{x^3 + 125}.$

43. $\dfrac{2x^2 + 9x + 4}{x^3 + 64} \div \dfrac{2x^2 - 7x - 4}{4x^2 - 1}.$

44. $\dfrac{x^2 - 9}{x^3 + 27} \div \dfrac{x^3 - 27}{x^2 + 3x + 9}.$

6.3. Addition of Fractions

In order to be able to add two or more algebraic fractions, they must have the same denominator just as in addition of arithmetic fractions. Any expression that has each denominator of several fractions as a factor is

called a *common denominator*. The common denominator of lowest degree*
is called the *lowest common denominator* and is abbreviated as LCD. We can
get a common denominator by multiplying all the denominators together,
but this is not necessarily the lowest common denominator; in fact, it is not
the lowest if two or more denominators have a factor in common. The
denominators should be factored before finding the lowest common denominator. Then the LCD is the product of all distinct factors that occur in any
denominator with each factor entering to the highest power to which it
enters in any denominator.

EXAMPLE

1. If the denominators are $(2x - 1)(3x + 2)^2$, $(2x - 1)^3(x + 3)$, and $(3x + 2)$
$(x + 1)$, then the common denominator is

$$(2x - 1)^3(3x + 2)^2(x + 3)(x + 1),$$

since that is the product of all distinct linear factors that occur in any denominator
and each linear factor is raised to the highest power to which it enters in any
given denominator.

If we are to add two or more fractions that do not have a common
denominator, our first task is to reduce them to a common denominator.
We then make use of the fact that the sum of several fractions with a common
denominator is a fraction whose denominator is the common denominator
and whose numerator is the sum of the numerators of the separate fractions.

EXAMPLE

2. Find the sum indicated by

$$\frac{2}{3} + \frac{a + b}{2a} + \frac{b - ab}{a^2} + \frac{2 - b}{b}.$$

SOLUTION

The common denominator is $(3)(2)(a^2)(b) = 6a^2b$, and all the fractions will
have that denominator if we multiply each denominator by the number of times it
goes into the common denominator. Thus we multiply 3 by $6a^2b \div 3 = 2a^2b$, the
second denominator by $6a^2b \div 2a = 3ab$, the third by $6a^2b \div a^2 = 6b$, and the last
by $6a^2b \div b = 6a^2$. The numerator of each fraction must be multiplied by the same
quantity as the denominator so as to leave the fraction unaltered in value. Now
multiplying each numerator and denominator by the factor needed to reduce the

---

* The degree of an algebraic expression is the largest of the numbers obtained by adding
the exponents of all letters representing variables that appear in a term. Thus the degree of
$2x^3y^2 + 7x^4 + 3x^2y^2$ is 5.

fractions to a common denominator, we have

$$\frac{2}{3} + \frac{a+b}{2a} + \frac{b-ab}{a^2} + \frac{2-b}{b}$$

$$= \frac{2}{3} \frac{2a^2b}{2a^2b} + \frac{a+b}{2a} \times \frac{3ab}{3ab} + \frac{b-ab}{a^2} \frac{6b}{6b} + \frac{2-b}{b} \times \frac{6a^2}{6a^2}$$

$$= \frac{4a^2b + 3a^2b + 3ab^2 + 6b^2 - 6ab^2 + 12a^2 - 6a^2b}{6a^2b}$$

$$= \frac{12a^2 + a^2b - 3ab^2 + 6b^2}{6a^2b}.$$

3. In adding

$$\frac{1}{x^2 - y^2}, \quad \frac{3}{x^2 + 3xy + 2y^2}, \quad \text{and} \quad \frac{2}{x^2 - 2xy + y^2},$$

we must get a common denominator; hence, we factor the denominators. In factored form, they are

$$x^2 - y^2 = (x - y)(x + y),$$
$$x^2 + 3xy + 2y^2 = (x + 2y)(x + y), \quad \text{and}$$
$$x^2 - 2xy + y^2 = (x - y)^2.$$

Consequently the lowest common denominator is $(x - y)^2(x + y)(x + 2y)$. Now multiplying each denominator by the quotient obtained by dividing it into the common denominator and multiplying each numerator by the same expression the denominator is multiplied by, we have

$$\frac{1}{x^2 - y^2} + \frac{3}{x^2 + 3xy + 2y^2} + \frac{2}{x^2 - 2xy + y^2}$$

$$= \frac{1}{(x - y)(x + y)} \times \frac{(x - y)(x + 2y)}{(x - y)(x + 2y)} + \frac{3}{(x + y)(x + 2y)} \times \frac{(x - y)^2}{(x - y)^2}$$

$$+ \frac{2}{(x - y)^2} \times \frac{(x + y)(x + 2y)}{(x + y)(x + 2y)}$$

$$= \frac{1(x^2 + xy - 2y^2) + 3(x^2 - 2xy + y^2) + 2(x^2 + 3xy + 2y^2)}{(x - y)^2(x + y)(x + 2y)}$$

$$= \frac{6x^2 + xy + 5y^2}{(x - y)^2(x + y)(x + 2y)}, \quad \text{removing parentheses and collecting.}$$

6.4. Equations That Involve Fractions

If an equation that involves fractions contains the unknown in the denominator it is called a *fractional equation*. The first step in solving any equation that involves fractions is to eliminate the fractions. This can be done by multiplying both members of the equation by the lowest common denominator of the fractions which occur in the equation. This procedure is called *clearing of fractions* and may produce an equation with roots that the original did not have and may produce an equation that is not linear. If the equation that is produced by clearing of fractions is linear, we can solve it as in Art. 3.2 but must check the value we get for the unknown to see if it is a root of the given equation.

EXAMPLES

1. Solve

$$\frac{2}{x-1} - \frac{3}{x(x-1)} = \frac{1}{x}.$$

SOLUTION

The lowest common denominator is $x(x-1)$, and multiplying through by it gives

$$2x - 3 = 1(x-1) = x - 1$$
$$x = 2.$$

We must check this possible solution by substituting it in the given equation. If we do this, the left member becomes

$$\frac{2}{2-1} - \frac{3}{2(2-1)} = \frac{2}{1} - \frac{3}{2} = \frac{1}{2},$$

and the right member is also $\frac{1}{2}$; hence $x = 2$ is a solution since it makes the two members equal.

2. Solve

$$\frac{x}{x+1} + \frac{4}{3x-4} = 1.$$

SOLUTION

The LCD is $(x+1)(3x-4)$, and if we multiply through by it, we get

$$x(3x-4) + 4(x+1) = (x+1)(3x-4).$$

Hence removing parentheses gives

$$3x^2 - 4x + 4x + 4 = 3x^2 - x - 4,$$
$$x = -8, \text{ collecting like terms.}$$

If we substitute -8 for x in the left member, we have

$$\frac{-8}{-8+1} + \frac{4}{3(-8)-4} = \frac{8}{7} + \frac{4}{-28}$$

$$= \frac{8}{7} - \frac{1}{7} = 1.$$

Therefore, the possible solution checks since
the two members are equal for $x = -8$.

3. Solve

$$\frac{2}{x+1} - \frac{1}{x+2} = \frac{2}{(x+2)(x+1)}.$$

SOLUTION

If we multiply through by the LCD, $(x+1)(x+2)$, we get

$$2(x+2) - 1(x+1) = 2$$
$$2x + 4 - x - 1 \quad = 2, \quad \text{removing parentheses,}$$
$$x = -1, \quad \text{solving for } x.$$

Consequently, if the given equation has a solution, it is $x = -1$. We shall now see whether this possibility is a root by substituting -1 for x in the given equation. Thus, we get

$$\frac{2}{-1+1} - \frac{1}{-1+2} \quad \text{from the left member and}$$

$$\frac{2}{(-1+2)(-1+1)} \quad \text{from the right member.}$$

Therefore, we can not admit -1 as a root since $1 - 1 = 0$ and division by zero is not defined.

Exercise 6.2

Combine the fractions in each of Problems 1 through 20 into a single fraction.

1. $\dfrac{2}{x} + \dfrac{1}{3x} + \dfrac{3}{x-2}.$

2. $\dfrac{1}{x} - \dfrac{3}{2x} + \dfrac{2}{3x-1}.$

3. $\dfrac{1}{2x} + \dfrac{2}{x-3} + \dfrac{1}{x(x-3)}.$

4. $\dfrac{3}{x} - \dfrac{1}{2x+1} + \dfrac{2}{x(2x+1)}.$

5. $\dfrac{2}{x+2} - \dfrac{3}{2x-1} + 1.$

6. $\dfrac{5}{2x+3} - \dfrac{4}{x-1} + 3.$

7. $\dfrac{3}{4x+1} - \dfrac{2}{3x+2} - 1.$

8. $\dfrac{2}{4x-3} - \dfrac{1}{2x+3} + 2.$

9. $\dfrac{1}{x+1} + \dfrac{2}{(x+1)^2} - \dfrac{2}{2x+5}.$

10. $\dfrac{3}{3x-1} - \dfrac{2}{(3x-1)^2} - \dfrac{1}{x-3}.$

11. $\dfrac{1}{3x+2} - \dfrac{1}{2x-1} + \dfrac{2}{(2x-1)^2}.$

12. $\dfrac{4}{(3x-2)^2} - \dfrac{3}{3x-2} + \dfrac{2}{2x-3}.$

13. $\dfrac{1}{x+1} + \dfrac{2}{x-1} - \dfrac{3}{2x-1}.$

14. $\dfrac{2}{3x-1} - \dfrac{1}{x+2} + \dfrac{4}{2x+3}.$

15. $\dfrac{3}{2x-1} + \dfrac{2}{2x+3} - \dfrac{5}{2x-3}.$

16. $\dfrac{1}{3x-5} - \dfrac{2}{5x-3} + \dfrac{1}{3x+5}$

17. $\dfrac{x}{x-1} + \dfrac{2x-1}{x+2} + \dfrac{1}{x-2}.$

18. $\dfrac{2x-3}{x+3} + \dfrac{x+3}{2x-3} + \dfrac{1}{x-3}.$

19. $\dfrac{3x-1}{4x+1} - \dfrac{2}{x+4} + \dfrac{x}{x+3}.$

20. $\dfrac{3x+2}{x+4} + \dfrac{1}{3x-2} + \dfrac{2x}{x-3}.$

Solve and check the following equations.

21. $\dfrac{1}{x} + \dfrac{2}{3x} - \dfrac{5}{2x+1} = 0.$

22. $\dfrac{2}{x} + \dfrac{1}{2x} - \dfrac{5}{x+2} = 0.$

23. $\dfrac{3}{x} + \dfrac{1}{3x} - \dfrac{10}{2x+3} = 0.$

24. $\dfrac{2}{x} - \dfrac{1}{4x} - \dfrac{7}{x-3} = 0.$

25. $\dfrac{1}{x} + \dfrac{3}{x+2} - \dfrac{10}{x(x+2)} = 0.$

26. $\dfrac{3}{x} - \dfrac{5}{3x-1} - \dfrac{1}{x(3x-1)} = 0.$

27. $\dfrac{2}{x+1} + \dfrac{1}{x+3} - \dfrac{1}{(x+1)(x+3)} = 0.$

28. $\dfrac{3}{x-2} - \dfrac{2}{2x-3} - \dfrac{11}{(x-2)(2x-3)} = 0.$

29. $\dfrac{3}{3x+1} - \dfrac{1}{(3x+1)^2} - \dfrac{2}{2x+1} = 0.$

30. $\dfrac{6}{2x-1} - \dfrac{5}{(2x-1)^2} - \dfrac{3}{x+2} = 0.$

31. $\dfrac{-2}{8x+5} - \dfrac{5}{(8x+5)^2} + \dfrac{1}{4x-5} = 0.$

32. $\dfrac{2}{4x-7} + \dfrac{1}{(4x-7)^2} - \dfrac{3}{6x-11} = 0.$

33. $\dfrac{1}{x+1} + \dfrac{2}{x} - \dfrac{3}{x-1} = 0.$

34. $\dfrac{1}{2x-1} + \dfrac{1}{x+2} - \dfrac{3}{2x+1} = 0.$

35. $\dfrac{1}{2x+1} + \dfrac{1}{x+2} - \dfrac{3}{2x+2} = 0.$

36. $\dfrac{1}{x+5} + \dfrac{2}{x+2} - \dfrac{3}{x+4} = 0.$

37. $\dfrac{2x}{3x+1} + \dfrac{1}{x} - \dfrac{4x+4}{6x-1} = 0.$

38. $\dfrac{2x}{x-1} - \dfrac{4x+3}{2x+1} - \dfrac{3}{2x-1} = 0.$

39. $\dfrac{2x}{3x+2} - \dfrac{2x+1}{3x-1} + \dfrac{1}{x} = 0.$

40. $\dfrac{2x}{3x-2} - \dfrac{4x-8}{6x-7} - \dfrac{1}{x+1} = 0.$

7

THE TRIGONOMETRIC FUNCTIONS

7.1. Introduction

Trigonometry is a branch of mathematics which deals with six fractions or ratios that are determined by angles. These fractions are the basis for much work in engineering, physics, and mathematics. In particular, they are used in finding other parts of a triangle when certain ones are given and in finding relations between angles.

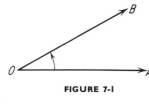

FIGURE 7-1

In our work with trigonometry, we say that an *angle* is formed by drawing two rays in the same direction from the same point and rotating one of them as indicated in Fig. 7-1. The common point of the two rays is called the *vertex* of the angle, the stationary side is called the *initial side* of the angle, and the final position of the rotating side is known as the *terminal side* of the angle. In the figure, OA is the initial side, OB is the terminal side, and O is the vertex.

The angle is called *positive* if the rotation is in the counterclockwise direction and *negative* if the rotation is clockwise. The angle AOB in Fig. 7-1 is positive.

The *size* of the angle is determined by the amount of rotation in bringing the moving ray from the initial to the terminal side. If it is rotated through one-fourth of a revolution, the resulting angle is a right angle, a degree is $\frac{1}{90}$ of a right angle, a minute is $\frac{1}{60}$ of a degree, and a second is $\frac{1}{60}$ of a minute.

7.2. The Relation Between the Coordinates and Radius Vector

In Art. 2.6, we discussed the reactangular coordinate system. In particular, we saw that the directed distance from the Y axis to a point is called the abscissa of the point and that the distance and direction from the X axis to the point is called the ordinate. Furthermore, the distance from the origin to the point is known as the *radius vector*, and we shall never consider it to be negative. In Fig. 7-2 the abscissa of P is $OQ = x$, its ordinate $QP = y$, and its radius vector is $OP = r$. The coordinates and radius vector of P are the same as the lengths of the sides of the right triangle OQP or differ

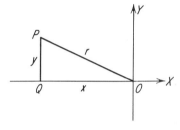

FIGURE 7-2

from them only in aglebraic sign, since the sides of the triangle are distances only and the coordinates of P are those same distances with a directional aspect added. Therefore, by use of the Pythagorean theorem, we have

(1) $$x^2 + y^2 = r^2,$$

since the square of a number and the square of the negative of that number are equal. This may be put in words: *the sum of the squares of the abscissa and ordinate of a point is equal to the square of the radius vector.*

This relationship enables us to find two possible values for the third one of the abscissa, ordinate, and radius vector if two of them are given. If the coordinates are given there is only one value for the radius vector since it is never negative; if, however, the radius vector and one coordinate are given, there is a positive value for the other coordinate and also a negative one. If the quadrant in which the point lies is given, then there is only one possible value for the coordinate that is to be determined.

EXAMPLES

1. If the coordinates of A are (5, 6), then the use of relation (1) above gives

$$5^2 + 6^2 = r^2$$
$$61 = r^2, \text{ squaring and adding;}$$
$$r = \sqrt{61}.$$

2. If $x = 7$ and $r = \sqrt{74}$, then relation (1) becomes

$$7^2 + y^2 = (\sqrt{74})^2,$$
$$y^2 = 25, \text{ squaring and solving for } y^2;$$

therefore,

$$y = \pm 5.$$

3. If the point $A(5, 12)$ is on a ray through the origin and if $B(10, y)$ is also on it, find the ordinate and radius vector of B.

SOLUTION

Since A and B are on the same ray through the origin as shown in Fig. 7-3, we can construct similar triangles by dropping perpendiculars to the X axis from A and B. Now, using the fact that corresponding sides of similar triangles are proportional, we see that

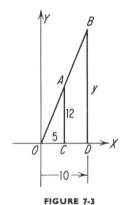

FIGURE 7-3

$$\frac{y}{12} = \frac{10}{5};$$

hence, multiplying both members by 12, we have

$$y = 12\left(\frac{10}{5}\right) = 24.$$

Finally, by use of the relation between the coordinates and radius vector of a point, we obtain

$$r^2 = 10^2 + 24^2$$
$$= 676, \text{ squaring and adding.}$$

Therefore the radius vector of B is $r = \sqrt{676} = 26$.

Exercise 7.1

Construct the angles in each of Problems 1 through 8 and indicate the amount and direction of rotation by an arrow.

1. $30°$, $150°$, $330°$.

2. $45°$, $225°$, $315°$.

3. $90°$, $180°$, $270°$.

4. $60°$, $120°$, $240°$.

5. $-45°$, $-135°$, $405°$.

6. $-60°$, $-240°$, $420°$.

7. $-150°$, $-210°$, $390°$.

8. $-180°$, $-270°$, $450°$.

Find the value or values of the one of x, y, and r that is missing in each of Problems 9 through 20 where x and y are the coordinates of a point and r is its radius vector.

9. $x = 5, y = 12$.   **10.** $x = 15, y = -8$.

11. $x = -4, y = 3$.   **12.** $x = -7, y = -24$.

13. $x = 3, r = 5$.   **14.** $y = -15, r = 17$.

15. $y = 7, r = 25$.   **16.** $x = -24, r = 25$.

17. $x = 12, r = 13$, the point in the fourth quadrant.

18. $y = -15, r = 17$, the point in the third quadrant.

19. $y = 7, r = 25$, the point in the second quadrant.

20. $x = 4, r = 5$, the point in the first quadrant.

In each of Problems 21 through 28, the points A and B have the indicated coordinates and are on the same line through the origin. Draw a figure in each case and find the quantities called for by use of similar triangles. In these problems, r represents the radius vector of B.

21. If $A(5, 12)$ and $B(10, y)$, find y and r.

22. If $A(3, -4)$ and $B(6, y)$, find y and r.

23. If $A(-8, 15)$ and $B(4, y)$, find y and r.

24. If $A(24, -7)$ and $B(8, y)$, find y and r.

25. If $A(12, -5)$ and $B(x, 10)$, find x and r.

26. If $A(-7, 24)$ and $B(x, 12)$, find x and r.

27. If $A(-9, -12)$ and $B(x, 4)$, find x and r.

28. If $A(10, 24)$ and $B(x, -12)$, find x and r.

29. The abscissa and radius vector of A are 8 and 10 and A is the first quadrant. Find the ordinate and radius vector of B if its abscissa is 4.

30. The abscissa and radius vector of A are -14 and 50 and A is in the second quadrant. Find the ordinate and radius vector of B if its abscissa is 7.

31. The ordinate and radius vector of A are -12 and 15 and A is in the third quadrant. Find the abscissa and radius vector of B if its ordinate is 8.

32. The ordinate and radius vector of A are -30 and 34 and A is in the fourth quadrant. Find the abscissa and radius vector of B if its ordinate is -15.

7.3. The Definition of the Functions of an Angle

We shall give a definition of the functions of an angle which makes use of the angle being in a definite position. We say that an angle is in *standard position* if its vertex is at the origin and its initial side is along the positive ray of the X axis. Both angles in Fig. 7-4 are in standard position.

If two angles are in standard position and have the same terminal side, they are *coterminal angles*. The two angles in Fig. 7-4 are coterminal. As a convenience in language we say that an angle is in the quadrant in which its terminal side lies. Thus we would say that 215° is a third-quadrant angle.

Before giving the definition of the trigonometric functions of an angle, we shall consider the six fractions or ratios formed by dividing the abscissa, ordinate, or radius vector of *P* by either of the other two along with the

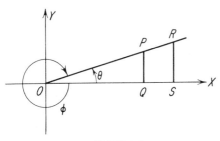

same ratios formed by using any other point *R* on the terminal side. We shall drop perpendiculars from *P* and *R* to the *X* axis and then have the similar triangles *OQP* and *OSR*, as shown in Fig. 7-4. Now, making use of the fact that corresponding parts of similar triangles are proportional, we see that each fraction formed from the abscissa, ordinate, and radius

FIGURE 7-4

vector of *P* is equal to the corresponding fraction formed from the coordinates and radius vector of *R*. The value of each of these fractions depends on an angle θ and is determined if a definite value is assigned to θ. These fractions are called *trigonometric functions* and each is given a name as shown in the following definition which should be memorized. See Fig. 7-5.

DEFINITION. *If θ is any angle in standard position, not an integral multiple of 90°, and if x, y, and r are the abscissa, ordinate, and radius vector of any point P on its terminal side, then the trigonometric functions of the angle and their abbreviations are*:

$$sine \ \theta = \frac{y}{r} = sin \ \theta,$$

$$cosine \ \theta = \frac{x}{r} = cos \ \theta,$$

$$tangent \ \theta = \frac{y}{x} = tan \ \theta,$$

$$cosecant \ \theta = \frac{r}{y} = csc \ \theta,$$

$$secant \ \theta = \frac{r}{x} = sec \ \theta,$$

$$cotangent \ \theta = \frac{x}{y} = cot \ \theta.$$

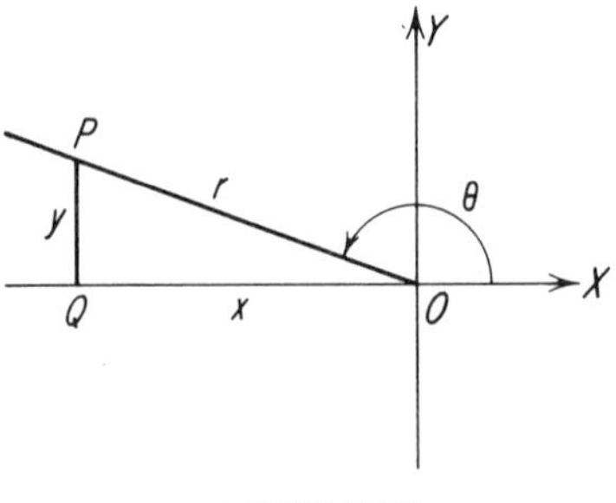

FIGURE 7-5

EXAMPLE

1. If $P(4, -3)$ is on the terminal side of an angle as shown in Fig. 7-6, then the radius vector is given by

$$r^2 = 4^2 + (-3)^2 = 25$$

and is $r = 5$. Consequently, using the definition of the trigonometric functions, we get

$$\sin \theta = \frac{-3}{5}, \qquad \csc \theta = \frac{5}{-3},$$

$$\cos \theta = \frac{4}{5}, \qquad \sec \theta = \frac{5}{4},$$

$$\tan \theta = \frac{-3}{4}, \qquad \cot \theta = \frac{4}{-3}.$$

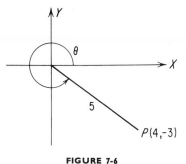

FIGURE 7-6

It may be easier to learn the definition of the six functions if we notice that they can be associated in pairs by the relations

$$\sin \theta \csc \theta = \frac{y}{r} \times \frac{r}{y} = 1,$$

$$\cos \theta \sec \theta = \frac{x}{r} \times \frac{r}{x} = 1,$$

$$\tan \theta \cot \theta = \frac{y}{x} \times \frac{x}{y} = 1.$$

The two functions in each pair are *reciprocals* since their product is one.

The algebraic sign of a trigonometric function of an angle can be determined readily by making use of the definition of the function and of the algebraic signs of the abscissa and ordinate of a point in the quandrant in which the angle lies.

EXAMPLE

2. If $P(-5, 12)$ is on the terminal side of an angle in standard position then, by using the relation $x^2 + y^2 = r^2$, we find that $r = 13$; consequently,

$$\sin \theta = \frac{12}{13}, \qquad \csc \theta = \frac{13}{12},$$

$$\cos \theta = \frac{-5}{13}, \qquad \sec \theta = \frac{13}{-5},$$

$$\tan \theta = \frac{12}{-5}, \qquad \cot \theta = \frac{-5}{12}.$$

7.4. Given One Function, to Find the Others

If we know the value of one function of an angle, we can find the value or values of the other five by using the Pythagorean relation between the coordinates and radius vector of a point along with the definition of the functions. By use of the Pythagorean relation, we find two values for the unknown if it is a coordinate and one value if it is the radius vector. We use as many of these values as are consistent with the conditions of the problem. We can use both values of a coordinate if only a function of an angle is given but can use only one if the quadrant in which the angle lies is also given.

EXAMPLES

1. Find the value of each of the other functions if $\cos A = \frac{5}{13}$.

SOLUTION

Since, as pointed out in Art. 7.3, the value of a function of an angle does not depend on the particular point chosen on the terminal side, we shall choose the one with $x = 5$ since then $r = 13$ and y also turns out to be an integer. Using these values for x and r, we have

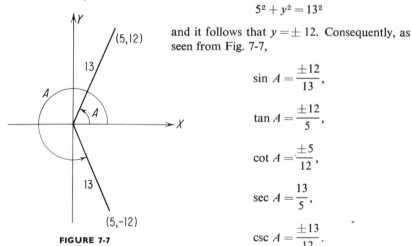

$$5^2 + y^2 = 13^2$$

and it follows that $y = \pm\, 12$. Consequently, as seen from Fig. 7-7,

$$\sin A = \frac{\pm 12}{13},$$

$$\tan A = \frac{\pm 12}{5},$$

$$\cot A = \frac{\pm 5}{12},$$

$$\sec A = \frac{13}{5},$$

$$\csc A = \frac{\pm 13}{12}.$$

FIGURE 7-7

2. Find the other functions if $\tan A = \frac{7}{24}$ and A is the third quadrant.

SOLUTION

The coordinates of any point on the terminal side of the angle are both negative since the angle is in the third quadrant. Using the definition of the tangent of an angle and the value given in this problem, we have

$$\tan A = \frac{y}{x} = \frac{7}{24};$$

hence,

$$y = \frac{7}{24} x$$

and, if we choose the point for which $x = -24$, we find that $y = -7$. Furthermore,

$$r^2 = (-24)^2 + (-7)^2 = 625,$$

and consequently $r = 25$. Therefore,

$$\sin A = \frac{-7}{25}, \qquad \csc A = \frac{25}{-7},$$

$$\cos A = \frac{-24}{25}, \qquad \sec A = \frac{25}{-24},$$

$$\tan A = \frac{7}{24} \text{ as given,} \quad \cot A = \frac{24}{7}.$$

Exercise 7.2

Draw an angle in standard position that has the point in each of Problems 1 through 12 on its terminal side and find the values of the functions of the angle in each case.

1. (5, 12).   **2.** (4, 3).   **3.** (8, 15).

4. (24, 7).   **5.** (12, −5).   **6.** (−7, 24).

7. (−8, −15).   **8.** (−3, 4).   **9.** $(-\sqrt{5}, 2)$.

10. $(5, -\sqrt{11})$.   **11.** $(\sqrt{21}, -2)$.   **12.** $(-3, -\sqrt{7})$.

If the given functions of an angle satisfy the conditions given in each of Problems 13 through 20, find the quadrant in which the angle is located.

13. The sine is negative and the cosine is positive.

14. The sine and cosine are both negative.

15. The sine and tangent are both negative.

16. The cosine is positive and the tangent is negative.

17. The cotangent is negative and the secant is positive.

18. The secant and cosecant are negative.

19. The secant is negative and the cosecant is positive.

20. The cosecant is negative and the cotangent is positive.

Construct all possible angles between 0 and 360° in standard position that satisfy the conditions in each of the following problems and find the value or values of the other functions of the angle.

21. $\sin A = \dfrac{3}{5}$.

22. $\cos A = \dfrac{5}{13}$.

23. $\sec A = \dfrac{25}{7}$.

24. $\csc A = \dfrac{17}{15}$.

25. $\cos A = -\dfrac{8}{17}$.

26. $\sec A = -\dfrac{13}{12}$.

27. $\csc A = -\dfrac{5}{4}$.

28. $\sin A = -\dfrac{24}{25}$.

29. $\tan A = \dfrac{7}{24}$.

30. $\cot A = \dfrac{12}{5}$.

31. $\cot A = -\dfrac{4}{3}$.

32. $\tan A = -\dfrac{15}{8}$.

33. $\cot A = \dfrac{12}{5}$, A in the third quadrant.

34. $\sin A = \dfrac{8}{17}$, A in the second quadrant.

35. $\cos A = \dfrac{-7}{25}$, A in the third quadrant.

36. $\tan A = -\dfrac{4}{3}$, A in the fourth quadrant.

7.5. Functions of 45°, 30°, 60° and Their Multiples

In order to find the values of the functions of 45°, we shall consider a 45°

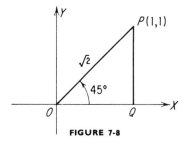

FIGURE 7-8

angle in standard position as shown in Fig. 7-8. We shall select any point P on the terminal side and drop a perpendicular PQ from it to the X axis. This forms a 45° right triangle; hence, it is an isosceles triangle. Consequently, the abscissa and ordinate of P are equal and we shall choose each of them to be one; then, by use of the relation between radius vector and coordinates of a point, we find that $OP = \sqrt{2}$. Now, using the definition of the functions gives

$$\sin 45° = \frac{1}{\sqrt{2}}, \qquad \csc 45° = \sqrt{2},$$

$$\cos 45° = \frac{1}{\sqrt{2}}, \qquad \sec 45° = \sqrt{2},$$

$$\tan 45° = 1, \qquad \cot 45° = 1.$$

In order to find the values of the functions of 30°, we shall make use of the theorem from plane geometry which states that the side opposite the 30° angle in a 30° right triangle is one-half as long as the hypotenuse. Therefore, if we put the 30° angle in standard position as shown in Fig. 7-9, and take P as the point on the terminal side with ordinate 1, then $OP = 2$, and by use of the relation between the coordinates and radius vector of a point, we have $x^2 + 1^2 = 2^2$ and $x = \sqrt{3}$ as shown in the figure. Now, the definition of the functions gives

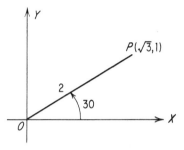

FIGURE 7-9

$$\sin 30° = \frac{1}{2}, \qquad \csc 30° = 2,$$

$$\cos 30° = \frac{\sqrt{3}}{2}, \qquad \sec 30° = \frac{2}{\sqrt{3}},$$

$$\tan 30° = \frac{1}{\sqrt{3}}, \qquad \cot 30° = \sqrt{3}.$$

In order to find the values of the functions of 60°, we shall draw a 60° angle in standard position as shown in Fig. 7-10 and select any point P on the terminal side. We now drop a perpendicular PQ from P to the X axis. We thus have a 30° right triangle with the 30° angle at P. Consequently, if we let $OQ = 1$, we have $OP = 2$ and can find that $QP = \sqrt{3}$ by use of the relation between the coordinates and radius vector of P. Now, the definition of the functions gives

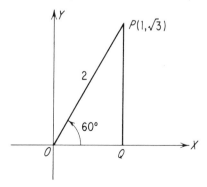

FIGURE 7-10

$$\sin 60° = \frac{\sqrt{3}}{2}, \qquad \csc 60° = \frac{2}{\sqrt{3}}.$$

$$\cos 60° = \frac{1}{2}, \qquad \sec 60° = 2,$$

$$\tan 60° = \sqrt{3}, \qquad \cot 60° = \frac{1}{\sqrt{3}}.$$

The values of the functions of multiples of 30° and 45° that are not multiples of 90° can be found in a manner similar to that already used in this article.

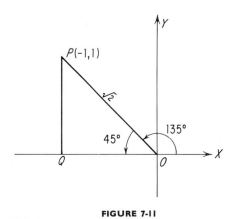

FIGURE 7-11

EXAMPLE

1. Find the values of the functions of 135°.

SOLUTION

We shall draw an angle of 135° in standard position as shown in Fig. 7-11. Then the angle formed by the terminal side of the 135° angle and the negative ray of the X axis is 45°. Hence, if P is any point on the terminal side of the 135° angle and if we drop a perpendicular PQ to the X axis, we then have an isosceles right triangle, OQP. Consequently the coordinates and radius vector of P are as shown and we can get the values of the functions of 135° by applying the definition. They are

$$\sin 135° = \frac{1}{\sqrt{2}}, \qquad \csc 135° = \sqrt{2},$$

$$\cos 135° = \frac{-1}{\sqrt{2}}, \qquad \sec 135° = -\sqrt{2},$$

$$\tan 135° = -1, \qquad \cot 135° = -1.$$

7.6. Functions of Quadrantal Angles

If an angle is in standard position and if its terminal side is along a coordinate axis, we say that we have a *quadrantal angle*. Since the terminal side is along a coordinate axis it follows that the ordinate or abscissa of a point on the terminal side is zero and the other one is the same size as the radius vector. Consequently, in attempting to find the functions of a quadrantal angle, we must bear in mind that *zero divided by any nonzero member is zero* and that *we do not divide by zero.* Therefore, two of the functions of a quandrantal angle are zero and there is no number to represent two others; furthermore, the other two are 1 or −1, since the abscissa or ordinate is the same size as the radius vector.

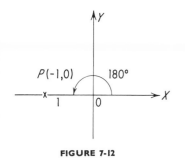

FIGURE 7-12

EXAMPLE

1. In order to find the values of the functions of 180°, we draw the angle in standard position and select a point *P* on the terminal side that is one unit from the origin. The coordinates of the point are (−1, 0), since it is one unit to the left of the *Y* axis and on the *X* axis. Therefore,

$$\sin 180° = \frac{0}{1} = 0, \qquad \csc 180° = \frac{1}{0} = \text{no number,}$$

$$\cos 180° = \frac{-1}{1} = -1, \qquad \sec 180° = \frac{1}{-1} = -1,$$

$$\tan 180° = \frac{0}{-1} - 0, \qquad \cot 180° = \frac{-1}{0} = \text{no number.}$$

Exercise 7.3

Find the values of the six trigonometric functions of each angle in Problems 1 through 8.

1. 135°, 240°.   **2.** 210°, 90°.   **3.** 225°, 270°.

4. 120°, 330°.   **5.** 150°, 0°.   **6.** 315°, 300°.

7. −30°, −240°.   **8.** −90°, −315°.

Evaluate each of the following.

9. $\sin 60° \cos 30° + \cos 30° \sin 60°$.

10. $\cos 60° \cos 45° - \sin 60° \sin 45°$.

11. $\sin 45° \cos 30° - \cos 315° \sin 150°$.

12. $\cos 300° \cos 315° - \sin 120° \sin 225°$.

13. $\sin 120° \cos 315° + \cos 240° \sin 225°$.

14. $\cos 120° \cos 45° + \sin 300° \sin 135°$.

15. $\dfrac{\tan 60° - \tan 225°}{1 - \tan 300° \tan 315°}.$

16. $\dfrac{\cot 210° + \tan 300°}{1 - \tan 240° \cot 330°}.$

Prove the statement in each of Problems 17 through 32 by making use of the values of the functions.

17. $\cos 60° = 2 \cos^2 330° - \sin^2 45° - \cos^2 225°.$

18. $\cos 150° = 2(\cos 45° \cos 30° - \sin 45° \sin 30°)^2 - \tan 60° \cot 240°.$

19. $\sin 150° = \sin 90° \cos 300° + \cos 270° \sin 240°.$

20. $\sin 270° = 2(\sin 90° \sin 45° + \cos 270° \cos 315°)(\cos 180° \cos 45° - \sin 0° \sin 215°).$

21. $2 \sin^2 30° = 1 - \cos 300°.$

22. $2 \sin^2 60° = 1 + \cos 300°.$

23. $2 \cos^2 120° = \tan 45° \cot 225° + \sin 210°.$

24. $2 \cos^2 300° = \sec 30° \cos 30° + \sin 210°.$

25. $\tan 210° = \csc 120° - \cot 240°.$

26. $\sec 330° - \cot 300° = \tan 240°.$

27. $\sin 120° = \cot 210° (\tan 225° + \sin 330°).$

28. $\sin 240° = (\csc 300° \sin 240° + \cos 60°) \cot 300°.$

29. $\sin 450° + \cos 360° = \cos 270° - \csc 210°.$

30. $\sin 90° + \csc 90° = 1 + \tan 225° \cot 45°.$

31. $\sin 180° = 2 \sin 300° \cos 630°.$

32. $\sin 270° = 2 \sin 225° \cos 315°.$

8

RIGHT TRIANGLES

8.1. Tables of Values of Trigonometric Functions

We studied the functions of 30°, 45°, and their multiples in Art. 7.5, but calculation of the values of functions of angles in general is beyond the scope of this book. We shall have a considerable need for the values of the functions and can find them in Table I. The angles are listed at intervals of ten minutes and the values of sines, cosines, tangents, and cotangents are given to four significant digits. There are tables which give the functions to a higher degree of accuracy, and they are available in many of the mathematical handbooks but will not be needed for our work.

The angles from 0° to 45° are listed on the left of the page and angles from 45° to 90° are listed on the right. The function names to be used with the angles on the left of the page are given at the top and the function names to be used with the angles on the right are given at the bottom of the page. In order to find a specified function of a given angle, we first locate the angle and then look across from it and below or above the function name according as the angle is on the left or the right.

EXAMPLES

1. In order to find the value of tan 22°40′ we first locate 22°40′. It is on the left of the third page of Table I; hence the function name to use is at the top of the page. Finally, across from 22°40′ and below **Tan**, we find that tan 22°40′ = .4176.

2. If we want the value of cos 71°30′, we make use of the fact that 71°30′ is on the right of the page and look above **Cos**; thus we find that cos 71°30′ = .3173.

If we know the value of a trigonometric function of an angle and want to find the angle, we must look below and, if necessary, above the function name until we find the given value; then the angle is on the left or the right according as the function name is at the top or bottom of the page.

EXAMPLES

3. If we know that cos A = .7826 and want to find A, we look above and below **Cos** until we locate .7826. It is on the fifth page of Table I, and since it is below **Cos** the angle is on the left. It is readily seen to be 38°30′.

4. If cot A = .4522, we look above and below **Cot** until we locate .4522 on the third page of the table. It is above Cot; hence the angle is on the right and is 65°40′.

8.2. Interpolation

In the last article, we saw how to find the functions of angles that are listed in the table and how to find the angle if one of its functions is given. We shall now present a method for finding functions of angles that are not in the table and angles whose functions are not listed. The method is known as *linear interpolation* and will be presented by means of two examples.

EXAMPLES

1. If we want the value of sin 28°23′, we begin by finding sin 28°20′ and sin 28°30′ since these are in the table and the two angles are nearer the angle we are interested in than are any other angles in the table. We find that sin 28°20′ = .4746 and sin 28°30′ = .4772; furthermore, we know that 28°23′ is $\frac{3}{10}$ of the way from 28°20′ toward 28°30′ since it is 10′ from 28°20′ to 28°30′ and 3′ from 28°20′ to 28°23′. We assume that sin 28°23′ is that same part, $\frac{3}{10}$, of the way from sin 28°20′ to sin 28°30′. By using the values of the functions, we know that sin 28°30′ − sin 28°20′ = .4772 − .4746 = .0026; hence $\frac{3}{10}$ of this difference is $\frac{3}{10}$(.0026) = .00078 and we use .0008 to four decimal places. Finally, we start at sin 28°20′ = .4746 and go .0008 toward sin 28°30′ = .4772 and thus see that sin 28°23′ = .4746 + .0008 = .4754. The essential parts of this work and discussion can be put in a diagram as follows:

$$\left.\begin{array}{l} 3 \\ 10 \end{array}\right. \left(\begin{array}{l} \sin 28°20' = .4746 \\ \sin 28°23' = \\ \sin 28°30' = .4772 \end{array}\right) .0026$$

$$\frac{3}{10}(.0026) = .0008$$

$$\sin 28°23' = .4746 + .0008 = .4754.$$

2. Find A, if $\tan A = .6647$.

SOLUTION

If we look in the table above and below **Tan**, we see that .6647 is not there and that $.6619 = \tan 33°30'$ and $.6661 = \tan 33°40'$ are nearer it than are any other entries. The remainder of the solution in diagram form is:

$$10' \left(\begin{array}{l} \tan 33°30' = .6619 \\ \tan A \quad\ = .6647 \\ \tan 33°40' = .6661 \end{array}\right) \begin{array}{l} .0028 \\ .0042 \end{array}$$

$$\frac{.0028}{.0042}(10') = 7' \text{ to the nearest minute.}$$

Therefore, $A = 33°30' + 7' = 33°37'.$

Exercise 8.1

Find the value of the specified function of the angle in each of Problems 1 through 12.

1. $\sin 32°20'$.

2. $\cos 28°40'$.

3. $\tan 15°10'$.

4. $\cot 43°50'$.

5. $\cos 7°$.

6. $\tan 36°30'$.

7. $\cot 47°40'$.

8. $\sin 83°10'$.

9. $\tan 65°20'$.

10. $\cot 79°50'$.

11. $\sin 51°30'$.

12. $\cos 89°$.

Find the angle in each of Problems 13 through 24.

13. $\cot A = 38.19$.

14. $\sin A = .1822$.

15. $\cos A = .9283$.

16. $\tan A = .6088$.

17. $\cot A = 1.250$.

18. $\cos A = .7716$.

19. $\tan A = 1.280$.

20. $\cot A = .7177$.

21. sin $A = .9100.$ **22.** cos $A = .3611.$

23. tan $A = 4.843.$ **24.** sin $A = .9996.$

Find· the value of the specified function of the angle in each of Problems 25 through 36.

25. cot 8°13′.   **26.** tan 13°8′.   **27.** cos 26°17′.

28. sin 34°46′.   **29.** tan 37°52′.   **30.** cos 44°44′.

31. sin 47°31′.   **32.** cot 53°57′.   **33.** cos 61°25′.

34. sin 68°59′.   **35.** cot 78°36′.   **36.** cot 86°33′.

By use of interpolation, find the angle to the nearest minute in each of Problems 37 through 48.

37. tan $A = .0225.$ **38.** cot $A = 15.00.$

39. sin $A = .2170.$ **40.** cos $A = .9300.$

41. cot $A = 1.667.$ **42.** sin $A = .5757.$

43. cos $A = .7272.$ **44.** tan $A = 1.160.$

45. cos $A = .5233.$ **46.** tan $A = 3.000.$

47. sin $A = .9200.$ **48.** cot $A = .0821.$

8.3. Special Definitions

We learned the definition of functions of angles in standard position in Art. 7.3. We shall now see what this definition becomes if the angle is a

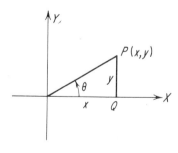

FIGURE 8-I

positive acute angle of a right triangle. If we put an acute angle in standard position as in Fig. 8-1 and then make a right triangle by dropping a perpendicular from any point $P(x, y)$ on the terminal side of the angle to the X axis, we see that x and y become the side adjacent to the angle and the side opposite it; furthermore, the radius vector becomes the hypotenuse of the triangle. Therefore, if applied to an acute angle of a right triangle, the definition becomes

$$\sin \theta = \frac{\text{side opposite}}{\text{hypotenuse}}, \qquad \tan \theta = \frac{\text{side opposite}}{\text{side adjacent}},$$

$$\cos \theta = \frac{\text{side adjacent}}{\text{hypotenuse}}, \qquad \cot \theta = \frac{\text{side adjacent}}{\text{side opposite}}.$$

The other two functions are not defined here since their values are not given in Table I. The reader must not forget that these definitions are applicable to acute angles and only to them but have the advantage that the angle does not have to be in any particular position.

We shall be consistent in using A and B for the acute angles and C for the right angle of a right triangle and the side opposite each angle shall be represented by the corresponding small letter as shown in Fig. 8-2.

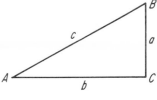

FIGURE 8-2

If we apply the special definitions to angles A and B, we get

$$\sin A = \frac{a}{c} = \cos B, \qquad \tan A = \frac{a}{b} = \cot B,$$

$$\cos A = \frac{b}{c} = \sin B, \qquad \cot A = \frac{b}{a} = \tan B.$$

The angles A and B are complementary, since their sum is $90°$; furthermore, each function on the left is called the *cofunction* of the one on the right, since the one can be obtained from the other by putting on or taking off the prefix "co". Consequently we can state the following theorem.

THEOREM. *Any trigonometric function of an acute angle is equal to the corresponding cofunction of the complementary angle.*

The fact stated in this theorem makes it possible to have the arrangement of functions and angles that is used in Table I. Each function name at the top of a column is the cofunction of the one at the bottom and each angle on the right of the page is the complement of the one directly across on the left; hence, each entry in the table is a function of an angle and also the corresponding cofunction of the complementary angle.

EXAMPLE

The value of sin 32°20′ is the same as that of cos 57°40′ since sine and cosine are cofunctions and 32°20′ and 57°40′ are complements.

8.4. Calculations with Numbers and Angles

Rounding off was discussed in Art. 1.12, and some discussion of procedures in calculating with approximate numbers was given in Art. 1.13. We shall now continue that discussion. The angles used should be to the degree of accuracy needed to have the desired function of the angle correct to the same number of significant digits as the numbers used. Hence, we want to determine a relation between degrees of accuracy for sides and angles of a triangle. Thus, if sin θ is given as .36, we know that it is between .355 and .365; hence, by use of Table I, we know that the angle is between 20°40' and 21°30'. Thus, information is sufficient to determine θ only approximately; hence, we take θ to be 21°. Similarly, if the value of the function of the angle is given to three significant digits, we can ordinarily get the value of the angle to the nearest multiple of 10'.

Although it may well be modified for some functions of angles near 0° and near 90°, as can be seen from an examination of Table I, we shall use the following table for determining comparable degrees of accuracy in sides and angles of a triangle.

Significant digits in sides	Value of the angle to the nearest
2	Degree
3	Multiple of 10 minutes
4	Minute

8.5. Solution of Right Triangles

The definition of each trigonometric function of an acute angle of a right triangle uses an angle and two sides of the triangle. Consequently, we can find the third of these three quantities if we know any two of them.

Every right triangle has three sides and three angles but one of the angles is always 90°; hence, there are five parts of a right triangle that may vary. If we know the length of a side and one other variable part, we can determine the other three parts. The procedure for finding any one of them is to *select the function and the angle so that the desired unknown and two known parts are used in the definition of the function of the angle.* We shall find each desired unknown to the degree of accuracy justified by the given data in keeping with the table of corresponding degrees of accuracy shown in Art. 8.4. If, in our work with triangles, an angle is expressed in degrees or in degrees and a multiple of ten minutes, we shall assume that it is correct to the nearest degree, the nearest ten minutes or the nearest minute according to whether the sides are given to two, three, or four significant figures.

EXAMPLES

1. Solve the right triangle in which $A = 38°40'$ and $c = 247$.

SOLUTION

In many cases the work progresses more smoothly if we draw a reasonably accurate sketch as the first step in solving. Figure 8-3 is such a sketch. If we decide to find a first, we must use $\sin A = a/c$ since then two known parts and the desired unknown are used in the definition. Thus, putting the given values in

$$\sin A = \frac{a}{c},$$

we have $$\sin 38°40' = \frac{a}{247}.$$

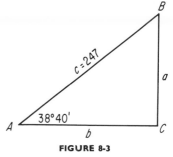

Multiplying through by 247 gives

$$a = 247 \sin 38°40'$$
$$= 247(.6248), \text{ using the table,}$$
$$= 154.$$

FIGURE 8-3

This value is given to three significant figures since that is in keeping with the given data. We can find B by subtracting A from $90°$ since A and B are the acute angles of a right triangle. Thus,

$$B = 90° - 38°40' = 51°20'.$$

Finally, we could find b by means of $\tan A = a/b$ but this would use a part that we have calculated, and consequently would carry over to b any error made in determining a. Therefore we will use $\cos A = b/c$. Hence, solving for b gives

$$b = c \cos A$$
$$= 247 \cos 38°40', \text{ putting in the given values,}$$
$$= 247(.7808), \text{ using the table,}$$
$$= 193 \text{ to three significant figures.}$$

2. Solve the right triangle in which $a = 34.16$ and $c = 49.23$.

SOLUTION

A sketch of the situation is shown in Fig. 8-4. We shall find A by use of

$$\sin A = \frac{a}{c} = \frac{34.16}{49.23} = .6939;$$

hence, $A = 43°56'$, interpolating.

We have no choice other than using a calculated part in determining B and b. Since B is the complement of A we have $B = 90° - 43°56' = 46°4'$. Now we can find b by use of

$$\tan B = \frac{b}{a}.$$

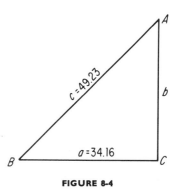

FIGURE 8-4

Multiplying by a gives

$$b = a \tan B$$
$$= (34.16)(\tan 46°4')$$
$$= (34.16)(1.038)$$
$$= 35.46 \text{ to four significant figures.}$$

Exercise 8.2

Solve the following triangles. Determine each angle to the degree of accuracy and each side to the number of significant digits justified by the given data.

1. $c = 283$, $A = 54°20'$.

2. $c = 572$, $A = 32°30'$.

3. $c = 41.4$, $B = 72°50'$.

4. $c = 739$, $B = 19°10'$.

5. $a = 307$, $c = 519$.

6. $a = .644$, $c = .976$.

7. $b = 1.03$, $c = 2.78$.

8. $b = .691$, $c = .994$.

9. $a = 439$, $b = 505$.

10. $a = 516$, $b = 427$.

11. $a = .0112$, $b = .0334$.

12. $a = 376$, $b = 299$.

13. $a = 23.9$, $A = 41°10'$.

14. $a = 537$, $A = 62°$.

15. $b = 428$, $B = 73°30'$.

16. $b = 1.03$, $B = 32°40'$.

17. $a = 17.3$, $B = 50°50'$.

18. $a = .386$, $B = 57°10'$.

19. $b = 2.77$, $A = 37°$.

20. $b = 404$, $A = 63°30'$.

21. $c = 8072$, $B = 62°13'$.

22. $c = 5941$, $B = 39°54'$.

23. $c = 949.8$, $A = 27°38'$.

24. $c = 3.772$, $A = 28°42'$.

25. $b = 71.24$, $c = 97.79$.

26. $b = .5807$, $c = .8570$.

27. $a = 4114$, $c = 5718$.

28. $a = 234.5$, $c = 876.5$.

29. $a = 2772$, $b = 3443$.

30. $a = .4996$, $b = .4134$.

31. $a = 7.381$, $b = 4.489$.

32. $a = 388.3$, $b = 356.7$.

33. $b = 6005$, $B = 52°36'$.

34. $b = 4718$, $B = 37°30'$.

35. $a = .2356$, $A = 23°56'$.

36. $a = 4817$, $A = 48°17'$.

37. $b = 60.13$, $A = 37°37'$.

38. $b = 3278$, $A = 63°44'$.

39. $a = .8117$, $B = 28°18'$.

40. $a = 1319$, $B = 78°33'$.

8.6. Angles of Elevation and Depression

There are times when it is not sufficient to be able to say that an object is above or below the observer; we shall now see how to be more exact. If an observer at O looks at object P, then OP is called the *line of sight* as shown in Figs. 8-5 and 8-6. If H is a point on the horizontal line through O and in the same vertical plane as OP, then the angle POH is called the *angle of elevation* if P is above OH and *angle of depression* if P is below OH.

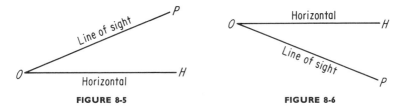

FIGURE 8-5   FIGURE 8-6

EXAMPLE

1. If the angle of elevation of the sun is 39°, how long on horizontal ground is the shadow of a vertical flagpole that is 43 feet tall?

SOLUTION

We shall begin by drawing a sketch of the situation as shown in Fig. 8-7. We shall use cot S since we know S and s and want to determine p. Thus

$$\cot 39° = p/43,$$
$$p = 43 \cot 39°$$
$$= 43(1.235)$$
$$= 53 \text{ to two digits}$$
$$\text{as justified.}$$

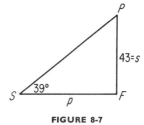

FIGURE 8-7

8.7. The Direction of a Line

A surveyor and an air navigator use different language in giving the direction of a line. We shall study that used by each. The direction of flight is often called *bearing* or *course* by a navigator and is the clockwise angle measured from due north to the line of flight. Courses of 55° and 240° are shown in Fig. 8-8.

In giving the direction of a line the surveyor tells the acute angle it makes with a north-south line and whether it is to the east or west of that line. Thus, to indicate that the line runs 23° east of north as shown by OA in Fig. 8-9, the surveyor says it runs N23°E. The directions of the other rays in the figure are as follows:

OB is S40°E,
OC is S70°W,
OD is N25°W.

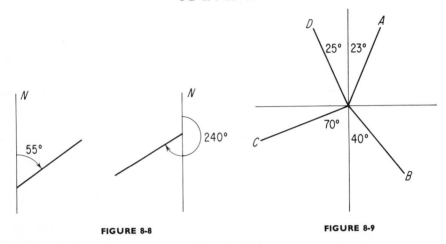

FIGURE 8-8   FIGURE 8-9

EXAMPLES

1. An airplane left a field F and flew 247 miles on a course of 130°20′ and landed at a field G. It then took off on a course of 220°20′ and landed at H which is due south of F. How far is it from H to F?

SOLUTION

A sketch of the situation is shown in Fig. 8-10. The angle of the triangle at F is $180° - 130°20′ = 49°40′$ and that at G is 90° since $360° - (220°20′ + 49°40′) = 90°$.

$$\cos 49°40′ = \frac{247}{g}.$$

$$g = \frac{247}{\cos 49°40′}$$

$$= \frac{247}{.6472}$$

$$= 382.$$

Therefore it is 382 miles from H to F.

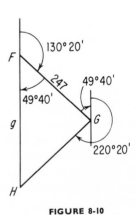

FIGURE 8-10

2. Find the length and direction of the third side of a tract of land that is bounded as follows: Beginning at an oak tree marked by an x, then S33°40′E for 427 feet, then S56°20′W for 671 feet, and then to the starting point.

SOLUTION

The situation in the problem is shown pictorially in Fig. 8-11. We can find the direction of the third side by subtracting B from $56°20'$; furthermore, B can be found by use of

$$\tan B = \frac{427}{671} = .6364.$$

Therefore, $B = 32°30'$, and, consequently, $\alpha = 56°20' - 32°30' = 23°50'$; hence, the direction from B to A is N23°50′E. In order to find the distance from B to A, we can use

$$\sin B = \sin 32°30' = \frac{427}{c};$$

$$c = \frac{427}{\sin 32°30'}$$

$$= \frac{427}{.5373}$$

$$= 795 \text{ to three figures.}$$

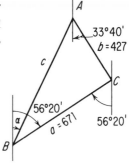

FIGURE 8-11

Exercise 8.3

1. Find the length of the diagonal of a rectangle that is 108 inches by 147 inches. Find the angle the diagonal makes with the shorter side.

2. Find the length and width of a rectangle if the 811 foot diagonal makes an angle of 28°20′ with the longer side.

3. The sides of a rhombus are 17.63 centimeters long and each acute angle is 43°34′. Find the lengths of the diagonals.

4. Find the angles of an isosceles triangle if its base is 42.8 centimeters and the altitude is 13.9 centimeters.

5. A boat is 123 feet from the bank of a lake. An observer in the boat finds the angle of elevation of the top of a tree on the bank is 21°30′. How tall is the tree?

6. The angle of elevation of the top of a 256 foot building is 41°20′ as seen by an observer who is in the same horizontal plane as the base of the building. How far is he from the building?

7. The angle of elevation of the top of an observation tower known to be 286 feet high is found to be 28°40′ by an observer in the same horizontal plane as the base of the building. How far is he from the building?

8. A house is 54.7 feet square and faces north. A water pipe runs under the house between a point 13.6 feet north of the southwest corner and one 18.3 feet south of the northeast corner. How long is the pipe?

9. A highway that ran down the west and across the south side of a parcel of land was rerouted to save distance. It now leaves the west boundary at a point 748 yards north of the southwest corner and proceeds S28°20′E until it intersects the south boundary. Find the distance saved.

10. A lot is bounded by three streets. One of them runs north and south, another east and west, and the third N38°W for a distance of 162 feet. Find the perimeter of the lot.

11. The angles of elevation of the top and bottom of a flagpole on top of a building are 70°50′ and 69°40′ as seen by an observer who is 112 feet from the base of the building and in the same horizontal plane as it. Find the length of the flagpole.

12. An observer in a second-story window finds that the angle of elevation of the top of a building across the street is 48° and that the angle of depression of its base is 34°20′. If the street is 63.2 feet wide, find the height of the building.

13. A pilot flew at 72° for 525 miles in going from *A* to *B* and then at 162° for 763 miles to *C*. Find the distance and direction from *C* to *A*.

14. Two planes left *A* at the same time. One flew at 170 miles per hour on a course of 38°40′ and the other at 145 miles per hour on a course of 308°40′. How far apart were the planes after 2 hours? What was the direction of the second from the first?

15. A pilot flew a course of 148°10′ for 536 miles from *F* to *G*, then a course of 58°10′ for 322 miles to *H*, and then back to *F*. Find the distance and direction of the last flight.

16. A pilot flew from Baton Rouge for 211 miles on a course of 322°40′ and then for 186 miles on a course of 52°40′. What course must he fly to return to Baton Rouge? How far must he fly?

17. Two walks intersect on a campus. The library is on one of the walks and is 225 feet due east of the intersection. The administration building is on the other and is 196 feet due north of the library. At what angle does the east-west walk intersect the line between the two buildings? How far is the administration building from the library?

18. A tract of land is described as follows: Begin at the Smith Oak, go N40°30′W for 324 feet, then S49°30′W for 267 feet and then to place of origin. Find the length and direction of the third side.

19. A surveyor's field notes describe the boundaries of a tract of land as follows: Beginning at a brass rod go due north for 1127 feet, then S69°48′W to a second marker, then S20°12′E to the starting point. Find the lengths of the last two sides.

20. Cheyenne, Wyoming is 425 miles due west of Omaha, Nebraska and Carlsbad, New Mexico is 550 miles due south of Cheyenne. Find the distance and direction from Omaha to Carlsbad.

9

INTRODUCTION TO POLAR COORDINATES

9.1. Some Definitions and Concepts

Beginning with Chapter 2 we encounter the rectangular coordinate system by means of which we are able to locate points and sketch the graph of a curve. There are, however, other types of coordinate systems that are often more desirable under some circumstances. We shall now present one known as the *polar coordinate system* or merely as *polar coordinates*. In this system, the frame of reference is a pair of perpendicular, directed lines. The horizontal line is called the *polar axis*, the vertical is known as the *normal axis* and their intersection is the *pole*. A point is located by giving its distance r from the pole and the angle θ from the positive end of the polar axis to the ray from the origin to the point as indicated in Fig. 9-1. The distance r is called the *radius vector* and the angle θ is called the *amplitude* or *vectorial angle*. As in trigonometry, the vectorial angle may be positive, zero, or negative. Distances measured along the radius vector from the pole are positive, and a negative sign before the radius vector indicates that the distance is to be measured along the extension of the radius vector through the pole.

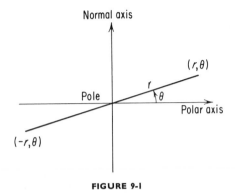

FIGURE 9-1

9.2. Pairs of Coordinates for a Point

In using rectangular coordinates, there is only one point for an ordered pair of numbers and only one ordered pair of numbers for a point. Thus, there is a one to one correspondence between points and ordered pairs of number in connection with the rectangular coordinate system. This is not the situation if the polar coordinate system is used. There is only one point for an ordered pair of numbers but a point may have an unlimited number of pairs of coordinates. If k is an integer, then $(r, \theta + k360°)$ represents the same point as (r, θ) since r is the radius vector in each case and the terminal sides of the angles θ and $\theta + k360°$ are the same. If we bear in mind that when r is preceded by a negative sign, then the distance is measured along the extension of the radius vector through the origin, we see $(-r, \theta + 180°)$ represents the same point as (r, θ); furthermore, $(-r, \theta + 180° + k360°)$ also represents that point for k an integer.

EXAMPLE

1. Locate the point $(3, 30°)$ and give three other pairs of coordinates for it.

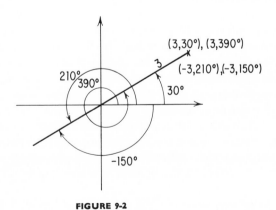

FIGURE 9-2

SOLUTION

The point is shown in Fig. 9-2 with three other pairs of coordinates for it. The point $(3, 390°)$ is the same as $(3, 30°)$ since $r = 3$ for each and $390°$ differs from $30°$ by an integral multiple of $360°$; furthermore, $(-3, 210°)$ and $(-3, -150°)$ also represent the same point as $(3, 30°)$ since $r = -3$ for each of them and the terminal side of each angle is the extension through the origin of the terminal side of $30°$.

9.3. The Reference Angle

In general, we shall need to express functions of angles in terms of the same function of a positive acute angle in some of our work with polar coordinates. In order to be able to do this, we shall state and prove a theorem that is called the reference angle theorem. By the *reference angle* of an angle in standard position, we mean the positive acute angle between the X axis and the terminal side of the angle if it is not a multiple of $90°$. If the given angle is zero or a multiple of $90°$ and is in standard position, then the reference angle is $0°$ or $90°$ according as the terminal side of the angle is along the X or Y axis.

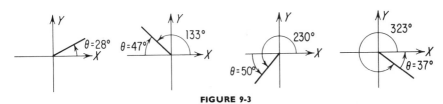

FIGURE 9-3

In Fig. 9-3, θ is the reference angle in each case since it is the positive acute angle between the X axis and the terminal side of the given angle.

To find the relations between the functions of an angle and those of the reference angle, we shall construct angles θ, $180° - \theta$, $180° + \theta$, and $360° - \theta$ in standard position since we then have an angle in each quadrant that has θ as reference angle. Instead of showing $360° - \theta$, we could have shown $-\theta$ since they have the same terminal side. The angles θ are equal in each of Figs. 9-4 (a), (b), (c), and (d); furthermore, by construction $OP = OP_2 = OP_3 = OP_4 = r$. Consequently, the triangles OMP, OM_2P_2, OM_3P_3 and OM_4P_4 are congruent. Therefore, x, x_2, x_3, and x_4 are the same size but x_2 and x_3 are negative, whereas x and x_4 are positive; furthermore, y and y_2 are equal and positive, whereas y_3 and y_4 are negative but the same size as y. Thus, $x_2 = x_3 = -x$ and $x_4 = x$; also, $y_3 = y_4 = -y$ and $y_2 = y$.

Now, by use of the definition of the sine of an angle, we have

$$\sin \theta = \frac{y}{r}, \qquad \sin(180° - \theta) = \frac{y_2}{r} = \frac{y}{r},$$

$$\sin(180° + \theta) = \frac{y_3}{r} = -\frac{y}{r}, \qquad \sin(360° - \theta) = \frac{y_4}{r} = -\frac{y}{r}.$$

Similarly, each of the other functions of $180° - \theta$, $180° + \theta$, and $360° - \theta$ is equal to plus or minus that function of θ. Consequently, we can now state

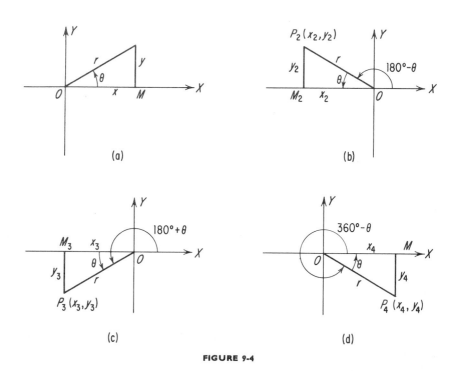

FIGURE 9-4

the reference angle theorem. It is: *Any trigonometric function of a given angle is numerically equal to the same function of the reference angle.* The algebraic sign of the function is determined by the quadrant in which the angle lies.

EXAMPLES

  1. The reference angle of 129° is 51°; hence, cos 129° is $\pm$ cos 51°. Since 129° is a second quadrant angle and since the cosine of a second quadrant angle is negative, we know that cos 129° = $-$ cos 51°.
  2. The reference angle of $-163°$ is 17°; hence, sin($-163°$) is $\pm$ sin 17°. Furthermore, $-163°$ is a third quadrant angle and the sine of a third quadrant angle is negative. Therefore, sin($-163°$) = $-$ sin 17°.
  3. tan 244° = tan 64° since 64° is the reference angle of 244°, and the tangent of a third quadrant angle is positive.

9.4. Functions of Negative Angles

The functions of negative angles can be found by use of the reference angle theorem, but it is often desirable to express a function of minus an angle in terms of the function of the angle. In order to be able to do this we shall consider the point $P(x, y)$ out a distance r on the terminal side of θ which is in standard position and the point $P_1(x_1, y_1)$ out a distance r on the terminal side of $-\theta$ which is also in standard position as shown in Fig. 9-5. Then $x_1 = x$ and $y_1 = -y$. Therefore,

$$\sin(-\theta) = \frac{y_1}{r} = -\frac{y}{r} = -\sin\theta$$

$$\cos(-\theta) = \frac{x_1}{r} = \frac{x}{r} = +\cos\theta$$

$$\tan(-\theta) = \frac{y_1}{x_1} = -\frac{y}{x} = -\tan\theta$$

$$\cot(-\theta) = \frac{x_1}{y_1} = -\frac{x}{y} = -\cot\theta$$

$$\sec(-\theta) = \frac{r}{x_1} = \frac{r}{x} = +\sec\theta$$

$$\csc(-\theta) = \frac{r}{y_1} = -\frac{r}{y} = -\csc\theta$$

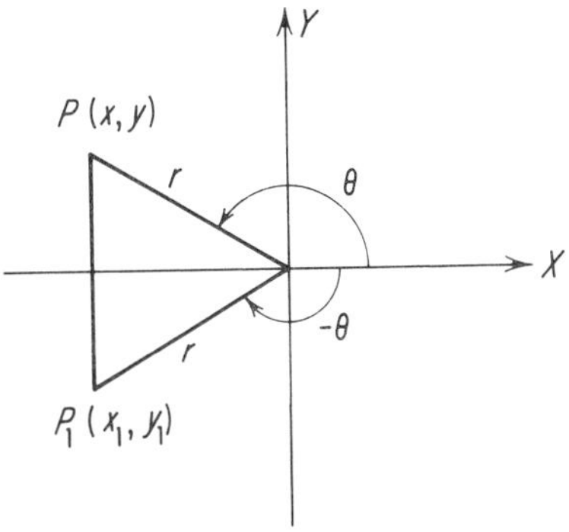

FIGURE 9-5

We now see that changing the sign of the angle does not change the value of the cosine or secant but that it does change the signs of the other four functions. This situation is sometimes expressed by stating that the cosine and secant of an angle are *even functions* and the other four are *odd functions*. These names are used because an even power of x is unchanged when x is replaced by $-x$, whereas an odd power is replaced by its negative if x is replaced by $-x$.

Exercise 9.1

Locate each point given in Problems 1 through 8 and give three other pairs of coordinates for each.

1. $(2, 45°)$, $(5, 140°)$, $(5, 215°)$, $(4, -70°)$.

2. $(3, 60°)$, $(-2, 130°)$, $(4, 230°)$, $(5, -40°)$.

3. $(4, 115°)$, $(3, -35°)$, $(6, 0°)$, $(2, -130°)$.

4. $(2, 180°), (4, 55°), (-3, 140°), (3, 280°)$.

5. $(5, 40°), (3, -45°), (-2, -60°), (5, 315°)$.

6. $(-2, 30°), (2, -30°), (4, 90°), (3, 240°)$.

7. $(-3, 270°), (7, -195°), (-4, -110°), (2, 210°)$.

8. $(5, 240°), (-3, 20°), (5, -70°), (-3, -50°)$.

Find the reference angle of each angle given in Problems 9 through 12.

9. $110°50', 215°, 290°, -35°$.

10. $140°, 220°, -310°, 310°20'$.

11. $340°, 260°, -300°30', 160°$.

12. $-305°10', -125°, 155°, 35°$.

Express the function of the given angle in each of Problems 13 through 24 in terms of the function of the reference angle.

13. $\sin 135°$.

14. $\cos 220°30'$.

15. $\tan 305°20'$.

16. $\cot(-320°)$.

17. $\sec 80°$.

18. $\csc 140°40'$.

19. $\cos 230°20'$.

20. $\tan 310°50'$.

21. $\cot 215°50'$.

22. $\sec(-40°)$.

23. $\csc(-280°)$.

24. $\sin 160°$.

25. $\tan 165°10'$.

26. $\cot(-135°)$.

27. $\sec(-70°)$.

28. $\csc 245°40'$.

29. $\sin 130°30'$.

30. $\cos 35°30'$.

31. $\tan 330°$.

32. $\cot 285°$.

Find the value of each of the following by use of the reference angle theorem and Table 1.

33. $\sin 220°20'$.

34. $\cos 130°50'$.

35. $\cot(-20°30')$.

36. $\tan 340°10'$.

37. $\cos(-309°20')$.

38. $\sin 330°20'$.

39. $\tan(-140°40')$.

40. $\cot 178°40'$.

9.5. Symmetry and Intercepts

This article will be devoted to a concept which, if understood and used, makes the sketching of a curve simpler than it otherwise would be.

We say that *two points P_1 and P_2 are located **symmetrically with respect to a third point** P if P is the midpoint of the segment that joins them.*

The points P_1 and P_2 in Fig. 9-6 are located symmetrically with respect to P since P is the midpoint of the segment between P_1 and P_2.

Two points P_1 and P_2 are said to be located symmetrically with respect to a line if the line is the perpendicular bisector of the segment that joins them.

The points P_1 and P_2 in Fig. 9-6 are symmetrically located with respect to the line L since it is the perpendicular bisector of the segment between P_1 and P_2.

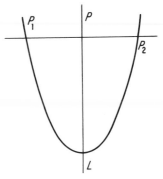

FIGURE 9-6

A curve is symmetrical with respect to a point P if for each point P_1 on the curve there is a point P_2 on it such that P_1 and P_2 are symmetrically located with respect to the point P. The point is called a center of symmetry.

A circle is symmetrical with respect to its center.

A curve is symmetrical with respect to a line if, for each point P_1 on the curve, there is a point P_2 on it such that P_1 and P_2 are symmetrically located with respect to the line. The line is called an axis of symmetry.

The curve in Fig. 9-6 is symmetrical with respect to the line L since corresponding to each point on the curve there is a second point on it such that L is the perpendicular bisector of the segment between the two points.

We shall be interested primarily in determining whether a curve is symmetrical with respect to the origin or a coordinate axis and shall give three theorems that furnish appropriate tests.

If a polar equation is unchanged or multiplied by -1

(a) *by replacing θ by $180° + \theta$, the curve is symmetrical with respect to the pole;*

(b) *by replacing θ by $-\theta$, the curve is symmetrical with respect to the polar axis;*

(c) *by replacing θ by $180 - \theta$, the curve is symmetrical with respect to the normal axis.*

If we look at Fig. 9-7, we see that 0 is the midpoint of P_2P since OP and OP_2 are each r in length; hence, test (a). Furthermore, by use of the congruent triangles, the polar axis is the perpendicular bisector of

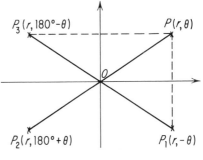

FIGURE 9-7

the segment PP_1; hence, test (b). Finally, the normal axis is the perpendicular bisector of PP_3; therefore, we have test (c).

As pointed out in Art. 9.2, a point has an infinite number of pairs of coordinates in the polar coordinate system; hence, it should not come as a surprise that there are other tests for symmetry besides those given above. For example, a curve is symmetrical with respect to the normal axis if its polar equation is unchanged by replacing r by $-r$ and θ by $-\theta$ simultaneously even though any symmetry with respect to the normal axis may fail to show up when the test (c) is applied. The curious and capable student can devise other tests for symmetry if he makes use of the various pairs of coordinates that P_1, P_2, and P_3 of Fig. 9-7 may have.

EXAMPLES

1. Test $f(r, \theta) = r - 3 \cos 2\theta = 0$ for symmetry.

SOLUTION

In testing for symmetry with respect to the pole, we shall see if the equation is changed by replacing θ by $180° + \theta$ as suggested in (a). Thus, we have

$$f(r, 180° + \theta) = r - 3 \cos 2(180° + \theta)$$
$$= r - 3 \cos(360° + 2\theta)$$
$$= r - 3 \cos 2\theta \text{ since } 360° + 2\theta \text{ and } 2\theta \text{ are coterminal}$$
$$= f(r, \theta)$$

Consequently, the curve is symmetrical with respect to the pole.

As stated in (b), we can test for symmetry with respect to the polar axis by replacing θ by $-\theta$. Thus, we get

$$f(r, -\theta) = r - 3 \cos 2(-\theta)$$
$$= r - 3 \cos 2\theta \text{ since the cosine is an even function}$$
$$= f(r, \theta)$$

Therefore, the curve is symmetrical with respect to the polar axis.

To test for symmetry with respect to the normal axis, we shall use the test (c) and have

$$f(r, 180° - \theta) = r - 3 \cos 2(180° - \theta)$$
$$= r - 3 \cos(360° - 2\theta)$$
$$= r - 3 \cos(-2\theta) \text{ since } 360° - 2\theta \text{ and } -2\theta \text{ are coterminal}$$
$$= r - 3 \cos 2\theta \text{ since the cosine is an even function.}$$

Consequently, the curve is symmetrical with respect to the normal axis.

The intersections of $f(r, \theta) = 0$ and the coordinate axes are called the *intercepts* of $f(r, \theta) = 0$. Consequently, we obtain the r coordinates of the intercepts by solving the equations in r that are obtained by putting $\theta = 0°, 90°, 180°, 270°, \ldots$ in $f(r, \theta) = 0$.

2. Find the intercepts of $f(r, \theta) = r - 3 \cos 2\theta = 0$.

SOLUTION

$f(r, 0°) = r - 3 \cos 0° = r - 3 = 0$ for $r = 3$; hence, $(3, 0°)$ is an intercept.

$f(r, 90°) = r - 3 \cos 180° = r + 3 = 0$ for $r = -3$; hence, $(-3, 90°)$ is an intercept.

$f(r, 180°) = r - 3 \cos 360° = r - 3 = 0$ for $r = 3$; hence, $(3, 180°)$ is an intercept.

$f(r, 270°) = r - 3 \cos 540° = r + 3 = 0$ for $r = -3$; hence, $(-3, 270°)$ is an intercept.

9.6. Construction of the Graph

If we have an equation $f(r, \theta) = 0$, we can sketch its graph by assigning a set of values to θ, evaluating each corresponding value of r, locating the points (r, θ), and drawing a smooth curve through them. The amount of labor is usually reduced if we make use of the intercepts and any symmetry that may exist.

The problem of determining the set of values to assign to θ is often a troublesome one. It is ordinarily sufficient to begin with $\theta = 0°$ and assign the multiples of $30°$ and $45°$ to θ until we have enough points to determine the graph or the desired part of it. At times, it may be advisable to assign values between the multiples of $30°$ and $45°$. This is likely to be the situation if a multiple of θ occurs in the given equation.

EXAMPLES

1. Sketch the graph of $f(r, \theta) = r - 3 \cos 2\theta = 0$.

SOLUTION

We found in the examples in Art. 9.5 that the curve is symmetrical with respect to both coordinate axis and the origin and that the intercepts are $(3, 0°)$, $(-3, 90°)$, $(3, 180°)$ and $(-3, 270°)$. We shall now continue collecting data on the curve by making a table of values of r and θ that has 2θ and $\cos 2\theta$ as intermediate steps between the values of θ and the corresponding values of r.

θ	0°	15°	22½°	30°	45°	60°	67½°	75°	90°
2θ	0°	30°	45°	60°	90°	120°	135°	150°	180°
$\cos 2\theta$	1	$\sqrt{3}/2$	$\sqrt{2}/2$	½	0	$-½$	$-\sqrt{2}/2$	$-\sqrt{3}/2$	-1
r	3	$3\sqrt{3}/2$	$3\sqrt{2}/2$	1.5	0	-1.5	$-3\sqrt{3}/2$	$-3\sqrt{3}/2$	-3

The half loops numbered 1 and 2 in Fig. 9-8 are obtained by locating the points (r, θ) indicated in the table. Those numbered 8 and 7 can be obtained from loops 1 and 2 by use of the symmetry with respect to the polar axis and the others from these four by use of symmetry with respect to the normal axis. We could have begun with half loops 1 and 2 and have obtained 5 and 6 by use of symmetry with respect to the pole, then the others could have been obtained by use of symmetry with respect to either axis. The half loops are numbered in the order in which they would have been obtained by using increasing values of θ. The curve is called a four-leaf rose.

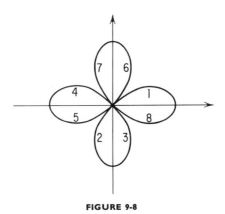

FIGURE 9-8

2. Test $f(r, \theta) = r - 2 \cos \theta - 3 = 0$ for symmetry, find its intercepts and sketch the curve.

SOLUTION

$$\text{Since } f(r, 180° + \theta) = r - 2 \cos(180° + \theta) - 3$$
$$= r + 2 \cos \theta - 3 \neq f(r, \theta),$$

we can only say that any symmetry that may exist with respect to the pole does not show up with the test used.

We shall now test for symmetry with respect to the polar axis by replacing θ by $-\theta$. Thus,

$$f(r, -\theta) = r - 2 \cos(-\theta) - 3$$
$$= r - 2 \cos \theta - 3 = f(r, \theta),$$

and the curve is symmetrical with respect to the polar axis. Consequently, if we get the graph for values of θ from 0° to 180°, the part for θ from 180° to 360° can be obtained by use of symmetry. Finally, replacing θ by $180° - \theta$, we have

$$f(r, 180° - \theta) = r - 2 \cos(180° - \theta) - 3$$
$$= r + 2 \cos \theta - 3 \neq f(r, \theta).$$

Consequently, we know that any symmetry that may exist relative to the normal axis does not show up with the test we have used.

We find the r coordinate of an intercept by assigning an integral multiple of 90° to θ and solving the resulting equation for r. Thus,

$$f(r, 0°) = r - 2 \cos 0° - 3 = r - 5 = 0 \text{ for } r = 5;$$

hence, $(5, 0°)$ is an intercept.

$$f(r, 90°) = r - 2 \cos 90° - 3 = r - 3 = 0 \text{ for } r = 3;$$

hence, $(3, 90°)$ is an intercept.

$$f(r, 180°) = r - 2 \cos 180° - 3 = r - 1 = 0 \text{ for } r = 1;$$

hence, $(1, 180°)$ is an intercept.

$$f(r, 270°) = r - 2 \cos 270° - 3 = r - 3 = 0 \text{ for } r = 3;$$

hence, $(3, 270°)$ is an intercept.

We shall now make a table of values of θ and r that gives $\cos \theta$ as an intermediate value.

θ	$0°$	$30°$	$45°$	$60°$	$90°$	$120°$	$135°$	$150°$	$180°$
$\cos \theta$	1	$\sqrt{3}/2$	$\sqrt{2}/2$	$\frac{1}{2}$	0	$-\frac{1}{2}$	$-\sqrt{2}/2$	$-\sqrt{3}/2$	-1
r	5	$\sqrt{3} + 3$	$\sqrt{2} + 3$	4	3	2	$-\sqrt{2} + 3$	$-\sqrt{3} + 3$	1

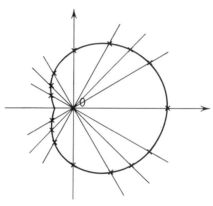

FIGURE 9-9

The upper half of the curve in Fig. 9-9 was obtained from the values in the table and the lower half from the symmetry with respect to the polar axis. The curve is called a cardioid.

Exercise 9.2

Find the intercepts, test for symmetry with respect to the pole and coordinate axes, and sketch the graph of the equation in each of the following problems.

1. $r = 4$.   **2.** $r = 3$.

3. $r = -2$.   **4.** $r = -5$.

5. $\theta = 45°$.   **6.** $\theta = 120°$.

7. $\theta = 240°$.

8. $\theta = 330°$.

9. $r = 4 \cos \theta$.

10. $r = 2 \sin \theta$.

11. $r = 6 \sin \theta$.

12. $r = 8 \cos \theta$.

13. $r = \sin 2\theta$.

14. $r = \sin 4\theta$.

15. $r = \cos 4\theta$.

16. $r = \cos 2\theta$.

17. $r = \cos 3\theta$.

18. $r = \cos 5\theta$.

19. $r = \sin 5\theta$.

20. $r = \sin 3\theta$.

21. $r = 2(\sin \theta - 1)$.

22. $r = 2(\cos \theta + 1)$.

23. $r = 3 \sin \theta - 4$.

24. $r = 3 \cos \theta + 2$.

25. $r = \sec \theta$.

26. $r = \tan \theta$.

27. $r = \cot \theta$.

28. $r = \csc \theta$.

29. $r^2 = \sin \theta$.

30. $r^2 = \cos \theta$.

31. $r^2 = \cos^2 2\theta$.

32. $r^2 = 4 \sin^2 2\theta$.

33. $r = \dfrac{2}{1 + \sin \theta}$.

34. $r = \dfrac{2}{1 - \sin \theta}$.

35. $r = \dfrac{3}{1 + \cos \theta}$.

36. $r = \dfrac{4}{1 - \cos \theta}$.

37. $r = \dfrac{4}{1 - 2 \cos \theta}$.

38. $r = \dfrac{6}{1 + 3 \sin \theta}$.

39. $r = \dfrac{1}{3 - \sin \theta}$.

40. $r = \dfrac{2}{2 + \cos \theta}$.

9.7. Relations between Rectangular and Polar Coordinates

We shall now use Fig. 9-10 in deriving four equations which can be used for changing an equation from rectangular or polar coordinates to the other. If we use the definition of the cosine and sine of an angle we have $\cos \theta = x/r$ and $\sin \theta = y/r$. Now solving for x and y, we have

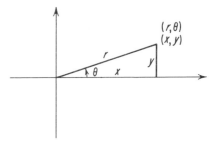

FIGURE 9-10

(1) $x = r \cos \theta$ and $y = r \sin \theta$.

Equations (1) can be used for changing an equation from the rectangular to the polar coordinate system. If we use the Pythagorean theorem and the definition of the tangent of an angle, we have $r^2 = x^2 + y^2$ and $\tan \theta = y/x$. Solving for r and θ, and using the usual symbolism, we have

(2) $$r = \sqrt{x^2 + y^2} \quad \text{and} \quad \theta = \arctan \frac{y}{x}$$

where the last equation is the symbolic form for θ *is an angle whose tangent is* y/x. Equations (2) can be used for changing from polar to rectangular form. Quite often the work of changing from one type of coordinate system to the other is simplified if a combination of equations (1) and (2) is used rather than just equations (1) or (2) alone.

EXAMPLES

1. Express $x^2 - 2y + y^2 = 0$ in terms of polar coordinates.

SOLUTION

If we replace y by $r \sin \theta$, as given in equations (1) and $x^2 + y^2$ by r^2 as given in equations (2), the given equation becomes $r^2 - 2r \sin \theta = 0$. Now, dividing by r, we see that the polar form of the given equation is $r - 2 \sin \theta = 0$. No part of the curve was lost in dividing through by r since the graph of $r - 2 \sin \theta = 0$ passes through the pole.

2. Express $r = \dfrac{2}{1 + \sin \theta}$ in rectangular form.

SOLUTION

We shall first clear of fractions by multiplying through by $1 + \sin \theta$. Thus we get $r + r \sin \theta = 2$. We can now replace r by $\sqrt{x^2 + y^2}$ and $r \sin \theta$ by y and have $\sqrt{x^2 + y^2} + y = 2$; hence, $\sqrt{x^2 + y^2} = 2 - y$. Now squaring each member, we have $x^2 + y^2 = 4 - 4y + y^2$ and, collecting terms, we find that

$$x^2 = 4 - 4y$$

is the rectangular form of the given equation.

Exercise 9.3

Express each of the following equations in terms of another type of coordinate system.

1. $x = 2$.   **2.** $r \cos \theta = 2$.

3. $r \sin \theta = 5$.   **4.** $y = 5$.

5. $r = 4$.

6. $x^2 + y^2 = 16$.

7. $x^2 + y^2 = 9$.

8. $r = 3$.

9. $r = \dfrac{1}{\cos \theta + 2 \sin \theta}$.

10. $x + 2y = 1$.

11. $2x - 3y = 5$.

12. $r = \dfrac{5}{2 \cos \theta - 3 \sin \theta}$.

13. $r \sin^2 \theta = 4 \cos \theta$.

14. $y^2 = 4x$.

15. $4x^2 + 9y^2 = 36$.

16. $4r^2 + 5r^2 \sin^2 \theta = 36$.

17. $\sin^2 \theta = r \cos^3 \theta$.

18. $y^2 = x^3$.

19. $(x - a)^2 + (y - b)^2 = a^2 + b^2$.

20. $r = 2a \cos \theta + 2b \sin \theta$.

21. $r^2(\cos^2 \theta - \sin^2 \theta) = 2 \sin \theta \cos \theta$.

22. $x^4 - y^4 = 2xy$.

23. $(x^2 + y^2)^2 = 2a^2xy$.

24. $r^2 = 2a^2 \sin \theta \cos \theta$.

25. $r \cos \theta = \sin^2 \theta$.

26. $(x^2 + y^2)x = y^2$.

27. $x(x^2 + y^2) = a(3x^2 - y^2)$.

28. $r \cos \theta = a(3 \cos^2 \theta - \sin^2 \theta)$.

10

MORE EXPONENTS AND
SOME RADICALS

10.1. Negative Exponents

We have studied positive integral and zero exponents and have found that

$$(1) \qquad a^m a^n = a^{m+n}.$$

$$(2) \qquad a^m/a^n = a^{m-n}, \qquad m \geq n, \qquad a \neq 0.$$

$$(3) \qquad (ab)^n = a^n b^n.$$

$$(4) \qquad (a/b)^n = a^n/b^n, \qquad b \neq 0.$$

$$(5) \qquad (a^m)^n = a^{mn}.$$

$$(6) \qquad a^0 = 1, \qquad a \neq 0.$$

We shall now decide on the meaning of a negative integral exponent in such a way that the laws given above for positive integral exponents hold.

We shall begin with a^{-n}, n positive, and multiply it by $a^n/a^n = 1$ and have

$$a^{-n} = a^{-n}\frac{a^n}{a^n}, \qquad a \neq 0$$

$$= \frac{a^{-n+n}}{a^n}, \text{ adding exponents,}$$

$$= \frac{a^0}{a^n} = \frac{1}{a^n}, \text{ since } a^0 = 1.$$

Consequently we shall define a^{-n} by the equation

$$\boldsymbol{a^{-n} = \frac{1}{a^n}, \qquad a \neq 0.}$$

We can now remove the restriction that requires m to be not smaller than n in division since we have a meaning for a number to a negative power. If an expression contains negative exponents we can get rid of them by multiplying by the base with the corresponding positive exponent and then dividing by the same quantity. This can be done before or after other indicated operations are performed.

EXAMPLES

1. $2^{-3} = 2^{-3}\dfrac{2^3}{2^3} = \dfrac{2^0}{2^3} = \dfrac{1}{2^3} = \dfrac{1}{8}.$

2. $3^{-2}3^{-4} = 3^{-2-4} = 3^{-6} = 3^{-6}\dfrac{3^6}{3^6} = \dfrac{3^0}{3^6} = \dfrac{1}{729}.$

3. $\dfrac{5^{-5}}{5^{-1}} = 5^{-5-(-1)} = 5^{-4} = 5^{-4}\dfrac{5^4}{5^4} = \dfrac{5^0}{5^4} = \dfrac{1}{625}.$

4. $(3^{-2})^{-3} = 3^{(-2)(-3)} = 3^6 = 729.$

5. $\dfrac{x^{-a}y^b}{w^{-c}} = \dfrac{x^{-a}y^b}{w^{-c}} \times \dfrac{x^a\,w^c}{x^a\,w^c} = \dfrac{x^0y^bw^c}{w^0x^a} = \dfrac{y^bw^c}{x^a}.$

We shall agree that an expression which involves negative exponents is simplified if all the reductions that can be made by use of the laws stated at the beginning of this article are made and if the result is expressed without zero or negative exponents. Furthermore, if the result is a fraction, all common factors should be removed from numerator and denominator.

EXAMPLES

6. Simplify $\dfrac{3^{-1} + 3^{-3}}{3^{-2}}.$

SOLUTION

We can eliminate all negative exponents by multiplying by $3^3/3^3$. Thus we have

$$\frac{3^{-1}+3^{-3}}{3^{-2}} = \frac{3^{-1}+3^{-3}}{3^{-2}} \times \frac{3^3}{3^3}$$

$$= \frac{3^{-1+3}+3^{-3+3}}{3^{-2+3}}$$

$$= \frac{3^2+3^0}{3} = \frac{9+1}{3} = \frac{10}{3}.$$

7. Simplify
$$\left(\frac{2a^{-3}b^2}{a^2c^{-2}}\right)^{-3}.$$

SOLUTION

Since we have a fraction with a numerator and denominator consisting of several factors, we must raise each factor to the power -3 and have

$$\left(\frac{2a^{-3}b^2}{a^2c^{-2}}\right)^{-3} = \frac{2^{-3}(a^{-3})^{-3}(b^2)^{-3}}{(a^2)^{-3}(c^{-2})^{-3}}$$

$$= \frac{2^{-3}a^9b^{-6}}{a^{-6}c^6}$$

$$= \frac{2^{-3}a^9b^{-6}}{a^{-6}c^6} \times \frac{2^3}{2^3} \times \frac{a^6}{a^6} \times \frac{b^6}{b^6}$$

$$= \frac{2^0a^{9+6}b^0}{2^3a^0b^6c^6} = \frac{a^{15}}{8b^6c^6}.$$

8. Simplify
$$\frac{x^{-1}+y^{-1}}{x^{-3}+y^{-3}}.$$

SOLUTION

We shall begin by multiplying numerator and denominator by x^3y^3, since that will get rid of the negative exponents. Thus

$$\frac{x^{-1}+y^{-1}}{x^{-3}+y^{-3}} = \frac{x^{-1}+y^{-1}}{x^{-3}+y^{-3}} \times \frac{x^3y^3}{x^3y^3} = \frac{x^2y^3+x^3y^2}{y^3+x^3}$$

$$= \frac{x^2y^2(y+x)}{(y+x)(y^2-yx+x^2)} = \frac{x^2y^2}{y^2-yx+x^2}.$$

Exercise 10.1

Simplify each of the following in keeping with the statement just after Example 5.

1. 2^{-3}.

2. 5^{-2}.

3. 6^{-1}.

4. 3^{-4}.

5. $5^{-1}5^{-2}$.

6. $3^{-2}3^{-3}$.

7. $6^3 6^{-5}$.

8. $2^{-6}2^4$.

9. $\dfrac{3^{-4}}{3^{-2}}$.

10. $\dfrac{4^{-1}}{4^{-3}}$.

11. $\dfrac{5^{-3}}{5^{-4}}$.

12. $\dfrac{2^{-5}}{2^{-2}}$.

13. $(2^{-1})^{-3}$.

14. $(3^2)^{-1}$.

15. $(4^{-1})^2$.

16. $(5^{-2})^{-2}$.

17. a^{-3}.

18. x^{-2}.

19. y^{-1}.

20. b^{-4}.

21. $a^{-2}a^{-4}$.

22. $b^{-1}b^0$.

23. $b^{-5}b^2$.

24. $c^{-3}c^{-5}$.

25. $\dfrac{x^{-2}}{x^{-3}}$.

26. $\dfrac{y^{-1}}{y^{-4}}$.

27. $\dfrac{a^{-2}}{a^3}$.

28. $\dfrac{b^2}{b^{-1}}$.

29. $\dfrac{a^{-2}b^2}{ab^{-3}}$.

30. $\dfrac{a^{-1}b^{-3}}{a^0b^{-1}}$.

31. $\dfrac{a^{-4}b^{-2}}{a^3b^{-1}}$.

32. $\dfrac{a^3b^{-2}}{a^{-2}b^0}$.

33. $(a^{-1}b^2)^{-4}$.

34. $(a^0b^{-3})^2$.

35. $(c^{-2}d)^{-2}$.

36. $(c^{-3}d^{-1})^3$.

37. $\left(\dfrac{a^2}{b^{-3}}\right)^{-3}$.

38. $\left(\dfrac{a^{-1}}{b^2}\right)^2$.

39. $\left(\dfrac{a^{-2}}{b}\right)^{-3}$.

40. $\left(\dfrac{a^{-1}}{b^0}\right)^3$.

41. $a^{-1} - b^{-1}$.

42. $ab^{-1} + a^{-1}b$.

43. $\dfrac{a}{b^{-1}} + \dfrac{b}{a^{-1}}$.

44. $2a^{-3} + \dfrac{3}{a^3}$.

45. $\dfrac{a^{-1} - b^{-2}}{b^{-1}}$.

46. $\dfrac{a^{-3} - b^{-2}}{a^{-1}}$.

47. $\dfrac{a^{-2} + b^{-1}}{a^{-3}}$.

48. $\dfrac{a^{-1} + b^{-3}}{a^{-2}}$.

49. $\dfrac{a^{-1} - b^{-1}}{a^{-2} - b^{-2}}$.

50. $\dfrac{a^{-2} + b^{-1}}{a^{-3} + b^{-1}a^{-1}}$.

51. $\dfrac{a^{-1} + b^{-1}}{(a-b)^{-1}}$.

52. $\dfrac{(a-b)^{-1}}{b^{-2} - a^{-2}}$.

10.2. Roots

Since $8^2 = (-8)^2 = 64$, we say that 8 and -8 are square roots of 64. We know that $(-4)^3 = -64$ and say that -4 is a cube root of -64; furthermore, 4 is called a cube root of 64 and a fifth root of 1024, since $4^3 = 64$ and $4^5 = 1024$. These examples illustrate the general statement that *a number a is called an nth root of b if and only if $a^n = b$*. Since the square of a number and of its negative are equal there are two square roots of any positive number. Furthermore, there is no positive nth root of a negative number because any integral power of a positive number is positive.

If there is a positive nth root of a number it is called the *principal* nth *root*. If there is a negative nth root of a number but no positive nth root of it, then the negative root is called the principal nth root.

EXAMPLES

1. The principal fourth root of 81 is 3, since 3 is positive and $3^4 = 81$.
2. The principal fifth root of -32 is -2, since there is no positive fifth root of -32 and $(-2)^5 = -32$.

The symbol $\sqrt[n]{a}$ is used to indicate the principal nth root of a. This symbol is called a *radical of order* **n**. The letter a is called the *radicand*, and n is the *index* of the radical. Since, by the definition of an nth root of a number, the nth power of an nth root gives the number, we know that

$$(\sqrt[n]{a})^n = a.$$

If there is a decimal number whose nth power gives a, then that number is an nth root of a, but if there is no terminating decimal number whose nth power gives a then the nth root cannot be expressed exactly as a terminating decimal and we must be content with a decimal approximation.

10.3. Extracting Square Roots

An approximation to an nth root can be found in appropriate tables or by use of logarithms. There is a table of square roots and cube roots in the back of this book. The use of logarithms will be explained in the chapter on logarithms. There is another procedure that can be used for finding an approximation to the square root of a number and we shall now present it.

EXAMPLES

1. Find a four-significant digit approximation to $\sqrt{713.465}$.

SOLUTION

We begin at the decimal point and mark off the digits in pairs in both directions as long as there is a pair to mark off. If there is an odd number of digits to the right of the decimal point, we add a zero so as to be able to mark off in pairs. If there is an odd number of digits to the left of the decimal point, we mark off in pairs as many times as possible and then leave the digit on the left by itself. If we do this to the given number and leave a space between each pair, we have 7 13. 46 50. We now

$$
\begin{array}{r}
2\ \ 6.\ \ 7\ \ 1 \\
\hline
\sqrt{7\ 13.\ 46\ 50} \\
4 \\
\hline
3\ 13 \\
2\ 76 \\
\hline
37\ \ 46 \\
36\ \ 89 \\
\hline
57\ 50 \\
53\ 41 \\
\hline
4\ 09
\end{array}
$$

$20(2) + 6 = 46$

$6(46) =$

$20(26) + 7 = 527$

$7(527) =$

$20(267) + 1 = 5341$

$1(5341) =$

find the largest number whose square is 7 or less. It is 2, we put it over 7, square it, subtract the square from 7 and get 3. Then we bring down the next pair of digits and have 313. We now multiply the 2 by 20 and add the largest possible number such that this number times $(40 + $ this number) is 313 or less. The number is 6, we put it above 13, add it to 40 and multiply 6 by $(40 + 6)$. The product is 276 and is placed below and subtracted from 313. Thus we get 37 and then bring down the next pair of digits. We now multiply the 26 by 20 and add the largest possible number such that this number times $(520 + $ this number) is 3746 or less The number is 7, we put it above 46, add it to 520 and multiply 7 by $(520 + 7)$. The product is 3689 and is placed below and subtracted from 3746. Thus we get 57 and then bring down the next pair of digits. We now multiply the 267 by 20 and add the largest possible number such that this number times $(5340 + $ this number) is 5750 or less. The number is 1, we put it above 50, add it to 5340 and multiply 1 by $(5340 + 1)$. The product is 5341 and is placed below and subtracted from 5750. The remainder is less than half of 5341. We are now through except for placing the decimal point. It is placed so that the square root has half as many decimal places as the number whose square root was taken. Thus we have

$$\sqrt{713.4650} = 26.71.$$

This result can be checked by squaring 26.71, $26.71 + .01 = 26.72$ and $26.71 - .01 = 26.70$ and seeing that 26.71^2 is nearer 713.465 than is the square of either of the others.

2. Find $\sqrt{76.3}$ to three significant figures.

SOLUTION

Since we want this square root to three figures we must take it to two decimal places; hence we must have four decimal places in the number whose square root we are taking. Consequently we must add three zeros and have 76.3000. We now mark this off in pairs both ways from the decimal point and continue as in the other example. The work is shown but the explanation is omitted.

$$
\begin{array}{r}
8 \quad 7 \quad 3 \\
\sqrt{76.\ 30\ 00} \\
64 \\
\hline
12\ 30 \\
11\ 69 \\
\hline
61\ 00 \\
52\ 29 \\
\hline
8\ 71
\end{array}
$$

$$20(8) + 7 = 167$$
$$7(167) =$$
$$20(87) + 3 = 1743$$
$$3(1743) =$$

The last subtrahend, 8 71, is less than half of 1743; hence,

$$\sqrt{76.3} = 8.73$$

Exercise 10.2

Find the indicated roots in Problems 1 through 12.

1. $\sqrt{9}$. **2.** $\sqrt{121}$. **3.** $\sqrt{289}$. **4.** $\sqrt{676}$.

5. $\sqrt[3]{27}$. **6.** $\sqrt[3]{343}$. **7.** $\sqrt[5]{243}$. **8.** $\sqrt[6]{64}$.

9. $\sqrt[3]{-64}$. **10.** $\sqrt[3]{-27}$. **11.** $\sqrt[5]{-32}$. **12.** $\sqrt[7]{-1}$.

Find the square root of each number in Problems 13 through 24 to three significant digits unless it comes out exactly as two digits.

13. 14.44 **14.** .2809. **15.** 5329. **16.** 428.49.

17. 76.23. **18.** 3.819. **19.** 281.31. **20.** 5872.3.

21. 7.623. **22.** 38.19. **23.** 28.131. **24.** 587.23.

Find the square root of the number in each of Problems 25 through 28 to four significant digits.

25. 1.53. **26.** 81.29. **27.** 608.713. **28.** 3.0219.

29. If the sides of a right triangle are 39 and 47, find the hypotenuse to three digits.

30. The hypotenuse and a side of triangle are 59 and 42, find the other side to two digits.

31. A guy wire is attached to a telephone pole 11.4 feet from the ground and also at a point 8.7 feet from the base of the pole. How long, to three digits, must the wire be if 2.8 feet are needed for connections?

32. Find to three digits the altitude of an isosceles triangle if the base is 42 and the equal sides are each 37.

33. The altitude and one of the equal sides of an isosceles triangle are 11.76 and 14.18, find the base to four digits.

34. The perimeter of an equilateral triangle is 63.6. Find the altitude to three digits.

35. Find the length of the diagonal of a square to three digits if its sides are 27.4 feet.

36. The sides of a rectangle are 21.7 and 36.4. Find the length of the diagonal to three digits.

10.4. Fractional Exponents

We shall now give an interpretation of fractional exponents. If we assume that

$$(1) \qquad (a^m)^n = a^{mn}$$

holds for $m = 1/n$, then it becomes

$$(a^{1/n})^n = a^{n/n} = a.$$

However, we found in Art. 10.2 that

$$(\sqrt[n]{a})^n = a;$$

consequently, we take

$$(2) \qquad a^{1/n} = \sqrt[n]{a}$$

as the definition of a fractional exponent with numerator one. If (1) holds for $m = 1/p$ and $n = q$, we have

$$(a^{1/p})^q = a^{q/p}$$

however, $\qquad\qquad a^{1/p} = \sqrt[p]{a};$

hence, $\qquad\qquad (\sqrt[p]{a})^q = a^{q/p}.$

Therefore we shall take, for positive a,

$$(3) \qquad a^{q/p} = (\sqrt[p]{a})^q$$

as the definition of a number to a fractional power. Furthermore,

$$a^{q/p} = (a^q)^{1/p} = \sqrt[p]{a^q} \qquad \text{for } a > 0.$$

Consequently, we have

(4) $$a^{q/p} = (\sqrt[p]{a})^q = \sqrt[p]{a^q}, \qquad a > 0$$

EXAMPLES

  1. $64^{1/3} = \sqrt[3]{64} = 4.$
  2. $16^{3/4} = (\sqrt[4]{16})^3 = 2^3 = 8.$

We can use (4) to change from either radical or fractional exponent form to the other as illustrated below.

EXAMPLES

  3. $\sqrt[3]{5x^9y^2} = 5^{1/3}x^{9/3}y^{2/3} = 5^{1/3}x^3y^{2/3}.$
  4. $(\sqrt[3]{3x^2})^2 = (3^{1/3}x^{2/3})^2 = 3^{2/3}x^{4/3}.$
  5. $5a^{1/2}y^{3/2} = 5(ay^3)^{1/2} = 5\sqrt{ay^3}.$
  6. $x^{3/5}y^{-2/3} = x^{3/5}y^{-2/3}\dfrac{y^{2/3}}{y^{2/3}} = \dfrac{x^{3/5}}{y^{2/3}} = \dfrac{\sqrt[5]{x^3}}{\sqrt[3]{y^2}}.$

10.5. Simplification of Exponential Expressions

As in Art. 10.1, we shall say that an expression that involves exponents is simplified if all reductions that can be made by means of the laws of exponents given in Art. 10.1 have been made and the result expressed without negative exponents anywhere, and we shall add the further requirement that there shall be no fractional exponents or radicals left in the denominator.

EXAMPLES

  1. Simplify

$$\frac{3x^3y^{-2}w^{-1/3}}{12t^{1/4}}.$$

SOLUTION

We shall first point out the things which need to be changed in order to have the expression in simplified form. The two negative exponents and the fractional one in the denominator must be eliminated; furthermore, we must be certain that we do not introduce some other objectionable feature in eliminating these. In order to get rid of y^{-2}, we multiply the numerator by y^2 and offset this by multiplying the denominator by it also. We could get rid of $w^{-1/3}$ by multiplying numerator and denominator by $w^{1/3}$, but this would introduce a fractional exponent in the denominator. We shall eliminate $w^{-1/3}$ by multiplying by w/w since this gets rid of the objectionable factor, $w^{-1/3}$, without bringing in another one. We shall multiply the denominator by $t^{3/4}$ since that gives t an integral exponent; hence we must multiply the numerator by $t^{3/4}$ also. If we do these things and remove the common factor 3, we have

$$\frac{3x^3y^{-2}w^{-1/3}}{12t^{1/4}} = \frac{x^3y^{-2}w^{-1/3}}{4t^{1/4}} \frac{y^2wt^{3/4}}{y^2wt^{3/4}}$$

$$= \frac{x^3y^{-2+2}w^{-1/3+1}t^{3/4}}{4t^{1/4+3/4}y^2w} = \frac{x^3y^0w^{2/3}t^{3/4}}{4t^1y^2w} = \frac{x^3w^{2/3}t^{3/4}}{4ty^2w}.$$

2. Simplify

$$\left(\frac{4s^2u^{-2}}{9t^{1/2}}\right)\left(\frac{3s^{-1/3}t^{2/3}}{2u^{-1}}\right)^3.$$

SOLUTION

$$\frac{4s^2u^{-2}}{9t^{1/2}}\left(\frac{3s^{-1/3}t^{2/3}}{2u^{-1}}\right)^3 = \frac{4s^2u^{-2}}{9t^{1/2}} \times \frac{27s^{-1}t^2}{8u^{-3}}, \text{ cubing,}$$

$$= \frac{108st^2u^{-2}}{72t^{1/2}u^{-3}}, \text{ multiplying,}$$

$$= \frac{3st^2u^{-2}}{2t^{1/2}u^{-3}} \times \frac{u^3t^{-1/2}}{u^3t^{-1/2}} = \frac{3st^{3/2}u}{2}.$$

We could have eliminated $t^{1/2}$ by multiplying by $t^{1/2}/t^{1/2}$, but this would have made it necessary to remove a factor t from numerator and denominator. We did not multiply by u^2 to get rid of u^{-2} since it was automatically eliminated in removing u^{-3}.

Exercise 10.3

Express each number in Problems 1 through 20 without exponents or radicals.

1. $9^{1/2}$ **2.** $8^{1/3}$.   **3.** $81^{1/4}$.   **4.** $32^{1/5}$.

5. $16^{3/2}$.   **6.** $27^{2/3}$.   **7.** $64^{5/6}$.   **8.** $81^{3/4}$.

9. $\left(\frac{16}{25}\right)^{3/2}$.   **10.** $\left(\frac{27}{8}\right)^{2/3}$.   **11.** $\left(\frac{81}{16}\right)^{5/4}$.   **12.** $\left(\frac{32}{3125}\right)^{3/5}$.

13. $8^{-1/3}$.   **14.** $25^{-1/2}$.   **15.** $243^{-1/5}$.   **16.** $32^{-1/5}$.

17. $125^{-2/3}$.   **18.** $36^{-3/2}$.   **19.** $32^{-3/5}$.   **20.** $81^{-5/4}$.

Simplify the following according to the statement made at the beginning of this article.

21. $a^{1/4}a^{1/3}$.   **22.** $b^{1/2}b^{1/5}$.   **23.** $c^{1/3}c^{1/5}$.   **24.** $d^{1/2}d^{1/7}$.

25. $\dfrac{a^{2/3}}{a^{1/2}}$.

26. $\dfrac{c^{3/4}}{c^{2/3}}$.

27. $\dfrac{d^{2/5}}{d^{1/4}}$.

28. $\dfrac{b^{3/5}}{b^{1/3}}$.

29. $a^{-3/4}$.

30. $a^{-2/3}$.

31. $a^{-1/2}$.

32. $a^{-3/5}$.

33. $\dfrac{a^{-1/2}}{b^{2/3}}$.

34. $\dfrac{a^{-2/3}}{b^{3/4}}$.

35. $\dfrac{a^{-3/4}}{b^{1/2}}$.

36. $\dfrac{a^{-2/5}}{b^{1/3}}$.

37. $(81a^{4/3}b^{4/5})^{1/4}$.

38. $(27a^{3/2}b^{3/4})^{1/3}$.

39. $(25x^{2/3}y^{2/5})^{1/2}$.

40. $(32a^{5/3}b^{5/2})^{1/5}$.

41. $(125a^{3/2}b^{-3/4})^{2/3}$.

42. $(16x^{4/5}y^{-4/3})^{3/4}$.

43. $(64a^{2/3}b^{2/5})^{-1/2}$.

44. $(243a^{5/3}b^{-5/4})^{-3/5}$.

45. $\dfrac{4^{1/2}x^{2/3}y^{-2/3}}{8^{1/3}x^{-1/3}y^{1/3}}$.

46. $\dfrac{9^{-1/2}a^{2/3}b^{-1/3}}{27^{-1/3}a^{1/3}b^{2/3}}$.

47. $\dfrac{16^{-3/4}a^{-1}b^{2/3}}{8^{-1/3}a^{1/2}b^{-1/3}}$.

48. $\dfrac{81^{-3/4}a^{-1/5}b^{1/3}}{27^{-1/3}a^{2/5}b^{-1/2}}$.

49. $\left(\dfrac{a^{2}b^{-3}}{a^{-1}b}\right)^{1/3}$.

50. $\left(\dfrac{8a^{-3}b^{3/2}}{c}\right)^{-2/3}$.

51. $\left(\dfrac{81a^{4}b^{-4/3}}{c}\right)^{-1/4}$.

52. $\left(\dfrac{243a^{-10}b^{5}}{c^{-2}}\right)^{-3/5}$.

53. $\dfrac{8a^{2}b^{-3}}{3c^{1/2}}\left(\dfrac{3a^{-1}b}{2c^{1/3}}\right)^{3}$.

54. $\dfrac{4a^{-1}b^{2}}{c^{1/3}}\left(\dfrac{2a^{2}b^{-1}}{c^{3/2}}\right)^{2}$.

55. $\left(\dfrac{4a^{3}b^{-2}}{c}\right)^{1/2}\dfrac{b^{1/3}c^{-1}}{a^{-1/2}}$.

56. $\left(\dfrac{27a^{-2}b^{3}}{c}\right)^{1/3}\dfrac{c^{-1/3}a^{-1/2}}{b^{1/2}}$.

$STOP$

10.6. Multiplication and Division of Radicals

There are rules or fomulas for use in multiplication and division of radicals, and we could develop them but shall not since we can do anything without them that we can by using them. Instead of deriving and using rules in connection with radicals, we shall change from radical to exponential form by making use of the relation

$$\sqrt[p]{a^{q}} = (\sqrt[p]{a})^{q} = a^{p/q}, \qquad a > 0,$$

perform the operations in terms of exponents, and then put the factors with fractional exponents back in radical form. We shall say a radical expression is simplified if the corresponding exponential expression is simplified and no factor of the radicand has an exponent as large as the index of the radical. The work will be simplified at times by making use of the fact that, for a positive,

$$\sqrt[n]{a^n} = a.$$

EXAMPLES

1. $\sqrt[3]{24} = \sqrt[3]{2^3 3} = (2^3 3)^{1/3} = 2(3^{1/3}) = 2\sqrt[3]{3}.$

2. $\sqrt{25a^3b^4} = (5^2a^3b^4)^{1/2} = 5a^{3/2}b^2 = 5a\,a^{1/2}b^2 = 5ab^2\sqrt{a}.$

3. $\sqrt{xy^2z^3}\,\sqrt[3]{x^3y^2z^{-1}} = (xy^2z^3)^{1/2}(x^3y^2z^{-1})^{1/3}$

$$= x^{1/2}yz^{3/2}xy^{2/3}z^{-1/3}$$
$$= xyx^{1/2}y^{2/3}z^{3/2-1/3}$$
$$= xyx^{1/2}y^{2/3}z^{7/6}$$
$$= xyzx^{1/2}y^{2/3}z^{1/6}$$
$$= xyz\sqrt{x}\sqrt[3]{y^2}\sqrt[6]{z}$$

4. $\dfrac{\sqrt[3]{x^2y^{-1}}}{\sqrt[4]{xt^3}} = \dfrac{(x^2y^{-1})^{1/3}}{(xt^3)^{1/4}} = \dfrac{x^{2/3}y^{-1/3}}{x^{1/4}t^{3/4}}$

$$= \frac{x^{2/3}y^{-1/3}}{x^{1/4}t^{3/4}} \times \frac{x^{-1/4}}{x^{-1/4}} \times \frac{y}{y} \times \frac{t^{1/4}}{t^{1/4}}$$

$$= \frac{x^{2/3-1/4}y^{-1/3+1}t^{1/4}}{x^{1/4-1/4}t^{3/4+1/4}y}$$

$$= \frac{x^{5/12}y^{2/3}t^{1/4}}{ty}$$

$$\text{since } \frac{2}{3} - \frac{1}{4} = \frac{8-3}{12} = \frac{5}{12} \text{ and } x^{1/4-1/4} = x^0 = 1$$

$$= \frac{\sqrt[12]{x^5}\sqrt[3]{y^2}\sqrt[4]{t}}{ty}$$

Exercise 10.4

Simplify the following radical expressions.

1. $\sqrt{25}.$ **2.** $\sqrt[3]{27}.$ **3.** $\sqrt[4]{256}.$ **4.** $\sqrt[5]{32}.$

5. $\sqrt[3]{24}.$ **6.** $\sqrt[4]{162}.$ **7.** $\sqrt[5]{96}.$ **8.** $\sqrt{147}.$

9. $\sqrt[5]{ab^5}.$ **10.** $\sqrt{a^3b^2}.$ **11.** $\sqrt[3]{a^3b^4}.$ **12.** $\sqrt[4]{a^7b^8}.$

13. $\sqrt{4a^2b^3}.$ **14.** $\sqrt{9a^4b^3}.$ **15.** $\sqrt{16a^3b^4}.$ **16.** $\sqrt{125a^5b^3}.$

17. $\sqrt[3]{24x^3y^4}$. **18.** $\sqrt[3]{81x^6y^4}$. **19.** $\sqrt[4]{32x^4y^6}$. **20.** $\sqrt[4]{162x^5y^7}$.

21. $\sqrt[4]{a^{-2}y^5}$. **22.** $\sqrt[5]{x^7y^{-1}}$. **23.** $\sqrt{x^{-1}y^3}$. **24.** $\sqrt[3]{a^5y^{-2}}$.

25. $\sqrt[5]{32x^{-4}}$. **26.** $\sqrt{8x^{-2}y^{-1}}$. **27.** $\sqrt[3]{54x^{-2}}$. **28.** $\sqrt[4]{a^{-3}b^{-5}}$.

29. $\sqrt{ab^3}\sqrt[3]{ab^4}$.   **30.** $\sqrt{a^3b^2}\sqrt[3]{a^4b^2}$.

31. $\sqrt[3]{a^5b}\sqrt[4]{a^3b^5}$.   **32.** $\sqrt[3]{a^2b^4}\sqrt[4]{a^9b^3}$.

33. $\sqrt{a^{-1}b}\sqrt[4]{a^3b^2}$.   **34.** $\sqrt{a^3b^{-1}}\sqrt[3]{a^2b^{-2}}$.

35. $\sqrt[3]{a^{-2}b^{-1}}\sqrt[5]{a^2b^{-2}}$.   **36.** $\sqrt[4]{a^{-3}b^5}\sqrt[5]{a^6b^{-1}}$.

37. $\dfrac{\sqrt{27x^{-1}y}}{\sqrt{3xy^{-3}}}$.   **38.** $\dfrac{\sqrt{98a^{-5}b}}{\sqrt{2a^{-3}b^{-5}}}$.

39. $\dfrac{\sqrt[3]{24a^4b^{-2}}}{\sqrt[3]{81a^{-1}b^2}}$.   **40.** $\dfrac{\sqrt[3]{54a^0b^{-4}}}{\sqrt[3]{16a^{-5}b}}$.

41. $\dfrac{\sqrt{9x^{-1}y^3}}{\sqrt[3]{27x^{-2}y^4}}$.   **42.** $\dfrac{\sqrt[3]{a^{-4}b^2}}{\sqrt{ab^{-1}}}$.

43. $\dfrac{\sqrt[5]{a^{-4}b^6}}{\sqrt[3]{a^4b^{-1}}}$.   **44.** $\dfrac{\sqrt[3]{ab^{-4}}}{\sqrt[5]{a^{-6}b^3}}$.

10.7. Addition of Radicals

Two radicals must have the same index and the same radicand if they are to be added. Radicals should be simplified before we decide whether they have the same radicand. If two radicals have the same index and the same radicand, we add them by adding their coefficients.

EXAMPLES

1. Simplify

$$\sqrt{50} + \sqrt{54} - \frac{3}{\sqrt{2}} - \sqrt{24}$$

SOLUTION

We shall first express each radical in terms of fractional exponents, then remove each factor that is a perfect square, rationalize the denominator of $3/2^{1/2}$, re-express in terms of radicals, and finally collect coefficients of like terms. Thus,

$$\sqrt{50} + \sqrt{54} - \frac{3}{\sqrt{2}} - \sqrt{24} = (5^2 2)^{1/2} + (3^2 6)^{1/2} - \frac{3}{2^{1/2}} - (2^2 6)^{1/2}$$

$$= 5(2^{1/2}) + 3(6^{1/2}) - \frac{3}{2^{1/2}} \frac{2^{1/2}}{2^{1/2}} - 2(6^{1/2})$$

$$= 5\sqrt{2} + 3\sqrt{6} - \frac{3}{2}\sqrt{2} - 2\sqrt{6}$$

$$= (5 - 1.5)\sqrt{2} + (3 - 2)\sqrt{6}$$

$$= 3.5\sqrt{2} + \sqrt{6}$$

2. Simplify

$$\sqrt{x^3 y} + \sqrt[3]{8xy^4} + \sqrt{4xy^3} - \sqrt[3]{x^4 y}$$

SOLUTION

1. We shall express each radical in terms of fractional exponents, take out each integral power of a factor, re-express in terms of radicals, and then collect coefficients of like terms. Thus,

$$\sqrt{x^3 y} + \sqrt[3]{8xy^4} + \sqrt{4xy^3} - \sqrt[3]{x^4 y} = (x^3 y)^{1/2} + (8xy^4)^{1/3} + (4xy^3)^{1/2} - (x^4 y)^{1/3}$$
$$= x^{3/2} y^{1/2} + 2x^{1/3} y^{4/3} + 2x^{1/2} y^{3/2} - x^{4/3} y^{1/3}$$
$$= xx^{1/2} y^{1/2} + 2x^{1/3} yy^{1/3} + 2x^{1/2} yy^{1/2} - xx^{1/3} y^{1/3}$$
$$= x\sqrt{xy} + 2y\sqrt[3]{xy} + 2y\sqrt{xy} - x\sqrt[3]{xy}$$
$$= (x + 2y)\sqrt{xy} + (2y - x)\sqrt[3]{xy}.$$

2. Some readers may prefer to make use of the fact that $\sqrt[n]{a^n} = a$ for $a > 0$ instead of changing from radical form to exponential and back again. If this is done, we have

$$\sqrt{x^3 y} + \sqrt[3]{8xy^4} + \sqrt{4xy^3} - \sqrt[3]{x^4 y} = \sqrt{x^2 xy} + \sqrt[3]{2^3 xy^3 y} + \sqrt{2^2 xy^2 y} - \sqrt[3]{x^3 xy}$$
$$= x\sqrt{xy} + 2y\sqrt[3]{xy} + 2y\sqrt{xy} - x\sqrt[3]{xy}$$
$$= (x + 2y)\sqrt{xy} + (2y - x)\sqrt[3]{xy}.$$

Exercise 10.5

Simplify

1. $\sqrt{12} + \sqrt{27} + \sqrt{48}$.   **2.** $\sqrt{20} + \sqrt{45} - \sqrt{80}$.

3. $\sqrt{72} - \sqrt{8} + \sqrt{18}$.   **4.** $\sqrt{75} + \sqrt{108} - \sqrt{243}$.

5. $\sqrt{8} - \sqrt{12} - \sqrt{18} + \sqrt{27}$.

6. $\sqrt{20} + \sqrt{162} - \sqrt{50} - \sqrt{45}$.

7. $\sqrt{12} + \sqrt{98} - \sqrt{108} - \sqrt{32}$.

8. $\sqrt{108} + \sqrt{54} - \sqrt{48} - \sqrt{24}$.

9. $\sqrt{72} - \sqrt[3]{81} - \sqrt{8} - \sqrt[3]{192}$.

10. $\sqrt{75} + \sqrt[3]{54} - \sqrt{147} - \sqrt[3]{128}$.

11. $\sqrt{150} + \sqrt[3]{32} - \sqrt{216} + \sqrt[3]{256}$.

12. $\sqrt[3]{108} + \sqrt{108} + \sqrt[3]{500} - \sqrt{500}$.

13. $\sqrt{xy^5} + \sqrt{x^3y^3} - \sqrt{x^5y}$. $\qquad$ **14.** $\sqrt{xy^3} + \sqrt{4x^3y} + \sqrt{x^3y^3}$.

15. $\sqrt{x^4y} + \sqrt{9x^2y^3} + \sqrt{4y^5}$. $\qquad$ **16.** $\sqrt{x^5y} + \sqrt{x^3y^3} - \sqrt{4xy^4}$.

17. $\sqrt{xy^4} - \sqrt{x^4y} + \sqrt{xy^2} - \sqrt{x^2y}$.

18. $\sqrt{x^5y^3} - \sqrt{9x^3y^2} + \sqrt{9xy^5} + \sqrt{4x^5y^4}$.

19. $\sqrt[3]{8x^4y^2} + \sqrt[3]{27x^5y^4} - \sqrt{4x^2y} - \sqrt{9xy^2}$.

20. $\sqrt[3]{16x^4y^3} + \sqrt[3]{24x^3y^4} - \sqrt{16x^4y^3} + \sqrt{24x^3y^4}$.

21. $\sqrt{\dfrac{x^3y}{2}} - \sqrt{\dfrac{xy^3}{8}} + \sqrt{\dfrac{2y^3}{x}}$.

22. $\sqrt{\dfrac{27x}{y}} + \sqrt{\dfrac{12x}{y^3}} - \sqrt{\dfrac{y^3}{3x}}$.

23. $\sqrt[3]{\dfrac{x}{y}} - \sqrt[3]{\dfrac{8y^2}{x^2}} - \sqrt[3]{\dfrac{1}{64x^2y}}$.

24. $\sqrt[3]{\dfrac{y}{x}} - \sqrt[3]{\dfrac{x^2}{27y^2}} - \sqrt[3]{\dfrac{8}{xy^2}}$.

25. $\sqrt{xy^{-1}} - \sqrt{4x^{-1}y} + \sqrt[3]{xy^{-2}} - \sqrt[3]{x^{-2}y^4}$.

26. $\sqrt{2xy^{-2}} + \sqrt{8y^2x^{-1}} + \sqrt[3]{2xy^{-3}} - \sqrt[3]{16x^4y^{-6}}$.

27. $\sqrt[3]{3xy^{-3}} + \sqrt[3]{9^{-1}x^{-2}y^3} + \sqrt{3^{-2}x^{-1}}$.

28. $\sqrt{16x^{-1}y} + \sqrt[4]{16x^{-1}y} + \sqrt{9xy^{-1}}$.

11

LOGARITHMS

11.1. Introduction

There are times when we are required to do a considerable amount of computation; hence, anything that will reduce the amount of labor involved is
worth while. If we use the concept of the logarithm of a number to a base,
we can replace raising to a power by multiplication, multiplication by addition, and division by subtraction. There is a close relation between logarithms
and exponents as is shown by the following definition: *The logarithm of a
positive number **N** to the base **b** where **b** is positive and different from **1** is
the exponent **L** that the base must have in order to produce the number.*

BASIC
DEF.

This definition can be put in symbolic form as

(1) $\log_b N = L$ *if and only if* $b^L = N$. $(N > 0, b > 0, \neq 1)$

These two equations are the logarithmic and exponential forms of the same
relation between N, b, and L. From the second equation in (1), it is clear that
L the logarithm is the exponent that b the base must have to produce N.

EXAMPLES

1. $\log_4 64 = 3$, since $4^3 = 64$.
2. $\log_2 64 = 6$, since $2^6 = 64$.
3. $\log_{64} 64 = 1$, since $64^1 = 64$.
4. $\log 4096 = \frac{1}{2}$, since $4096^{1/2} = 64$.
5. Find N if $\log_6 N = 3$.

SOLUTION

If we make use of the two equivalent equations in (1) and put the equation of this problem in exponential form, we have $6^3 = N$; hence, $N = 216$.

6. Find b if $\log_b 64 = 2$.

SOLUTION

Changing the given equation to exponential form, we get $b^2 = 64 = 8^2$; hence, $b = 8$.

Exercise 11.1

By use of (1), change the statements in Problems 1 through 12 to exponential form and those in Problems 13 through 24 to logarithmic form.

1. $\log_5 125 = 3$.   **2.** $\log_3 81 = 4$.

3. $\log_8 64 = 2$.   **4.** $\log_4 64 = 3$.

5. $\log_4 32 = 2.5$.   **6.** $\log_9 27 = 1.5$.

7. $\log_{27} 81 = \frac{4}{3}$.   **8.** $\log_{16} 128 = 1.75$.

9. $\log_3 \frac{1}{9} = -2$.   **10.** $\log_4 \frac{1}{64} = -3$.

11. $\log_2 \frac{1}{32} = -5$.   **12.** $\log_4 \frac{1}{128} = -3.5$.

13. $2^6 = 64$.   **14.** $3^4 = 81$.

15. $7^3 = 343$.   **16.** $4^5 = 1024$.

17. $9^{1.5} = 27$.   **18.** $16^{1.25} = 32$.

19. $32^{.8} = 16$.   **20.** $243^{.2} = 3$.

21. $2^{-3} = \frac{1}{8}$.   **22.** $5^{-4} = \frac{1}{625}$.

23. $6^{-3} = \frac{1}{216}$.   **24.** $11^{-2} = \frac{1}{121}$.

Find the value of the letter in each of Problems 25 through 48.

25. $\log_5 N = 2$.   **26.** $\log_5 N = 3$.

27. $\log_7 N = 0$.   **28.** $\log_2 N = 5$.

29. $\log_{16} N = \frac{1}{2}$.   **30.** $\log_8 N = \frac{1}{3}$.

31. $\log_9 N = \frac{3}{2}$.   **32.** $\log_{16} N = \frac{3}{4}$.

33. $\log_4 N = -1$.   **34.** $\log_3 N = -3$.

35. $\log_5 N = -4$.   **36.** $\log_7 N = -3$.

37. $\log_2 8 = L$.   **38.** $\log_3 27 = L$.

39. $\log_5 625 = L$.   **40.** $\log_7 49 = L$.

41. $\log_4 8 = L$.   **42.** $\log_8 16 = L$.

43. $\log_{25} 3125 = L$.   **44.** $\log_{16} 32 = L$.

45. $\log_b 125 = 3$.   **46.** $\log_b 64 = 5$.

47. $\log_b 81 = 4$.   **48.** $\log_b 49 = 2$.

49. $\log_b 27 = \frac{3}{2}$.   **50.** $\log_b 4 = \frac{2}{3}$.

51. $\log_b 8 = \frac{3}{4}$.   **52.** $\log_b 4 = \frac{2}{5}$.

11.2. The Common or Briggs System

If the base 10 is used in determining the logarithm of a number, we say that we have the _Common_ or _Briggs_ logarithm of the number. Consequently, the common logarithm of 1000 is 3 since $10^3 = 1000$. It is customary to omit the base if working with common logarithms; hence, we write log 1000 = 3 instead of writing $\log_{10} 1000 = 3$. Hereafter, we shall omit the base and, thereby, indicate that the common logarithm is being used.

11.3. Characteristic and Mantissa

We know that log 100 = 2 since $10^2 = 100$ and that log 1000 = 3 but do not know log 347 since we do not know the power to which 10 must be raised to produce 347. It seems reasonable that log 347 should be between 2 and 3 since 347 is between 10^2 and 10^3. The value of log 347 cannot be expressed exactly, but to two decimal places it is 2.54. In exponential form, this is

$$10^{2.54} = 347;$$

hence, $3470 = 347(10) = 10^{2.54} \, 10^1 = 10^{3.54}$.

Consequently,

$$\log 3470 = 3.54.$$

Similarly,

$$\log 34.7 = \log \frac{347}{10} = \log (347)(10^{-1})$$

$$= \log (10^{2.54} \times 10^{-1}) = \log 10^{1.54} = 1.54.$$

If the logarithm of a number is expressed as an integer plus a positive fraction, then the integer is called the *characteristic* and the positive fraction is called the *mantissa* of the logarithm. Thus the characteristic of log 347 is 2 and the mantissa is .54 since log 347 = 2.54.

The mantissa can be found in a table such as Table III in this book. In order to find the mantissa of the logarithm of a three-digit number, we look in the column of Table III which is headed by N for the first two digits and then look in line with this and in the column headed by the third digit of the number and find the mantissa.

EXAMPLE

1. In order to find the mantissa of log 514, we first locate 51 in the column headed by N; then, in line with this and in the column headed by 4 we find 7110. We must supply a decimal to the left of each mantissa; hence, the mantissa of log 514 is .7110. Furthermore, the mantissa of log 668 is .8248.

In order to see how to determine the characteristic we need the following definition: *The __reference position__ for the decimal point in a number is immediately to the right of the first nonzero digit in the number.*

EXAMPLE

2. If we use this definition, we see that the decimal point is in reference position in 2.37, is three places to the right of reference position in 2370, and one place to the left of reference position in .237.

If n is an integer and we multiply a number by 10^n, the effect is to move the decimal point n places to the right if n is positive and n places to the left if n is negative. If the decimal point is in reference, the number is between $1 = 10^0$ and $10 = 10^1$; hence its logarithm is zero plus a positive fraction, and consequently the characteristic of the logarithm is zero. Therefore we have the following method for determining the characteristic of the logarithm of a number: *The characteristic of the logarithm of N is numerically equal to the number of places the decimal point is removed from reference position. It is positive or negative according as the decimal point is to the right or the left of reference position.*

E XAMPLE

3. The characteristic of log 85.3 is 1, since the decimal is one place to the right of reference position; furthermore, the characteristic of log .00372 is −3 since the decimal is 3 places to the left of reference position. It is customary to write a negative characteristic as the proper positive integer minus 10. In accordance with this we would write −3 as 7 − 10. Finally, log .234 = 9.3692 − 10 since the characteristic is −1 = 9 − 10 and the mantissa is .3692.

11.4. Given log N, to Find N

If we are given the value of log N and want to find N, we must locate the mantissa in the body of the tables if it is there. We then find the first two digits of N to the left of the mantissa and the third digit above it. The decimal point is then placed in keeping with the value of the characteristic. If the mantissa of log N is not in the body of the table, we locate the entry, if any, that is nearer it than any other and proceed as though the mantissa of log N were in the table. If the mantissa is midway between two entries, we use the three-digit number that corresponds to the larger one.

E XAMPLES

1. Find N if log $N = 2.6712$.

SOLUTION

We look for the mantissa .6712 in the body of the tables and find it across from 46 and under 9; hence, the sequence of digits in N is 469. Furthermore, the decimal point must be two places to the right of reference position since the characteristic is two; consequently, $N = 469$.

2. Find N if log $N = 9.2609 − 10$.

SOLUTION

The mantissa .2609 is not in the table but .2601 is there and is nearer the given mantissa than any other entry. Furthermore, .2601 is across from 18 and under 2; hence, the sequence of digits in N is 182. Finally, the decimal point belongs one place to the left of reference position since the characteristic is $9 − 10 = −1$; hence, $N = .182$.

3. Find N if log $N = .8817$.

SOLUTION

The mantissa .8817 is not in the table but it is half way between the two entries .8814 and .8820. Therefore, we use the sequence of digits that corresponds to .8820. Thus, the sequence of digits is 761 and $N = 7.61$ since the characteristic is zero.

Exercise 11.2

Find the logarithm of each number given in Problems 1 through 24.

1. 326.	**2.** 71.5.	**3.** 8.62.	**4.** 93.8.
5. 1.74.	**6.** 583.	**7.** 24.7.	**8.** 3.15.
9. 23.8.	**10.** 9.06.	**11.** 876.	**12.** 739.
13. .605.	**14.** .0177.	**15.** .00707.	**16.** .0383.
17. .00211.	**18.** .835.	**19.** .0724.	**20.** .00500.
21. .0552.	**22.** .00119.	**23.** .678.	**24.** .974.

If log N is as given in the following problems, find N in each case.

25. 1.1206.	**26.** 2.4425.	**27.** 0.8710.	**28.** 1.9763.
29. 0.8585.	**30.** 1.6075.	**31.** 2.6866.	**32.** 0.7679.
33. $9.9609 - 10$.		**34.** $8.1931 - 10$.	
35. $8.2833 - 10$.		**36.** $7.5502 - 10$.	
37. 1.7184.	**38.** 2.8856.	**39.** 0.8410.	**40.** 1.5480.
41. 2.2424.	**42.** 0.9262.	**43.** 1.6773.	**44.** 2.5127.
45. 0.2956.	**46.** 1.5334.	**47.** 2.8907.	**48.** 0.9901.

11.5. Computation Theorems

We shall make use of the laws of exponents in order to derive some properties of logarithms that are used in numerical computation.

If we are given that

(1)　　　　　　　$\log_b M = m$　and　$\log_b N = n,$

then, by use of the definition of the logarithm of a number to a base as given in (1) of Art. 11.1, we know that

(2)　　　　　　　$M = b^m$　and　$N = b^n.$

Therefore,

$$\log_b MN = \log_b b^m b^n$$

$$= \log_b b^{m+n}, \text{ adding exponents,}$$

$$= m + n, \text{ by (1) of Art. 11.1,}$$

$$= \log_b M + \log_b N, \text{ by (1) above.}$$

Hence, we have the following theorem.

THEOREM 1. *The logarithm of the product of two positive numbers is equal to the sum of the logarithms of the numbers.*

EXAMPLE

1. $\log 15 = \log (3 \times 5) = \log 3 + \log 5$
                        $= .4771 + .6990 = 1.1761$

This theorem can be extended to the product of three or even more numbers as follows.

$$\log_b MNP = \log_b M(NP)$$

$$= \log_b M + \log_b NP$$

$$= \log_b M + \log_b N + \log_b P.$$

We shall now derive another computation theorem. Consider

$$\log_b \frac{M}{N} = \log_b \frac{b^m}{b^n}, \text{ by (2) above,}$$

$$= \log_b b^{m-n}, \text{ subtracting exponents,}$$

$$= m - n, \text{ by (1) of Art. 11.1,}$$

$$= \log_b M - \log_b N, \text{ by (1) above.}$$

Consequently, we have the following theorem.

THEOREM 2. *The logarithm of the quotient of two positive numbers is equal to the logarithm of the dividend minus that of the divisor.*

EXAMPLE

2. $$\log \frac{5}{3} = \log 5 - \log 3$$

$$= .6990 - .4771 = .2219.$$

In order to get a third computation thereoem, we shall consider $M = b^m$ and raise each member to the kth power and have

$$M^k = (b^m)^k = b^{km}, \text{ multiplying exponents.}$$

Consequently,

$$\log_b M^k = \log_b b^{km}$$

$$= km, \text{ by (1) of Art. 11.1,}$$

$$= k \log_b M, \text{ by (1) of this article.}$$

Now we have the following theorem.

THEOREM 3. *The logarithm of a power of a positive number is equal to the exponent of the power multiplied by the logarithm of the number.*

EXAMPLE

3.
$$\log 5^3 = 3 \log 5$$
$$= 3(0.6990) = 2.0970.$$

If we make use of the fact that

$$\sqrt[q]{a^p} = a^{p/q}, \qquad a > 0,$$

we can use Theorem 3 to extract roots of positive members.

EXAMPLES

4.
$$\log \sqrt{6} = \log 6^{1/2}$$

$$= \frac{1}{2} \log 6 = \frac{1}{2}(.7782) = .3891;$$

hence, using the table, we have

$$\sqrt{6} = 2.45.$$

5. Use the computation theorems to express

$$\log \frac{ab^r}{c}$$

as the sum and difference of logarithms.

SOLUTION

$$\log \frac{ab^r}{c} = \log ab^r - \log c, \text{ by Theorem 2,}$$

$$= \log a + \log b^r - \log c, \text{ by Theorem 1,}$$
$$= \log a + r \log b - \log c, \text{ by Theorem 3.}$$

6. Use the computation theorems to express $\log M - A \log N + \log P$ as the logarithm of a single expression.

SOLUTION

$$\log M - A \log N + \log P = \log MP - \log N^A, \text{ Theorems 1 and 3,}$$

$$= \log \frac{MP}{N^A}, \text{ Theorem 2.}$$

11.6. Logarithmic Computation

We shall further illustrate the use of logarithms in computation by means of several examples.

EXAMPLES

1. By use of logarithms, find

$$N = (23.7)(386)(.0519).$$

SOLUTION

Taking the logarithm of each member, we have

$$\log N = \log(23.7)(386)(.0519)$$
$$= \log 23.7 + \log 386 + \log .0519, \text{ Theorem 1.}$$

Now looking up the logarithm of each number and furnishing each characteristic, we get

$$\log 23.7 = 1.3747$$
$$\log 386 = 2.5866$$
$$\log .0519 = 8.7152 - 10$$
$$\overline{\log N = 2.6765,} \qquad \text{adding.}$$

The mantissa that is nearest .6765 is .6767 and is across from 47 and under 5; hence,

$$N = 475$$

since the characteristic is 2.

2. Use logarithms to find the value of

$$N = \frac{(37.6)(.859)}{4.21}.$$

SOLUTION

If we take the logarithm of each member, we get

$$\log N = \log \frac{(37.6)(.859)}{4.21}$$

$$= \log 37.6 + \log .859 - \log 4.21, \text{ Theorems 1 and 2.}$$

We must now look up the value of each of these logarithms and perform the indicated operations. Thus

$$\log 37.6 = 1.5752$$
$$\log .859 = 9.9340 - 10$$
$$\overline{\text{sum} = 1.5092,} \qquad \text{adding,}$$
$$\log 4.21 = 0.6243$$
$$\overline{\text{difference} = 0.8849,} \qquad \text{subtracting.}$$

Now we know log $N = .8849$ and can find N from the table. Since the entry nearest .8849 is across from 76 and under 7, we know that 767 is the sequence of digits in N; furthermore, the characteristic is zero and

$$N = 7.67.$$

3. Evaluate

$$N = \frac{2.79}{36.8}.$$

SOLUTION

$$\log N = \log \frac{2.79}{36.8}$$
$$= \log 2.79 - \log 36.8$$

Use of the table gives log $2.79 = .4456$ and log $36.8 = 1.5658$; hence, log 36.8 is larger than log 2.79 as should be expected since 36.8 is greater than 2.79; consequently, the subtraction required by the problem leads to a negative number including a negative fraction unless we change the form of log $2.79 = .4456$. The fractions in the table are all positive, and we can avoid getting a negative one by adding 10 to log 2.79 and also subtracting 10. Thus, we obtain

$$\log 2.79 = 10.4456 - 10$$
$$\underline{\log 36.8 = 1.5658}$$
$$\log N = 8.8798 - 10 \quad \text{subtracting,}$$
$$N = .0758$$

since the entry nearest .8798 is across from 75 and under 8 and the characteristic is $8 - 10 = -2$.

4. Evaluate

$$N = 2.71^3$$

by using logarithms.

SOLUTION

$$\log N = \log 2.71^3$$
$$= 3 \log 2.71, \quad \text{Theorem 3 of Art. 11.5,}$$
$$= 3(.4330) = 1.2990;$$

hence, $\qquad N = 19.9$

to three significant figures.

5. Use logarithms to evaluate

$$N = \sqrt[3]{.273}.$$

SOLUTION

Since

$$\sqrt[3]{.273} = .273^{1/3}$$

we know that

$$\log N = \log .273^{1/3}$$

$$= \frac{1}{3} \log .273$$

$$= \frac{1}{3} (9.4362 - 10).$$

If we performed the indicated operations, we would get a negative number including a negative fraction, but this can be avoided by adding and subtracting the same properly chosen number. It should be chosen so that the total amount subtracted is divisible by 3; hence, we shall add and subtract 20 and have

$$\log N = \frac{1}{3} (29.4362 - 30) = 9.8121 - 10.$$

Therefore, $N = .649.$

Exercise 11.3

Express the logarithm of the number in each of Problems 1 through 16 as the sum or difference of logarithms.

1. $\log ab.$

2. $\log cd.$

3. $\log par.$

4. $\log a(x - 3).$

5. $\log \dfrac{a}{b}.$

6. $\log \dfrac{k}{t}.$

7. $\log \dfrac{k}{a + 2}$

8. $\log \dfrac{b - 5}{b + 7}.$

9. $\log a^b.$

10. $\log c^k.$

11. $\log(x - 1)^p.$

12. $\log A^{B-1}.$

13. $\log \dfrac{bd}{s}.$

14. $\log \dfrac{a}{st}.$

15. $\log \dfrac{am}{bt}.$

16. $\log \dfrac{a(s - 2)}{(t + 3)x}.$

17. $\log a^t y.$

18. $\log b^3 k^a.$

19. $\log s\sqrt[3]{t}.$

20. $\log \sqrt[a]{b} \sqrt[5]{a}.$

Express the combination of logarithms in each of Problems 21 through 32 as the logarithm of a single term.

21. $\log a + \log k.$

22. $\log b + \log 5.$

23. $\log a + \log b + \log c.$

24. $\log M + \log N + \log K.$

25. $\log A - \log S.$

26. $\log B - \log 3.$

27. $\log A + \log K - \log T$.

28. $\log S - \log A - \log P$.

29. $a \log M - \log N$.

30. $\log P - a \log M$.

31. $2 \log a - 3 \log(b - 1)$.

32. $4 \log K - 5 \log(L + 3)$.

Evaluate the combination of numbers in each of the following problems by means of logarithms.

33. $(5.72)(31.6)$.

34. $(48.3)(5.03)$.

35. $(27.1)(.314)$.

36. $(80.7)(.773)$.

37. $\dfrac{279}{38.4}$.

38. $\dfrac{27.9}{384}$.

39. $\dfrac{78.9}{2.63}$.

40. $\dfrac{58.4}{.907}$.

41. $\dfrac{(37.6)(7.32)}{82.9}$.

42. $\dfrac{(48.5)(73.1)}{976}$.

43. $\dfrac{84.7}{(.748)(91.1)}$.

44. $\dfrac{399}{(60.3)(8.73)}$.

45. 17.3^2.

46. 3.74^3.

47. $\sqrt[3]{.809}$.

48. $\sqrt[7]{.697}$.

49. $\dfrac{\sqrt{7.47}}{\sqrt[5]{183}}$.

50. $\dfrac{30.1\sqrt{.418}}{\sqrt[3]{.778}}$.

51. $\dfrac{\sqrt[7]{.813}}{2.75}$.

52. $\dfrac{\sqrt{.892}}{\sqrt[3]{.403}}$.

11.7. Interpolation

The process of interpolation was discussed in Art. 8.2 in connection with tables of trigonometric functions of angles. The procedure here will be the same as there and will be illustrated by three examples.

EXAMPLES

1. Find N to four figures if
$$\log N = 2.5518.$$

SOLUTION

The mantissa .5518 is not in the table; hence, we look up the two that are nearer it than any others. We then have

$$10 \left(\begin{array}{l} \text{man log } 3560 = .5514 \\ \text{man log } N \quad\; = .5518 \\ \text{man log } 3570 = .5527 \end{array} \right) \; 4 \; \bigg) \; 13$$

Hence .5518 is $\frac{4}{13}$ of the way from .5514 toward .5527; consequently, we assume that N is $\frac{4}{13}$ of the way from 3560 toward 3570. The zero was added to 356 and to 357 so as to have a four-digit number, since N is to be computed to four figures. We now get

$$\frac{4}{13}(10) = 3, \text{ to the nearest unit,}$$

and go that much from 3560 toward 3570; hence, the sequence of digits in N is 3563. Therefore,

$$N = 356.3,$$

since the characteristic of log N is 2.

2. If we want to find log 47.62 or of any other four-digit number, we must interpolate as indicated below.

$$10 \left(2 \left(\begin{array}{l} \text{man log } 4760 = .6776 \\ \text{man log } 4762 = \\ \text{man log } 4770 = .6785 \end{array} \right) .0009 \right.$$

Since 4762 is $\frac{2}{10}$ of the way from 4760 toward 4770 we then assume that man log 4762 is $\frac{2}{10}$ of the way from man log 4760 toward man log 4770. Now, we take

$$\frac{2}{10}(.0009) = .0002$$

and go that much from man log 4760 toward man log 4770 and see that the mantissa of log 4762 is .6778; hence,

$$\log 47.62 = 1.6778.$$

3. Evaluate

$$N = (2.374)(354.8)$$

by using logarithms and interpolation.

SOLUTION

$$\begin{aligned} \log N &= \log (2.374)(354.8) \\ &= \log 2.374 + \log 354.8. \end{aligned}$$

We now interpolate and find that

$$\begin{aligned} \log 2.374 &= 0.3755 \\ \underline{\log 354.8 = 2.5500} \\ \log N = 2.9255, \quad \text{adding.} \end{aligned}$$

We must interpolate again in order to obtain N. Thus, we get

$$N = 842.4.$$

Exercise 11.4

Find either N or log N in Problems 1 through 16 by use of interpolation.

1. log $N = 1.2588$.

2. log $N = .3476$.

3. log $N = 2.8039$.

4. log $N = 3.5991$.

5. log $N = 9.7228 - 10$.

6. log $N = 8.4235 - 10$.

7. log $N = 7.1782 - 10$.

8. log $N = 6.2784 - 10$.

9. $N = 38.42$.

10. $N = 6.728$.

11. $N = 4997$.

12. $N = 943.9$.

13. $N = .7883$.

14. $N = .01454$.

15. $N = .002585$.

16. $N = .5846$.

By use of logarithms and interpolation perform the computations indicated in the following problems.

17. $(2.359)(327.4)$.

18. $(.5842)(78.33)$.

19. $(64.71)(5.903)$.

20. $(.8037)(.7126)$.

21. $\dfrac{162.6}{81.77}$.

22. $\dfrac{59.63}{28.44}$.

23. $\dfrac{2.713}{857.9}$

24. $\dfrac{71.25}{932.2}$.

25. $\dfrac{(38.41)(67.29)}{885.3}$.

26. $\dfrac{(4.937)(581.2)}{676.7}$.

27. $\dfrac{47.28}{(386.3)(.9177)}$.

28. $\dfrac{8059}{(9286)(.9425)}$.

29. $3.142^{2.72}$.

30. $2.718^{3.14}$.

31. $\sqrt{81.96}$.

32. $\sqrt[5]{7824}$.

33. $\sqrt[3]{.6885}$.

34. $\sqrt[7]{.1776}$.

35. $\sqrt{22.37}\,(38.41)$.

36. $\sqrt[3]{28.03}\,(71.08)$.

37. $\sqrt[3]{128.3}\,\sqrt[6]{.6034}$.

38. $\sqrt[5]{37.37}\,\sqrt[7]{.2639}$.

39. $\dfrac{\sqrt[5]{3.806}}{\sqrt[3]{.7777}}$.

40. $\dfrac{\sqrt[4]{.6223}}{\sqrt[7]{28.32}}$.

12

LAWS OF SINES AND COSINES

12.1. Introduction

In Chapter 8, we saw how to find the unknown parts of a right triangle and shall now develop formulas for solving oblique triangles—an oblique triangle is solved if the unknown sides and angles are determined. As proved in plane geometry, we can determine the unknown parts if three parts including a side are known. These known parts may be:

A. Two angles and a side;

B. Two sides and the angle opposite one of them;

C. Two sides and the included angle;

D. Three sides.

Since the use of logarithms reduces the amount of physical labor involved in computation, we shall see how to use them in connection with the trigonometric functions of angles before developing formulas for solving oblique triangles.

12.2. Logarithms of Trigonometric Functions

In order to find the value of the logarithm of a trigonometric function of an angle, we could find the value of the function and then look up the logarithm of that number. It is, however, not necessary to do all this work, since there are tables which give the logarithms of the functions of the angles. Table IV in this book is such a table. It contains the angles from $0°$ to $90°$ at intervals of $10'$ and gives the values of log sin, log tan, log cot and log cos for each angle listed except that 10 must be subtracted from all entries but those in the columns headed L. Cot. These tens could have been printed but a considerable amount of space was saved by ommitting them.

The angles from $0°$ to $45°$ are listed on the left side of the page at intervals of $10'$ and the corresponding function names are given at the top. Thus log tan $18°40' = 9.5287 - 10$ since, except for the $- 10$, this is the entry in the table that is across from $18°40'$ and below L. Tan. The angles from $45°$ to $90°$ are listed on the right and the corresponding function names are given at the bottom of the page. If we look across from $56°30'$ and above log sin, we find 9.9211; hence, log sin $56°30' = 9.9211 - 10$.

If we want the logarithm of a function of an angle between $0°$ and $90°$ that is not listed in the table, we can find it by resorting to interpolation. Furthermore, if the logarithm of a function of an angle is given but is not in the table, we can find the angle by interpolation.

EXAMPLES

1. Find the value of log cos $23°17'$.

SOLUTION

The angle $23°17'$ is not in the table; hence, we look up the value of log cos $23°\,10'$ and log cos $23°20'$, since these angles are nearer the one we want than any others that are in the table. If we make the usual array for interpolation, we have

$$10' \left(7' \begin{array}{l} \text{log cos } 23°10' = 9.9635 - 10 \\ \text{log cos } 23°17' = \\ \text{log cos } 23°20' = 9.9629 - 10 \end{array} \right) .0006$$

Since $23°17'$ is $\frac{7}{10}$ of the way from $23°10'$ toward $23°20'$, we assume that log cos $23°17'$ is $\frac{7}{10}$ of the way from log cos $23°10'$ toward log cos $23°20'$. Hence, we perform the following operations:

$$\frac{7}{10}(.0006) = .0004,$$

log cos $23°17' = 9.9635 - 10 - .0004 = 9.9631 - 10$.

2. Find A if log tan $A = 10.3889 - 10$.

SOLUTION

This value of log tan A is not in the table; hence, we look up the two entries that are closer to it than any other entries and have

$$10' \left(\begin{array}{l} \text{log tan } 67°40' = 10.3864 - 10 \\ \quad\text{log tan } A = 10.3889 - 10 \\ \text{log tan } 67°50' = 10.3900 - 10 \end{array} \right) .0025 \right) .0036$$

$$\frac{.0025}{.0036}(10') = 7'$$

Therefore,

$$A = 67°47'$$

Exercise 12.1

Find the value called for in each of Problems 1 through 12.

1. log sin 15°30′.   **2.** log cos 18°40′.

3. log tan 25°50′.   **4.** log cot 28°10′.

5. log cos 39°20′.   **6.** log tan 41°.

7. log cot 84°10′.   **8.** log sin 63°20′.

9. log tan 48°40′.   **10.** log cot 51°30′.

11. log sin 73°30′.   **12.** log cos 87°50′.

Find the angle determined by the equation in each of Problems 13 through 24.

13. log cos $A = 9.9991 - 10$.   **14.** log tan $A = 9.1194 - 10$.

15. log cot $A = 0.5873$.   **16.** log sin $A = 9.5673 - 10$.

17. log tan $A = 9.7701 - 10$.   **18.** log cot $A = .1629$.

19. log sin $A = 9.8699 - 10$.   **20.** log cos $A = 9.8140 - 10$.

21. log cot $A = 9.8153 - 10$.   **22.** log sin $A = 9.9672 - 10$.

23. log cos $A = 9.4314 - 10$.   **24.** log tan $A = 1.1739$.

Use interpolation to find the value called for in each of Problems 25 through 36.

25. log tan 8°23′.   **26.** log cot 17°34′.

27. log sin 21°38′.   **28.** log cos 29°47′.

29. log cot 36°36′.   **30.** log sin 42°42′.

31. log cos 51°51′.   **32.** log tan 65°29′.

33. log sin 68°57′.   **34.** log cos 73°33′.

35. log tan 80°4′.   **36.** log cot 88°25′.

Find the angle to the nearest minute in each of Problems 37 through 48.

37. log cot $A = 0.3742$.   **38.** log sin $A = 9.7348 - 10$.

39. log cos $A = 9.3025 - 10$.   **40.** log tan $A = 9.0274 - 10$.

41. log sin $A = 9.4776 - 10$.   **42.** log cos $A = 9.8029 - 10$.

43. log tan $A = 0.7184$.   **44.** log cot $A = 9.3841 - 10$.

45. log cos $A = 9.8847 - 10$.   **46.** log tan $A = 9.7887 - 10$.

47. log cot $A = 1.4735$.   **48.** log sin $A = 9.5566 - 10$.

12.3. The Law of Sines

We shall now consider a general triangle whose angles are A, B, and C and shall call the sides opposite them a, b, and c, respectively. In order to be able to derive a relation between the sides and angles, we shall drop a

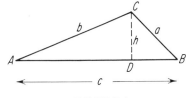

FIGURE 12-1   FIGURE 12-2

perpendicular from either vertex to the opposite side or to the opposite side produced. If we call the foot of the perpendicular D and its length h, we have Fig. 12-1 if the perpendicular strikes the opposite side and Fig. 12-2 if it strikes the side produced. If we use either figure, we have

$$\sin A = \frac{h}{b};$$

hence,

(1) $$h = b \sin A.$$

From Fig. 12-1, we also have

$$\sin B = \frac{h}{a} \text{ or}$$

(2) $$h = a \sin B.$$

Furthermore, in Fig. 12-2, angle *ABC* is angle *B* and its reference angle is *DBC*. Therefore

$$\sin B = \sin DBC = \frac{h}{a} \quad \text{and}$$

(2') $$h = a \sin B.$$

Now equating the expressions for *h* as given by (1) and by (2) or (2'), we have

$$a \sin B = b \sin A;$$

hence, dividing by sin *A* sin *B*, we get

$$\frac{a}{\sin A} = \frac{b}{\sin B}.$$

We can prove in a similar manner that

$$\frac{a}{\sin A} = \frac{c}{\sin C}.$$

Consequently,

(1) $$\frac{a}{\sin A} = \frac{b}{\sin B} = \frac{c}{\sin C}.$$

This relation between the sides and angles of a triangle is called the *law of sines* and can be put in words as:

In any triangle, the three fractions obtained by dividing a side by the sine of the opposite angle are equal.

We can use the law of sines to solve for the fourth part provided three of the four parts involved in any two of the fractions are known. Since the four parts used in any two of the fractions are two sides and the two angles opposite them we can use the law of sines provided the known parts are:

(1) any two angles and any side, or,

(2) any two sides and an angle opposite either of them.

We shall illustrate (1) now by means of an example and then discuss (2) in the next article.

EXAMPLES

1. Solve the triangle in which $A = 38°20'$, $B = 67°30'$, and $c = 43.7$.

SOLUTION

Since *c* is the given side we must have *C* and can find it by subtracting the sum of *A* and *B* from 180°. Thus,

$$C = 180° - (38°20' + 67°30')$$
$$= 180° - 105°50' = 74°10'.$$

If we decide to find a next, we must use the law of sines in the form

$$\frac{a}{\sin A} = \frac{c}{\sin C}.$$

Substituting the known values in this gives

$$\frac{a}{\sin 38°20'} = \frac{43.7}{\sin 74°10'};$$

hence, multiplying by $\sin 38°20'$, we get

$$a = \frac{43.7 \sin 38°20'}{\sin 74°10'}.$$

Therefore,

$$\log a = \log 43.7 + \log \sin 38°20' - \log \sin 74°10'$$

$$
\begin{aligned}
\log 43.7 &= 1.6405 \\
\log \sin 38°20' &= 9.7926 - 10 \\
\hline
\text{sum} &= 11.4331 - 10 \\
\log \sin 74°10' &= 9.9832 - 10 \\
\hline
\log a &= 1.4499, \qquad \text{subtracting,} \\
a &= 28.2 \qquad \text{to three figures.}
\end{aligned}
$$

We now use

$$\frac{b}{\sin B} = \frac{c}{\sin C}$$

in order to find b. Putting in the known values, we have

$$\frac{b}{\sin 67°30'} = \frac{43.7}{\sin 74°10'},$$

hence, multiplying by $\sin 67°30'$,

$$b = \frac{43.7 \sin 67°30'}{\sin 74°10'}$$

and

$$\log b = \log 43.7 + \log \sin 67°30' - \log \sin 74°10'$$

$$
\begin{aligned}
\log 43.7 &= 1.6405 \\
\log \sin 67°30' &= 9.9656 - 10 \\
\hline
\text{sum} &= 11.6061 - 10 \\
\log \sin 74°10' &= 9.9832 - 10 \\
\hline
\log b &= 1.6229, \qquad \text{subtracting.} \\
b &= 42.0.
\end{aligned}
$$

2. One side of a triangular plot of land is 362.7 feet and the other sides meet this one at angles of 27°28′ and 85°56′. Find the other two sides.

SOLUTION

We shall first draw a figure and label the various parts as shown in Fig. 12-3. We then find the third angle by subtracting the sum of the first two from 180° and have

$$C = 180° - (27°28′ + 85°56′)$$
$$= 180° - 113°24′ = 66°36′.$$

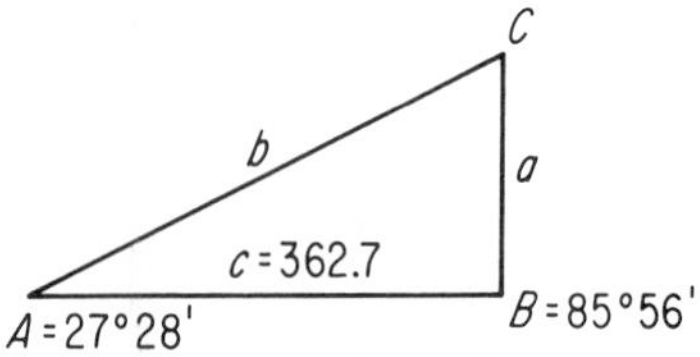

We now use the law of sines in the form

FIGURE 12-3

$$\frac{a}{\sin A} = \frac{c}{\sin C}$$

to find a, since A, c, and C are known. Putting the known values in, we have

$$\frac{a}{\sin 27°28′} = \frac{362.7}{\sin 66°36′};$$

hence,

$$a = \frac{362.7 \sin 27°28′}{\sin 66°36′}$$

and

$$\log a = \log 362.7 + \log \sin 27°28′ - \log \sin 66°36′.$$

We must interpolate to obtain each of these logarithms. Doing that gives

$$\begin{array}{rl}
\log 362.7 = & 2.5595 \\
\log \sin 27°28′ = & 9.6639 - 10 \\
\hline
\text{sum} = & 12.2234 - 10 \\
\log \sin 66°36′ = & 9.9627 - 10 \\
\hline
\log a = & 2.2607, \\
a = & 1823,
\end{array}$$

subtracting,
interpolating.

Similarly,

$$\frac{b}{\sin 85°56′} = \frac{362.7}{\sin 66°36′},$$

$$b = \frac{362.7 \sin 85°56′}{\sin 66°36′},$$

$$\log b = \log 362.7 + \log \sin 85°56′ - \log \sin 66°36′$$

$$\begin{array}{rl}
\log 362.7 = & 2.5595 \\
\log \sin 85°56′ = & 9.9989 - 10 \\
\hline
\text{sum} = & 12.5584 - 10 \\
\log \sin 66°36′ = & 9.9627 - 10 \\
\hline
\log b = & 2.5957, \\
b = & 394.2,
\end{array}$$

subtracting,
interpolating.

Exercise 12.2

Solve the triangle determined by the data in each of Problems 1 through 20. Obtain each part to the justified degree of accuracy.

1. $A = 28°30', B = 64°10', a = 257.$
2. $B = 37°40', C = 56°30', b = 146.$
3. $C = 57°10', A = 46°40', c = 60.3.$
4. $A = 48°20', B = 53°30', a = 5.14.$
5. $B = 23°30', C = 51°20', b = 27.5.$
6. $C = 34°40', A = 48°, c = 3.02.$
7. $A = 19°50', B = 62°30', a = .218.$
8. $B = 49°30', C = 38°10', b = .503.$
9. $C = 37°40', A = 58°40', b = 9.41.$
10. $A = 24°30', B = 49°20', c = 88.3.$
11. $B = 40°20', C = 57°50', a = .977.$
12. $C = 54°50', A = 73°30', b = 60.9.$
13. $A = 72°23', B = 39°42', a = 8.762.$
14. $B = 56°56', C = 47°47', b = 4714.$
15. $C = 49°35', A = 62°26', c = 3807.$
16. $A = 54°44' B = 67°39', a = 6.073.$
17. $B = 29°29', C = 55°57', a = 9.695.$
18. $C = 47°38', A = 34°43', b = 8976.$
19. $A = 35°53', B = 49°49', c = .9138.$
20. $B = 41°17', C = 47°36', c = 7.354.$

21. One angle of a rhombus is $58°40'$ and its sides are 24.1 centimeters long. Find the length of the longer diagonal.

22. A triangular tract of land is 547 varas on one side. Find the lengths of the other sides if they meet the given side at angles of $41°20'$ and $53°30'$.

23. The shortest side of a triangular lot is 106 feet. Find the longest side if two of the angles are $46°10'$ and $63°20'$.

24. In triangle ABC, $A = 58°26'$, $C = 45°32'$ and side AC is 76.92 inches. Find the length of the shortest side.

25. A range finder AC is 37.4 feet long. Lines drawn from a point B that is not on AC make angle $BAC = 77°30'$ and angle $BCA = 82°10'$. Find the distance between A and B.

26. Two trees on the west bank of a river are 139 feet apart. The lines of sight of a tree on the east bank as seen from the first two trees make angles of $40°40'$ and $58°30'$ with the line between the two trees on the west bank. How far from the third tree is each of the others?

27. A pilot flew at $159°$ with an air speed of 230 miles per hour. The wind was blowing from the west and his course was $154°$. Find the velocity of the wind and the ground speed of the plane.

28. A pilot flew at $78°$ from A to B and then at $212°$ to C. If C is 387 miles due south of A, how far did he fly?

12.4. The Ambiguous Case of the Law of Sines

We found in plane geometry that it is not always possible to form a triangle with two given lengths as sides and a given angle opposite one of them.

We also found that the lengths and angle may be such that there can be one or even two triangles. We shall now see how to determine, by use of a trigonometric treatment, the number of triangles that are possible.

If the given sides are a and b and if the given angle is A, then use of the law of sines gives

$$\sin B = \frac{b \sin A}{a}.$$

Furthermore, since a, b, and A are known, we can find the value of $\sin B$. Since there is no other alternative, the value of $\sin B$ must be greater than, equal to, or less than 1. We shall consider these three possibilities separately in order to determine the number of triangles in each case.

If $\sin B > 1$, there is no value of B, since the sine of an angle is never greater than one. Consequently, there is no triangle if $\sin B > 1$.

If $\sin B = 1$, then $B = 90°$. Consequently, there is a triangle if the given angle A is less than $90°$, and there is no triangle if A is equal to or greater than $90°$, since the sum of the angles of a triangle is $180°$.

If the $\sin B < 1$, then one value of B can be found in the table, and $B' = 180 - B$ is another since $\sin B$ and $\sin (180° - B)$ are equal. If neither B nor B' fits* into a triangle with the given angle, there is no solution; if only one of B and B' fits into a triangle with the given angle, there is one solution; and if both fit, then there are two solutions.

This discussion can be summarized as: *If the two sides **a** and **b** and an angle **A** are given, we have the ambiguous case. It can be handled by finding* sin **B**, *then finding all positive angles **B** that are less than* **180°** *and using as many of them as fit into a triangle with the given angle.*

The different situations that may arise are shown pictorially in Figs. 12-4 to 12-10.

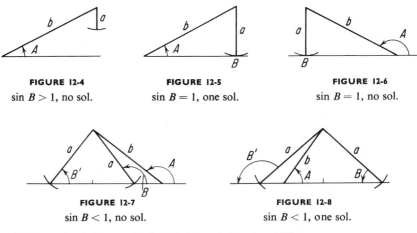

FIGURE 12-4   FIGURE 12-5   FIGURE 12-6
$\sin B > 1$, no sol. $\sin B = 1$, one sol. $\sin B = 1$, no sol.

FIGURE 12-7   FIGURE 12-8
$\sin B < 1$, no sol. $\sin B < 1$, one sol.

* Two angles fit into a triangle if their sum is less than $180°$.

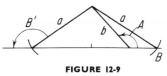

FIGURE 12-9   **FIGURE 12-10**

sin $B < 1$, one sol.   sin $B < 1$, two sols.

EXAMPLES

1. How many solutions are there if $a = 264$, $b = 314$, and $A = 64°10'$?

SOLUTION

If the given data are substituted in

$$\sin B = \frac{b \sin A}{a}$$

we have

$$\sin B = \frac{314 \sin 64°10'}{264} = \frac{314(.9001)}{264} = 1.07 ;$$

consequently, there is no angle B and no solution.

2. How many solutions are there if $a = 188$, $b = 176$, and $A = 63°40'$? Find each.

SOLUTION

If we substitute the given data in

$$\sin B = \frac{b \sin A}{a},$$

we obtain

$$\sin B = \frac{176 \sin 63°40'}{188} = \frac{176(.8962)}{188} = .8390.$$

Therefore,

$$B = 57°0'$$

and

$$B' = 180° - 57°0' = 123°0'.$$

The angles B and A fit into a triangle, since $A + B = 63°40' + 57°0' = 120°40'$ is less than 180°; however, A and B' do not fit into a triangle, since their sum $63°40' + 123°0' = 186°40'$ is greater than 180°. Consequently, there is only one solution. For it, $C = 180° - (A + B) = 180° - 120°40' = 59°20'$; hence, the third side c can be found by use of the law of sines in the form

$$\frac{c}{\sin C} = \frac{a}{\sin A}.$$

Substituting the known values in this gives

$$\frac{c}{\sin 59°20'} = \frac{188}{\sin 63°40'} ;$$

hence

$$c = \frac{188 \sin 59°20'}{\sin 63°40'}$$

$$\log c = \log 188 + \log \sin 59°20' - \log \sin 63°40'$$

$$
\begin{array}{rr}
\log 188 = & 2.2742 \\
\log \sin 59°20' = & 9.9346 - 10 \\
\hline
\text{sum} = & 12.2088 - 10 \\
\log \sin 63°40' = & 9.9524 - 10 \\
\hline
\log c = & 2.2564
\end{array}
$$

$$c = 180, \text{ to three figures.}$$

3. How many solutions are there if $a = 374$, $b = 669$, and $A = 32°10'$? Draw a figure for each solution.

SOLUTION

If we solve

$$\frac{a}{\sin A} = \frac{b}{\sin B}$$

for $\sin B$ and put in the known values, we have

$$\sin B = \frac{669 \sin 32°10'}{374} \,;$$

hence,

$$\log \sin B = \log 669 + \log \sin 32°10' - \log 374$$

$$
\begin{array}{rr}
\log 669 = & 2.8254 \\
\log \sin 32°10' = & 9.7262 - 10 \\
\hline
\text{sum} = & 12.5516 - 10 \\
\log 374 = & 2.5729 \\
\hline
\log \sin B = & 9.9787 - 10, \text{ subtracting,} \\
B = & 72°10', \qquad \text{using the table,} \\
B' = & 180° - 72°10' = 107°50'.
\end{array}
$$

Consequently, there are two solutions, since both B and B' fit into a triangle with A. The figures for the two triangles are shown in Figs. 12-11 and 12-12.

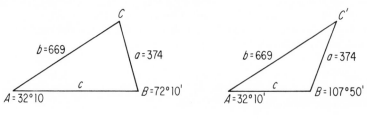

FIGURE 12-11   FIGURE 12-12

Exercise 12.3

Determine the number of solutions in each of Problems 1 through 16. Find each solution.

1. $A = 32°20'$, $b = 27.3$, $a = 11.4$. **2.** $A = 48°30'$, $b = 169$, $a = 117$.

3. $A = 134°40'$, $b = 14.2$, $a = 10.1$. **4.** $A = 129°50'$, $b = 267$, $a = 205$.

5. $C = 63°10'$, $a = 975$, $c = 870$. **6.** $B = 37°$, $a = 309$, $b = 186$.

7. $B = 41°50'$, $a = 727$, $b = 772$. **8.** $C = 63°40'$, $b = 473$, $c = 508$.

9. $A = 117°20'$, $b = 108$, $a = 150$. **10.** $B = 134°10'$, $c = 212$, $b = 269$.

11. $C = 108°50'$, $c = 236$, $b = 262$. **12.** $A = 119°30'$, $a = .119$, $b = .127$.

13. $A = 39°30'$, $a = 29.7$, $b = 32.8$. **14.** $B = 46°50'$, $b = 3.95$, $c = 4.18$.

15. $C = 57°40'$, $c = .503$, $a = .548$. **16.** $A = 72°20'$, $a = 72.2$, $b = 74.3$.

17. A pilot flew at 138° from A to B and then at 235° from B to C. Find the distance from B to C and the direction from A to C if A is 630 miles from B and 920 miles from C.

18. Due to an erroneous calculation, a pilot flew for 256 miles in a direction that was 20°30′ off course. He then turned through an acute angle toward his intended destination and reached it after flying a total of 402 miles. If the ground speed of the plane was 141 miles per hour for the entire trip, how much delay in reaching his destination was caused by the miscalculation?

19. Town C is 213 miles N27°20′E of town A and town B is N73°10′E of A and 392 miles from C. Find the distance from A to B and the direction from B to C.

20. A 32-foot ladder makes an angle of 38° with the vertical when it reaches a window. What angle will a 28-foot ladder make with the vertical when it reaches the same window?

12.5. The Law of Cosines

We shall now develop another formula for use in solving a triangle. This new formula is based on the Pythagorean theorem, and Figs 12-13, 12-14,

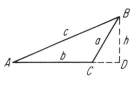

FIGURE 12-13

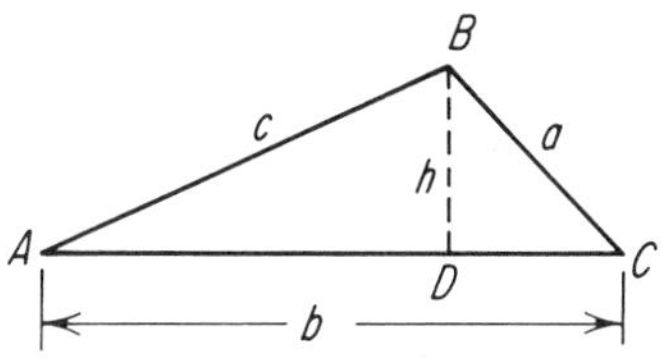

FIGURE 12-14

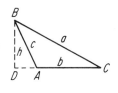

FIGURE 12-15

and 12-15 are used in obtaining it. Each of them is drawn by dropping a perpendicular from a vertex of a triangle to the opposite side or the opposite side produced. We call the length of the perpendicular h and its foot D. Now, applying the Pythagorean theorem, we have

(1) $$c^2 = h^2 + (AD)^2.$$

Since we want a relation between sides and angles of the triangle, we must express h and AD in terms of them. In each figure, we have

$$AD = b + CD$$

$$= b - DC, \text{ since } CD = -DC$$

$$= b - a \cos C, \text{ since } \cos C = \frac{DC}{a}$$

and

$$h = a \sin C, \quad \text{since} \quad \sin C = \frac{h}{a}.$$

Now, putting these expressions for AD and h in (1), we get

$$c^2 = (a \sin C)^2 + (b - a \cos C)^2$$

$$= a^2 \sin^2 C + b^2 - 2ab \cos C + a^2 \cos^2 C$$

$$= a^2(\sin^2 C + \cos^2 C) + b^2 - 2ab \cos C.$$

Finally, replacing $\sin^2 C + \cos^2 C$ by 1*, we have the relation

(2) $$c^2 = a^2 + b^2 - 2ab \cos C$$

between the three sides and an angle. Since the lettering is immaterial, we can state the following theorem: *The square of any side of a triangle is equal to the sum of the squares of the other two sides decreased by twice their product times the cosine of the angle between them.*

The two symbolic forms, besides (2), in which this can be put are

(3) $$b^2 = a^2 + c^2 - 2ac \cos B$$

and

(4) $$a^2 = b^2 + c^2 - 2bc \cos A.$$

Equations (2), (3), and (4) and the theorem are called the *law of cosines*. Since all three sides and one angle are used in each equation, we can use the law of cosines to find an additional part if we know

* We can see that $\sin^2 C + \cos^2 C = 1$ by dividing each member of the relation $x^2 + y^2 = r^2$ of Art. 7.2 by r^2 and then using the definition of the sine and cosine of an angle.

(i) any two sides and the included angle,

(ii) all three sides,

(iii) any two sides and an angle opposite one of them.

We shall not, however, use the law of cosines for (iii), since that situation is called the ambiguous case of the law of sines and was handled in Art. 12.4.

EXAMPLES

1. Find the smallest angle of the triangle in which $a = 42$, $b = 31$, and $c = 47$.

SOLUTION

The smallest angle is opposite the smallest side; hence, it is B. Consequently, we shall use

$$b^2 = a^2 + c^2 - 2ac \cos B$$

Substituting the given values, we get

$$31^2 = 42^2 + 47^2 - 2(42)(47) \cos B$$
$$961 = 1764 + 2209 - 3948 \cos B$$
$$\cos B = \frac{3012}{3948} = .7629$$

Therefore,

$$B = 40° \text{ to the nearest degree.}$$

2. Find side c if $a = 132$, $b = 213$, and $C = 53°20'$.

SOLUTION

Since it involves known quantities and the desired unknown, we shall use

$$c^2 = a^2 + b^2 - 2ab \cos C.$$

Substituting the known quantities, we have

$$c^2 = 132^2 + 213^2 - 2(132)(213) \cos 53°20'$$
$$= 17,424 + 45,369 - 33,582 = 29,211.$$

We shall use logarithms to solve for c and have

$$\log c^2 = \log 29,211$$
$$2 \log c = 4.4656$$
$$\log c = 2.2328$$
$$c = 171 \text{ to three figures.}$$

After one part has been determined by use of the law of cosines, the other unknown parts can be found by applying the law of sines. This will not be illustrated, since the reader became familiar with that procedure in Art. 12.3 and 12.4.

Exercise 12.4

Find the specified part in each of Problems 1 through 24.

1. If $a = 32$, $b = 41$, and $c = 37$, find A.

2. If $a = 19$, $b = 23$, and $c = 17$, find B.

3. If $a = 72$, $b = 63$, and $c = 58$, find C.

4. If $a = 49$, $b = 38$, and $c = 41$, find A.

5. If $a = 59$, $b = 93$, and $c = 64$, find B.

6. If $a = 63$, $b = 48$, and $c = 81$, find C.

7. If $a = 47$, $b = 21$, and $c = 36$, find A.

8. If $a = 23$, $b = 42$, and $c = 24$, find B.

9. If $a = 173$, $b = 207$, and $c = 355$, find C.

10. If $a = 805$, $b = 396$, and $c = 533$, find A.

11. If $a = 377$, $b = 773$, and $c = 619$, find B.

12. If $a = 809$, $b = 361$, and $c = 905$, find C.

13. If $a = 38$, $b = 32$, and $C = 72°$, find c.

14. If $b = 73$, $c = 61$, and $A = 58°$, find a.

15. If $c = 49$, $a = 58$, and $B = 43°$, find b.

16. If $a = 71$, $b = 60$, and $C = 64°$, find c.

17. If $b = 24$, $c = 33$, and $A = 103°$, find a.

18. If $c = 45$, $a = 39$, and $B = 97°$, find b.

19. If $a = 53$, $b = 61$, and $C = 93°$, find c.

20. If $b = 26$, $c = 35$, and $C = 118°$, find a.

21. If $c = 294$, $a = 347$, and $B = 44°$, find b.

22. If $a = 413$, $b = 508$, and $C = 68°$, find c.

23. If $b = 331$, $c = 417$, and $A = 109°$, find a.

24. If $c = 587$, $a = 605$, and $B = 113°$, find b.

25. If the sides and a diagonal of a parallelogram are 207, 318, and 423, respectively, find the angles.

26. If the diagonals of a parallelogram are 40 and 54 inches and if they meet at an angle of 53°, find the length of each side.

27. A pilot is flying at an airspeed of 215 miles per hour and a heading of 195°. The wind is blowing from due east at 25 miles per hour. Find the ground speed of the plane and the direction of flight.

28. If A is 427 feet N36°20′E of B and C is 541 feet from B and 843 feet from A, find the direction from B to C.

12.6. The Area of a Triangle

We shall now find two formulas for the area of a triangle. Each of them is based on the theorem that states the area is equal to one-half the product of the base and altitude. In Fig. 12-16, the base is represented by c and the altitude by h; furthermore, $h = b \sin A$.

Consequently,

$$K = \frac{1}{2} ch$$

FIGURE 12-16

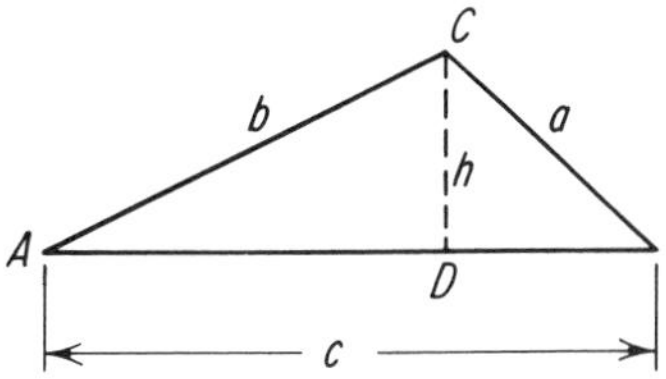

becomes

$$\text{(1)} \qquad K = \frac{1}{2} cb \sin A.$$

Finally, since the lettering is immaterial, this equation can be put in words: *The area of a triangle is one-half the product of any two sides and the sine of the included angle.*

EXAMPLE

1. Find the area of the triangle with $a = 239$, $b = 186$, and $C = 72°30′$.

SOLUTION

We must use the symbolic form

$$K = \frac{1}{2} ab \sin C$$

because of the parts that are known. Putting the given values in, we have

$$K = \frac{1}{2} (239)(186) \sin 72°30′$$

$$\log K = \log 239 + \log 186 + \log \sin 72°30′ - \log 2$$
$$\log 239 = 2.3784$$
$$\log 186 = 2.2695$$
$$\log \sin 72°30′ = 9.9794 - 10$$
$$\overline{\text{sum} = 4.6273}$$
$$\log 2 = 0.3010$$
$$\overline{\log K = 4.3263,} \qquad \text{subtracting,}$$
$$K = 2.12(10^4), \qquad \text{to three figures.}$$

In order to obtain a second formula for the area we shall put the expression obtained for b from

$$\frac{b}{\sin B} = \frac{c}{\sin C}$$

in (1). Thus

$$K = \frac{1}{2} cb \sin A$$

becomes

$$K = \frac{1}{2} c \frac{c \sin B}{\sin C} \sin A;$$

hence,

(2) $$K = \frac{c^2 \sin B \sin A}{2 \sin C}$$

is a formula for area and can be used if we know two angles and a side.

Since the lettering is immaterial, (2) can be put in words: *The area of a triangle is the square of any side times the sines of the adjacent angles and divided by twice the sine of the opposite angle.*

EXAMPLE

2. Find the area of a triangle if $a = 137$, $A = 52°30'$, and $B = 39°10'$.

SOLUTION

Equation (2) requires the use of all three angles, and we shall find the third one by subtracting the sum of the given two from 180°. Thus $C = 180° - (52°30' + 39°10') = 88°20'$. Since the known side is a, we must use the symbolic form

$$K = \frac{a^2 \sin B \sin C}{2 \sin A}.$$

Now putting the known values in this, we have

$$K = \frac{137^2 \sin 39°10' \sin 88°20'}{2 \sin 52°30'};$$

hence,

$$\log K = 2 \log 137 + \log \sin 39°10' + \log \sin 88°20' - \log 2 - \log \sin 52°30'$$

$$2 \log 137 = 2(2.1367) = 4.2734$$

$\log \sin 39°10' = 9.8004 - 10$	$\log 2 = 0.3010$
$\log \sin 88°20' = 9.9998 - 10$	$\log \sin 52°30' = 9.8995 - 10$
$\log$ numerator $= 24.0736 - 20$	$\log$ denominator $= 10.2005 - 10$
$\log$ denominator $= 10.2005 - 10$	

$$\log K = 13.8731 - 10, \quad \text{subtracting,}$$
$$= 3.8731$$
$$K = 7.47(10^3).$$

Exercise 12.5

Find the area of each triangle in Problems 1 through 24.

1. $a = 31, b = 42, C = 57°$.   **2.** $b = 17, c = 33, A = 29°$.

3. $c = 47, a = 53, B = 68°$.   **4.** $a = 83, b = 72, C = 75°$.

5. $b = 64, c = 37, A = 102°$.   **6.** $c = 77, a = 65, B = 141°$.

7. $a = 91, b = 83, C = 127°$.   **8.** $b = 49, c = 76, A = 134°$.

9. $c = 37.2, a = 48.9, B = 57°20'$.   **10.** $a = 278, b = 303, C = 39°40'$.

11. $b = 524, c = 611, A = 113°10'$.   **12.** $c = .707, a = .936, B = 138°50'$.

13. $a = 27, B = 54°, C = 69°$.   **14.** $a = 13, A = 46°, B = 73°$.

15. $a = 49, A = 63°, C = 71°$.   **16.** $b = 52, C = 34°, A = 68°$.

17. $b = 391, A = 37°10', B = 69°20'$.   **18.** $b = 487, B = 84°30', C = 52°20'$.

19. $c = 416, A = 60°20', B = 41°30'$.   **20.** $c = 793, B = 37°50', C = 69°40'$.

21. $c = 38.47, A = 41°14', C = 58°37'$.   **22.** $a = 8.123, A = 53°35', B = 66°42'$.

23. $b = 655.6, B = 72°31', C = 74°48'$.   **24.** $c = .7024, B = 54°44', C = 67°29'$.

13

QUADRATIC EQUATIONS

13.1. Solution by Factoring

We studied equations of degree one in Chapter 3 and shall study equations of degree two in this chapter. The general form of the equation is

$$ax^2 + bx + c = 0, \qquad a \neq 0,$$

and it is called a *quadratic equation*. Except for the requirement that a shall not be zero, the equation could reduce to one of the first degree.

We shall present several methods for solving quadratic equations. The first of these is solution by factoring and it is based on the principle that *the product of two or more factors is zero if and only if at least one of them is zero*. We can use this principle in solving $ax^2 + bx + c = 0$ if and only if we can express $ax^2 + bx + c$ as the product of two linear factors and can then solve each of the linear equations obtained by setting these factors equal to zero. We studied how to factor certain quadratic expressions in Art. 5.4 and how to solve a linear equation in Art. 3.2. We shall now need a combination of these two skills.

EXAMPLE

1. Solve

$$2x^2 + 5x - 3 = 0 \text{ for } x \text{ and check.}$$

SOLUTION

We shall factor $2x^2 + 5x - 3$ and have

$$2x^2 + 5x - 3 = (2x - 1)(x + 3);$$

consequently, in accordance with the principle stated above, we set each factor equal to zero and solve for x. Thus, we have

$$2x - 1 = 0 \quad \text{and} \quad x + 3 = 0.$$

Consequently, $x = \frac{1}{2}$ and $x = -3$. We shall check each value by substituting in the given equation. If we replace x by $\frac{1}{2}$, the left member becomes

$$2 \left(\frac{1}{2}\right)^2 + 5 \left(\frac{1}{2}\right) - 3 = 2 \left(\frac{1}{4}\right) + \frac{5}{2} - 3 = 0;$$

hence, this value checks, since the right member of the given equation is also zero. Substituting the other possible root, -3, for x in the left member of the equation gives

$$2(-3)^2 + 5(-3) - 3 = 2(9) - 15 - 3 = 0;$$

therefore, this value also checks. Consequently, the roots or solutions are $x = \frac{1}{2}$ and $x = -3$.

Exercise 13.1

Solve the following quadratic equations by factoring.

1. $x^2 - 4 = 0.$ **2.** $x^2 - 1 = 0.$

3. $x^2 - 25 = 0.$ **4.** $x^2 - 49 = 0.$

5. $9x^2 - 16 = 0.$ **6.** $4x^2 - 9 = 0.$

7. $16x^2 - 25 = 0.$ **8.** $49x^2 - 4 = 0.$

9. $x^2 - 3x + 2 = 0.$ **10.** $x^2 - 5x + 6 = 0.$

11. $x^2 + 2x - 8 = 0.$ **12.** $x^2 - 2x - 3 = 0.$

13. $x^2 - 3x - 10 = 0.$ **14.** $x^2 - 4x - 21 = 0.$

15. $x^2 + 9x + 20 = 0.$ **16.** $x^2 + 4x + 3 = 0.$

17. $3x^2 - 11x + 10 = 0.$ **18.** $2x^2 - x - 3 = 0.$

19. $3x^2 + 7x - 6 = 0.$ **20.** $4x^2 - 13x - 12 = 0.$

21. $2x^2 + 3x - 2 = 0$.

22. $3x^2 - 7x - 6 = 0$.

23. $2x^2 - 7x - 15 = 0$.

24. $5x^2 - 7x - 6 = 0$.

25. $6x^2 - 13x + 6 = 0$.

26. $12x^2 + 7x - 12 = 0$.

27. $10x^2 - 21x - 10 = 0$.

28. $14x^2 - 53x + 14 = 0$.

29. $12x^2 - 17x + 6 = 0$.

30. $20x^2 - 37x + 15 = 0$.

31. $10x^2 + 7x - 12 = 0$.

32. $28x^2 + 5x - 12 = 0$.

33. $x^2 - (a + b)x + ab = 0$.

34. $2x^2 + (b - 2a)x - ab = 0$.

35. $cx^2 - (bc + d)x + bd = 0$.

36. $acx^2 + (cd - ab)x - bd = 0$.

13.2. Solution by Completing the Square

We shall make use of the fact that the square of $x + a$ is $x^2 + 2ax + a^2$ in developing a second procedure for solving a quadratic equation. We shall notice that the first and third terms are perfect squares and that the second term is twice their product. Hence, in order to complete the square of a quadratic without a constant term and with one as coefficient of the second degree term, we add the square of half the coefficient of the linear term. If the leading coefficient is not 1, we factor out that coefficient so as to obtain a leading coefficient of 1. Thus, in order to complete the square of $x^2 + 6x$, we must add the square of one half of 6; thus, we get $x^2 + 6x + 9$. Furthermore, to complete the square of $2x^2 + 10x$, we write $2x^2 + 10x = 2(x^2 + 5x)$ and, consequently, must add $[\frac{1}{2}(5)]^2 = \frac{25}{4}$ in the parentheses. If, in working with an equation we add a number to one member, we must remember to add the same number to the other member.

EXAMPLES

1. The trinomial $x^2 - 8x + 16$ is a perfect square, since the first term is $1x^2$ and the third term 16 is the square of one-half the coefficient of x. The square root of the trinomial is $\pm(x - 4)$, according to whether $x - 4$ is positive or negative.

2. In order to solve

$$x^2 + 6x + 5 = 0,$$

we notice that 5 is not the square of one-half the coefficient of x; hence, we must add the proper number to $x^2 + 6x$ so as to make it a perfect square, but must first get the 5 out of the way. We do this by adding -5 to each member and obtaining

$$x^2 + 6x = -5.$$

We now take half the coefficient of x, square it, and add the square, 9, to the left member, since that makes it a perfect square. We must also add 9 to the right member to offset having added it to the left member. Thus, we have

$$x^2 + 6x + 9 = -5 + 9 = 4,$$
$$(x + 3)^2 = 4,$$
$$x + 3 = \pm 2,$$
$$x = -3 \pm 2$$
$$= -1, -5.$$

These values can be checked by substituting in the given equation. Substituting -1 gives $(-1)^2 + 6(-1) + 5 = 1 - 6 + 5 = 0$, and -1 is a root. The other value can be checked similarly.

3. Solve

$$2x^2 - 7x + 3 = 0.$$

SOLUTION

We begin the solution by dividing by the coefficient 2 of x^2 so as to have an expression with 1 as leading coefficient. Thus, we get

$$x^2 - \frac{7}{2}x + \frac{3}{2} = 0,$$

and adding $-\frac{3}{2}$ to each member gives

$$x^2 - \frac{7}{2}x = -\frac{3}{2}.$$

Now, adding $\frac{49}{16}$ to each member, since that is the square of half the coefficient of x, we have

$$x^2 - \frac{7}{2}x + \frac{49}{16} = -\frac{3}{2} + \frac{49}{16} = \frac{25}{16},$$

$$\left(x - \frac{7}{4}\right)^2 = \frac{25}{16},$$

$$x - \frac{7}{4} = \frac{\pm 5}{4},$$

$$x = \frac{7}{4} \pm \frac{5}{4} = 3, \frac{1}{2}.$$

It is quite possible that, after adding the square of half the coefficient of x to each member of the equation, we may have a negative number on the right. This requires the use of a new type of number, since the square of a positive or a negative number is positive. Any negative number can be thought of and written as a positive number N times (-1); hence, we have

$$\sqrt{-N} = \sqrt{(-1)N}$$
$$= \sqrt{-1}\sqrt{N},$$

and this is written in the form $i\sqrt{N}$, since it is customary in mathematics to represent $\sqrt{-1}$ by i. Since $i = \sqrt{-1}$, it follows that $i^2 = -1$, and we will need to use this fact in checking possible solutions.

EXAMPLE

    4. Solve

$$x^2 + 4x + 8 = 0.$$

SOLUTION

    Since the coefficient of x^2 is already 1, we begin by adding -8 to each member. Thus,

$$x^2 + 4x = -8,$$
$$x^2 + 4x + 4 = -8 + 4, \text{ completing the square,}$$
$$(x + 2)^2 = -4,$$
$$x + 2 = \pm\sqrt{-4} = \pm 2i,$$
$$x = -2 \pm 2i.$$

    We shall check $-2 + 2i$ and leave the checking of $-2 - 2i$ to the reader. Substituting $-2 + 2i$ for x in the given equation, we get

$$(-2 + 2i)^2 + 4(-2 + 2i) + 8 = 4 - 8i + 4i^2 - 8 + 8i + 8$$
$$= 4 + 4i^2, \text{ collecting,}$$
$$= 4 - 4, \, i^2 = -1,$$
$$= 0,$$

and $x = -2 + 2i$ is a root.

Exercise 13.2

    Solve the following quadratic equations by completing the square.

1. $x^2 - 5x + 6 = 0.$ **2.** $x^2 - x - 6 = 0.$

3. $x^2 - 5x + 4 = 0.$ **4.** $x^2 - 2x - 15 = 0.$

5. $x^2 + 2x - 8 = 0.$ **6.** $x^2 + 2x - 15 = 0.$

7. $x^2 - 11x + 28 = 0.$ **8.** $x^2 + x - 30 = 0.$

9. $2x^2 - 5x + 2 = 0.$ **10.** $3x^2 - 11x + 6 = 0.$

11. $4x^2 - 21x + 5 = 0.$ **12.** $3x^2 + 5x - 2 = 0.$

13. $12x^2 - 25x + 12 = 0.$ **14.** $15x^2 - 31x + 10 = 0.$

15. $15x^2 + 19x + 6 = 0.$ **16.** $6x^2 - 5x - 4 = 0.$

17. $x^2 - 4x + 1 = 0.$ **18.** $x^2 - 6x + 5 = 0.$

19. $x^2 - 2x - 4 = 0$.   **20.** $x^2 + 2x - 2 = 0$.

21. $x^2 - 4x - 3 = 0$.   **22.** $x^2 - 6x + 6 = 0$.

23. $x^2 - 4x + 2 = 0$.   **24.** $x^2 + 4x + 1 = 0$.

25. $x^2 - 2x + 2 = 0$.   **26.** $x^2 - 4x + 5 = 0$.

27. $x^2 - 2x + 5 = 0$.   **28.** $x^2 - 6x + 10 = 0$.

29. $2x^2 - 2x + 1 = 0$.   **30.** $9x^2 - 12x + 5 = 0$.

31. $9x^2 - 12x + 20 = 0$.   **32.** $4x^2 - 8x + 13 = 0$.

33. $x^2 + (b - a)x - ab = 0$.   **34.** $x^2 - (2a + b)x + a(a + b) = 0$.

35. $x^2 - (a + a^2)x + a^3 = 0$.   **36.** $x^2 - 2ax + a^2 - b^2 = 0$.

13.3. Solution by Formula

We shall now develop a formula for use in solving the quadratic equation

$$ax^2 + bx + c = 0, \qquad a \neq 0$$

In order to do this, we shall complete the square. Consequently, we begin by adding $-c$ to each member and then dividing by the coefficient of x^2. Thus, we get

$$x^2 + \frac{b}{a}x = -\frac{c}{a}$$

and continue the solution by adding the square of $\frac{1}{2} \cdot \frac{b}{a}$ to each member. This gives

$$x^2 + \frac{b}{a}x + \left(\frac{1}{2} \cdot \frac{b}{a}\right)^2 = -\frac{c}{a} + \frac{1}{4} \cdot \frac{b^2}{a^2},$$

$$\left(x + \frac{1}{2} \cdot \frac{b}{a}\right)^2 = \frac{b^2 - 4ac}{4a^2},$$

$$x + \frac{b}{2a} = \pm \frac{\sqrt{b^2 - 4ac}}{2a}$$

Now, solving for x we see that

$$x = \frac{-b \pm \sqrt{b^2 - 4ac}}{2a}$$

are the two solutions of $ax^2 + bx + c = 0$, $a \neq 0$. These expressions for x are called the *quadratic formula* and can be used for solving any quadratic.

EXAMPLES

1. Solve

$$2x^2 + 3x - 5 = 0.$$

SOLUTION

This equation is in the form $ax^2 + bx + c = 0$. The coefficient of x^2 is $a = 2$, that of x is $b = 3$, and the constant term is $c = -5$. Putting these in the quadratic formula gives

$$x = \frac{-3 \pm \sqrt{3^2 - 4(2)(-5)}}{2(2)} = \frac{-3 \pm \sqrt{9 + 40}}{4}$$

$$= \frac{-3 \pm 7}{4} = 1, -2.5$$

as the solutions. The 1 was obtained by using the plus sign with the 7, and -2.5 came from using -7.

2. Solve

$$9x^2 - 12x + 8 = 0.$$

SOLUTION

In this equation, $a = 9$, $b = -12$, and $c = 8$; hence, the quadratic formula becomes

$$x = \frac{-(-12) \pm \sqrt{(-12)^2 - 4(9)8}}{2(9)}$$

$$= \frac{12 \pm \sqrt{144 - 288}}{18} = \frac{12 \pm \sqrt{-144}}{18}$$

$$= \frac{12 \pm 12i}{18} = \frac{2 \pm 2i}{3}.$$

Exercise 13.3

Solve the following equations by use of the quadratic formula.

1. $x^2 - 5x + 6 = 0.$ **2.** $x^2 - 7x + 12 = 0.$

3. $x^2 - 7x + 10 = 0.$ **4.** $x^2 - 2x - 3 = 0.$

5. $x^2 - 3x - 10 = 0.$ **6.** $x^2 + 4x - 21 = 0.$

7. $x^2 + 8x + 12 = 0.$ **8.** $x^2 + 8x + 15 = 0.$

9. $3x^2 - 7x + 2 = 0.$ **10.** $2x^2 - 7x + 3 = 0.$

11. $4x^2 - 5x + 1 = 0.$ **12.** $3x^2 - 10x + 3 = 0.$

13. $3x^2 + 10x - 8 = 0.$ **14.** $4x^2 + 5x - 6 = 0.$

15. $5x^2 - 13x - 6 = 0.$ **16.** $5x^2 - 22x - 15 = 0.$

17. $6x^2 - 13x + 6 = 0.$ **18.** $12x^2 + 7x - 12 = 0.$

19. $14x^2 - 55x + 21 = 0.$ **20.** $20x^2 + x - 12 = 0.$

21. $21x^2 + 55x + 14 = 0.$ **22.** $20x^2 + 7x - 6 = 0.$

23. $12x^2 + 17x + 6 = 0.$ **24.** $4x^2 + 16x + 15 = 0.$

25. $x^2 - 2x - 1 = 0.$ **26.** $x^2 - 4x + 1 = 0.$

27. $x^2 - 4x - 1 = 0.$ **28.** $x^2 - 10x + 22 = 0.$

29. $2x^2 - 4x - 1 = 0.$ **30.** $9x^2 - 18x + 4 = 0.$

31. $9x^2 - 36x + 28 = 0.$ **32.** $2x^2 - 6x - 27 = 0.$

33. $x^2 + 4 = 0.$ **34.** $x^2 + 25 = 0.$

35. $x^2 + 16 = 0.$ **36.** $x^2 + 49 = 0.$

37. $x^2 - 6x + 11 = 0.$ **38.** $x^2 - 4x + 7 = 0.$

39. $x^2 + 2x + 6 = 0.$ **40.** $x^2 + 4x + 11 = 0.$

41. $x^2 + (a - b)x - ab = 0.$ **42.** $x^2 - bx - a^2 - ab = 0.$

43. $x^2 - ax + ab - b^2 = 0.$ **44.** $x^2 - 2ax + a^2 - b^2 = 0.$

13.4. Equations in Quadratic Form

The unknown in a quadratic equation can be any quantity. All that is necessary is for the unknown to enter to the second power. It may also be to the first power, and the equation may contain a constant. Thus,

$$2(x^2 + 3x)^2 - 3(x^2 + 3x) - 5 = 0$$

is a quadratic equation in which the unknown is $x^2 + 3x$. Hence, the equation can be solved for $x^2 + 3x$ by factoring, completing the square, or formula.

If, in thinking of an equation as a quadratic, the quantity considered as the unknown is linear or quadratic, then we can find the values of x which satisfy the given equation.

EXAMPLES

1. Solve

$$3(x^2 - 3x)^2 + 13(x^2 - 3x) + 14 = 0.$$

SOLUTION

As a first step in the solution we shall solve the equation for $x^2 - 3x$, since the given equation is a quadratic if this quantity is considered as the unknown. Thus,

$$x^2 - 3x = \frac{-13 \pm \sqrt{13^2 - 4(3)(14)}}{2(3)}$$

$$= \frac{-13 \pm \sqrt{169 - 168}}{6}$$

$$= \frac{-13 \pm 1}{6} = -2, -\frac{7}{3}.$$

We now have two quadratics in x and can get each in the form $ax^2 + bx + c = 0$. If we do this and multiply the second through by 3, the equations become
$$x^2 - 3x + 2 = 0 \quad \text{and} \quad 3x^2 - 9x + 7 = 0.$$
The solutions of the first one of these are

$$x = \frac{-(-3) \pm \sqrt{(-3)^2 - 4(1)2}}{2(1)}$$

$$= \frac{3 \pm \sqrt{9 - 8}}{2} = \frac{3 \pm 1}{2} = 2, 1,$$

and those of the second are

$$x = \frac{-(-9) \pm \sqrt{(-9)^2 - 4(3)7}}{2(3)}$$

$$= \frac{9 \pm \sqrt{81 - 84}}{6} = \frac{9 \pm i\sqrt{3}}{6}.$$

Consequently, the solutions of the given equations are 2, 1,

$$\frac{9 + i\sqrt{3}}{6}, \quad \text{and} \quad \frac{9 - i\sqrt{3}}{6}.$$

2. Solve

$$4x^4 - 21x^2 + 20 = 0.$$

SOLUTION

If we think of x^4 as $(x^2)^2$, the given equation is a quadratic with x^2 as the unknown. Writing it as such, we have

$$4(x^2)^2 - 21(x^2)^1 + 20 = 0.$$

The solutions of this quadratic are

$$x^2 = \frac{-(-21) \pm \sqrt{(-21)^2 - 4(4)20}}{2(4)}$$

$$= \frac{21 \pm \sqrt{441 - 320}}{8}$$

$$= \frac{21 \pm \sqrt{121}}{8} = \frac{21 \pm 11}{8} = 4, \frac{5}{4};$$

hence, $\qquad x = \pm 2 \quad \text{and} \quad \pm \frac{\sqrt{5}}{2}$

are the solutions of the given equation.

Exercise 13.4

Solve each of the following equations by considering it as a quadratic.

1. $x^4 - 5x^2 + 4 = 0.$ **2.** $x^4 - 10x^2 + 9 = 0.$

3. $x^4 - 8x^2 - 9 = 0.$ **4.** $x^4 + 12x^2 - 64 = 0.$

5. $x^6 - 9x^3 + 8 = 0.$ **6.** $x^6 + 26x - 27 = 0.$

7. $x^6 + 19x^3 - 216 = 0.$ **8.** $x^6 + 126x^3 + 125 = 0.$

9. $x^{-2} + 3x^{-1} - 10 = 0.$ **10.** $x^{-2} - x^{-1} - 12 = 0.$

11. $x^{-2} - 6x^{-1} - 7 = 0.$ **12.** $x^{-2} + 2x^{-1} - 15 = 0.$

13. $(x - 1)^2 + 2(x - 1) - 8 = 0.$ **14.** $(2x - 1)^2 + 2(2x - 1) - 3 = 0.$

15. $(3x + 2)^2 - (3x + 2) - 20 = 0.$ **16.** $(x + 2)^2 + (x + 2) - 2 = 0.$

17. $(x^2 + x)^2 + 8(x^2 + x) - 12 = 0.$ **18.** $(x^2 + 3x)^2 - 2(x^2 + 3x) - 8 = 0.$

19. $(x^2 - 5x)^2 + 10(x^2 - 5x) + 24 = 0.$

20. $(x^2 - 4x)^2 + 7(x^2 - 4x) + 12 = 0.$

21. $(2x^2 - 5x)^2 - (2x^2 - 5x) - 6 = 0.$

22. $(3x^2 - 2x)^2 - 6(3x^2 - 2x) + 5 = 0.$

23. $(2x^2 + 3x)^2 - 4(2x^2 + 3x) - 5 = 0.$

24. $(3x^2 - 5x)^2 - 4 = 0.$

25. $\left(\frac{2x - 1}{x - 2}\right)^2 + 4\left(\frac{2x - 1}{x - 2}\right) + 3 = 0$

26. $\left(\dfrac{x+6}{2x+1}\right)^2 + 5\left(\dfrac{x+6}{2x+1}\right) + 6 = 0.$

27. $\left(\dfrac{3x-2}{x+3}\right)^2 + 6\left(\dfrac{3x-2}{x+3}\right) + 8 = 0.$

28. $\left(\dfrac{4x+3}{x-3}\right)^2 + 4\left(\dfrac{4x+3}{x-3}\right) + 3 = 0.$

29. $2\left(\dfrac{x+1}{x-1}\right) - 5\left(\dfrac{x-1}{x+1}\right) - 3 = 0.$

30. $3\left(\dfrac{2x-1}{x-2}\right) + 2\left(\dfrac{x-2}{2x-1}\right) - 5 = 0.$

31. $5\left(\dfrac{3x+2}{2x-3}\right) + \dfrac{2x-3}{3x+2} - 6 = 0.$

32. $6\left(\dfrac{3x+1}{x+2}\right) - 20\left(\dfrac{x+2}{3x+1}\right) - 7 = 0.$

13.5. Radical Equations

If an equation contains a radical, it is called a *radical equation*. We shall now make use of two facts in solving radical equations. These facts are:

1. Any root of a given equation is also a root of the equation obtained by equating the squares of the two members of the given equation.

2. The equation obtained by squaring the members of the given equation may have roots that are not roots of the given equation.

The first of these statements is true since the squares of two equal quantities are equal. The second is true since the square of a quantity and the square of the negative of it are equal.

EXAMPLE

1. In order to solve

$$\sqrt{5x^2 - 1} = 2x,$$

we square each member and obtain

$$5x^2 - 1 = 4x^2.$$

Now adding $1 - 4x^2$ to each member gives

$$x^2 = 1;$$

hence, $x = \pm 1$

are the roots of the equation obtained by squaring the members of the given equation. If we substitute $x = 1$ in the left member of the given equation, it becomes $\sqrt{5(1)^2 - 1} = \sqrt{4} = 2$; consequently, $x = 1$ is a root of the given equation since the right-hand member is also 2. The other value cannot be a root since the right member is -2 for $x = -1$, whereas the left member is positive since it is the square root of a positive number.

A value of x that is a root of the equation obtained by squaring the members of the given equation but is not a root of the given equation is called an *extraneous root*.

The general procedure for solving a radical equation consists of:

(1) *rearranging terms so that one member of the equation contains only a radical;*

(2) *squaring each member of the rearranged equation;*

(3) *solving for x if the equation obtained in (2) does not contain a radical [repeating (1) and (2) if the equation obtained in (2) does contain a radical];*

(4) *substituting the possible roots in the given equation to see if they are roots or extraneous roots.*

EXAMPLE

2. Solve

$$\sqrt{x^2 + 3x + 6} - 3x + 2 = 0.$$

SOLUTION

Adding $3x - 2$ to each member isolates a radical in one member of the equation and gives

$$\sqrt{x^2 + 3x + 6} = 3x - 2.$$

Squaring each member, we have

$$x^2 + 3x + 6 = (3x - 2)^2$$
$$= 9x^2 - 12x + 4.$$

Collecting like powers leads to the quadratic

$$-8x^2 + 15x + 2 = 0.$$

We shall solve by use of the quadratic formula and have

$$x = \frac{-15 \pm \sqrt{15^2 - 4(-8)(2)}}{2(-8)}$$

$$= \frac{-15 \pm \sqrt{225 + 64}}{-16}$$

$$= \frac{-15 \pm \sqrt{289}}{-16} = \frac{-15 \pm 17}{-16} = -\frac{1}{8}, 2.$$

Substituting 2 for x in the left member of the given equation, we have

$$\sqrt{2^2 + 3(2) + 6} - 3(2) + 2 = \sqrt{16} - 6 + 2$$
$$= 4 - 4 = 0;$$

hence, $x = 2$ is a root of the given equation. It is obvious that a negative number cannot be a root since -3 times any negative number is positive as are $\sqrt{x^2 + 3x + 6}$ and 2; consequently, their sum cannot be zero as it would have to be for the equation to have a negative root.

3. Solve

$$\sqrt{x + 1} = \sqrt{4x - 3} - \sqrt{x - 2}.$$

SOLUTION

Since one member contains only a radical, we shall square each member as the equation stands and get

$$x + 1 = 4x - 3 - 2\sqrt{(4x - 3)(x - 2)} + x - 2.$$

Isolating the radical in the right member gives

$$-4x + 6 = -2\sqrt{(4x - 3)(x - 2)}$$

and, dividing through by -2, we have

$$2x - 3 = \sqrt{(4x - 3)(x - 2)}.$$

Squaring again, since there is still a radical and it is in a member by itself, gives

$$(2x - 3)^2 = (4x - 3)(x - 2),$$
$$4x^2 - 12x + 9 = 4x^2 - 11x + 6,$$
$$4x^2 - 4x^2 - 12x + 11x = 6 - 9,$$
$$-x = -3,$$
$$x = 3.$$

This is the only possible root. Substituting in the left member gives $\sqrt{3 + 1} = 2$, and putting $x = 3$ in the right member, we have

$$\sqrt{4(3) - 3} - \sqrt{3 - 2} = \sqrt{9} - \sqrt{1}$$
$$= 3 - 1 = 2;$$

hence, $x = 3$ is a root because the two members are equal.

Exercise 13.5

Solve the following equations.

1. $\sqrt{x^2 + 3x - 1} = \sqrt{4x + 1}.$ **2.** $\sqrt{x^2 - 3x + 3} = \sqrt{3x - 2}.$

3. $\sqrt{2x^2 - x + 1} = \sqrt{1 - 3x}.$ **4.** $\sqrt{x^2 + 4x + 5} = \sqrt{2x + 8}.$

5. $\sqrt{2x + 3} = 2x - 3.$

6. $\sqrt{14 - 2x} = x - 3.$

7. $\sqrt{11 + x} = 5 + x.$

8. $\sqrt{3x + 4} = 8 - x.$

9. $\sqrt{x^2 - 5x + 3} = 9 - x.$

10. $\sqrt{2x^2 + 2x + 1} = 1 - x.$

11. $\sqrt{3x^2 - x + 4} = x - 2.$

12. $\sqrt{2x^2 - x + 1} = 2x - 2.$

13. $\sqrt{4x^2 + 6x + 7} = 4x - 1.$

14. $\sqrt{9x^2 + 3x - 2} = 6x - 2.$

15. $\sqrt{2x^2 + 3x + 7} = 2x + 2.$

16. $\sqrt{3x^2 + 2x + 8} = 3x + 2.$

17. $\sqrt{2x^2 + 3x - 1} = 1 + x.$

18. $\sqrt{3x^2 + 5x + 1} = x + 2.$

19. $\sqrt{x^2 - x - 6} = 2x - 6.$

20. $\sqrt{3x^2 + 2x + 4} = 3x + 2.$

21. $\sqrt{3x - 2} + 2 = \sqrt{6x + 3}.$

22. $\sqrt{3x + 7} = 1 + \sqrt{x + 2}.$

23. $\sqrt{8x + 5} - 1 = \sqrt{2x + 3}.$

24. $\sqrt{6x - 1} - 2 = -\sqrt{9x - 2}.$

25. $\sqrt{x^2 - x + 2} + 1 = \sqrt{x^2 + x + 3}.$

26. $\sqrt{2x^2 + x + 3} - 2 = \sqrt{2x^2 + x - 1}.$

27. $\sqrt{3x^2 + 2x + 4} + 1 = \sqrt{3x^2 + x + 9}.$

28. $\sqrt{x^2 - x - 2} + 3 = \sqrt{x^2 + 5x + 1}.$

29. $\dfrac{\sqrt{2x + 3} + 1}{\sqrt{4x + 2} + 1} = 1.$

30. $\dfrac{\sqrt{3x + 2}}{\sqrt{6x + 5} + 1} = \dfrac{1}{2}.$

31. $\dfrac{\sqrt{x - 4} + 3}{\sqrt{x + 4} - 1} = 2.$

32. $\dfrac{\sqrt{2x + 1} + 1}{\sqrt{5 - x} + 1} = 2.$

33. $\sqrt{3x - 2} - \sqrt{4x + 1} = \sqrt{x - 1}.$

34. $\sqrt{2x + 7} + \sqrt{x + 4} = \sqrt{x + 7}.$

35. $\sqrt{3x + 1} + \sqrt{2x - 1} = \sqrt{7x + 2}.$

36. $\sqrt{x + 3} + \sqrt{2x + 5} = \sqrt{x + 6}.$

37. $\sqrt{5x + 4} - \sqrt{2x + 1} = \sqrt{3x + 1}.$

38. $\sqrt{3 - x} - \sqrt{x + 2} = \sqrt{2x - 3}.$

39. $\sqrt{4x + 3} - \sqrt{2x - 2} = \sqrt{4x - 2}.$

40. $\sqrt{3x + 5} + \sqrt{6x + 3} = \sqrt{6x + 11}.$

13.6. Problems That Lead to Quadratics

Many stated problems lead to quadratic equations. The procedure in obtaining these equations is the same as used in Art. 3.9. We must bear in mind, however, that the equation may have solutions that are not solutions of the problem.

EXAMPLES

1. The sum of the reciprocals of two consecutive integers is $\frac{11}{30}$. What are the numbers?

SOLUTION

If one of the integers is represented by x, then the next larger one must be represented by $x + 1$. Furthermore, the reciprocal of a number is 1 divided by the number; consequently, the reciprocals of the numbers are

$$\frac{1}{x} \text{ and } \frac{1}{x + 1}.$$

Therefore,
$$\frac{1}{x} + \frac{1}{x + 1} = \frac{11}{30}$$

is the desired equation, since each member is an expression for the sum of the reciprocals of the numbers.

In order to solve the equation, we multiply each term by the common denominator $30x(x + 1)$. Thus, we have

$$30x(x + 1) \frac{1}{x} + 30x(x + 1) \frac{1}{x + 1} = 30x(x + 1) \frac{11}{30},$$

$$30(x + 1) + 30x = x(x + 1)11, \text{ removing common factors,}$$
$$30x + 30 + 30x = 11x^2 + 11x, \text{ removing parentheses,}$$
$$-11x^2 + 49x + 30 = 0, \text{ collecting like terms,}$$
$$(x - 5)(-11x - 6) = 0, \text{ factoring.}$$

Hence,
$$x = 5, \frac{-6}{11}$$

are the solutions of the equation, but $-6/11$ is not a solution of the problem, since it calls for integers. Therefore, 5 is one of the integers, and $5 + 1 = 6$ is the other.

2. Find the length and width of a rectangular flower bed in a park if the perimeter is 146 feet and the area is 780 square feet.

SOLUTION

If we represent the length by L, then the width is $\frac{1}{2}(146) - L = 73 - L$; consequently, $L(73 - L)$ is one expression for the area and 780 is given as another.

Therefore, the desired equation is

$$L(73 - L) = 780;$$

hence,

$$73L - L^2 = 780, \text{ removing parentheses,}$$
$$-L^2 + 73L - 780 = 0, \text{ adding } -70 \text{ and rearranging,}$$
$$(L - 13)(-L + 60) = 0, \text{ factoring.}$$

Solving for L, we see that $L = 13$ or 60; hence, the width is 60 ft or 13 ft.

3. A pilot flew a round trip of 2100 miles in 6.5 hours. If there was a head wind of 20 miles an hour for the first half of the trip and a tail wind of 30 miles per hour for the return trip, find the speed in still air.

SOLUTION

If we represent his speed in still air by x, then $x - 20$ represents the speed against the 20-mile wind, and $x + 30$ represents the speed with the 30-mile wind. Since the round trip was 2100 miles, each half of it was 1050; furthermore, since distance divided by rate gives time, we see that

$$\frac{1050}{x - 20} \quad \text{is the time against the wind,}$$

$$\frac{1050}{x + 30} \quad \text{is the time with the wind.}$$

Therefore,

$$\frac{1050}{x - 20} + \frac{1050}{x + 30} = \frac{13}{2}$$

is the desired equation, since each member is an expression for the time required for the trip. Now, clearing of fractions, we have

$$1050(2)(x + 30) + 1050(2)(x - 20) = 13(x - 20)(x + 30),$$
$$2100x + 63{,}000 + 2100x - 42{,}000 = 13x^2 + 130x - 7800,$$
$$-13x^2 + 4070x + 28{,}800 = 0, \text{ collecting terms,}$$
$$(x - 320)(-13x - 90) = 0, \text{ factoring;}$$

hence,

$$x = 320, \ -\frac{90}{13}.$$

Consequently, the air speed is 320 miles per hour.

Exercise I3.6

1. The sum of the squares of two consecutive integers is 421. What are they?

2. Determine two consecutive odd integers whose product is 255.

3. What are the two consecutive even integers whose product exceeds their sum by 194?

4. Find two consecutive, positive, even integers such that the difference of their reciprocals is $\frac{1}{24}$.

5. Two positive integers differ by 5, and the difference of their squares is 105. What are they?

6. Find two positive numbers whose product and sum are 294 and 35, respectively.

7. Find two integers that differ by 4 and whose product is 396.

8. Two positive numbers differ by 7 and the sum of their squares is 337. Find them.

9. What is the positive integer that differs from its reciprocal by $\frac{143}{12}$?

10. Find a number that is 5 greater than 14 times its reciprocal.

11. What is the integer such that the sum of it and 8 times the reciprocal of the next larger integer is 8?

12. Find the number that exceeds 12 times its reciprocal by 4.

13. The area of a rectangle is 143 square feet. Find its dimensions if the length exceeds the width by 2.

14. The floor space of a rectangular room is 315 feet. Find the dimensions if the length is 9 less than twice the width.

15. The side of a square is 5 feet shorter than the length of a rectangle and 12 feet longer than the width of the rectangle. Find the dimensions of the rectangle if its area is half that of the square.

16. The outside dimensions of a picture with a mat on it are 12 inches and 10 inches. Find the width of the mat if the mat covers 1.5 times as much as it enclosed.

17. Two boys drive a motor boat upstream 30 miles and back in 4 hours. Find the rate of the boat in still water if the river current is 4 miles per hour.

18. A plane travelled 900 miles at a uniform rate. If it had gone 45 miles per hour slower, the trip would have required another hour. Find the rate travelled.

19. Jim drove 135 miles at a uniform rate and Joe drove 160 miles at a rate that was 5 miles per hour slower. Find the rate of each driver if Joe drove an hour longer than Jim did.

20. A cattle buyer paid $11,220 for each of two herds of steers. The second herd cost $8 per head more than the first, and there were 212 head in all. Find the price per head of the higher priced herd.

21. A rancher drove around a rectangular pasture in his jeep to inspect the fence and found that the mileage was 4.8 miles. If the pasture contained 864 acres, find the dimensions of the pasture. A square mile is 640 acres.

22. A square piece of carpet covered the floor of a room except for 4 feet at one end. What size piece of carpet is needed to finish covering the floor if the entire area is 192 square feet?

23. A driver figured that the daily cost of taking his car to work was $1.50. He divided this evenly between himself and each of his passengers. He obtained two more riders and was able to reduce the price per person by $.20 per day. How many riders did he have at the new rate?

24. A swimming pool holds 3000 cubic feet of water and can be drained at 10 cubic feet per minute faster than it can be filled. Find the rate of filling if 50 minutes more is required for filling than for draining.

14

INEQUALITIES

14.1. Definitions and Fundamental Theorems

We studied conditions under which two functions are equal in Chapters 3, 6, and 13. We shall now consider conditions under which two functions are unequal. In order to be more specific, we shall determine conditions under which one function is greater than another and conditions under which one is less than the other. If a is greater than b, we write $a > b$ and mean that $a - b > 0$. We indicate that a is less than b by writing $a < b$ and meaning that $a - b < 0$. In each case the inequality sign points toward the smaller of the two quantities. We say that two inequalities have the *same sense* or *opposite senses* according to whether the signs in them point in the same or opposite directions.

An inequality that is true for all real values of the variable is called an *absolute inequality*, and an inequality that is true for some real values of the variable and not true for others is called a *conditional inequality*. Thus, $x^2 + 7 > 0$ is an absolute inequality since it is true for all real values of x; furthermore, $3x > 12$ is a conditional inequality since it is true for $x > 4$ and not true for other real values of x.

The set of values of the variable for which the inequality is true is called the *solution*. The process of determining the solution is based on several theorems. The first is:

(1) *The sense of an inequality is not changed if the same number is added to or subtracted from each member.*

We shall prove this theorem for adding the same number to $a < b$. If $a < b$, then

$$a - b = n, \qquad n < 0$$

Therefore,

$$a - b + c - c = n, \qquad c - c = 0,$$

$$(a + c) - (b + c) = n, \qquad \text{rearranging terms and} \\ \text{inserting parentheses.}$$

Consequently,

$$a + c < b + c$$

By means of a similar argument, we can prove that $a + c > b + c$ if $a > b$. Since subtracting the same number from each member is equivalent to adding the negative of that number to each member, the theorem is true for subtraction as well as for addition.

The second theorem is:

(2) *The sense of an inequality is not changed if the members are multiplied by the same positive number.*

We shall prove this theorem for $a > b$. If $a > b$, then

$$a - b = n, \qquad n < 0$$

Therefore,

$$ka - kb = kn, \quad \text{multiplying each member by } k > 0,$$

Consequently, since $kn < 0$, we have

$$ka < kb$$

as was to be proved.

Since dividing by $d > 0$ is equivalent to mutliplying by $1/d$, it follows that if $a < b$ then $a/d < b/d$. A similar argument can be used to prove that if $a > b$, then for $k > 0$, $ka > kb$ and, for $d > 0$, $a/d > b/d$.

The third theorem is:

(3) *The sense of an inequality is reversed if the members are multiplied or divided by the same negative number.*

The proof of this theorem is essentially the same as that for the second theorem.

The fourth theorem is:

(4) *If* a *and* b *are two unequal positive numbers and if* n *is a positive integer, then* a^n *and* b^n *and also* $\sqrt[n]{a}$ *and* $\sqrt[n]{b}$ *are unequal in the same sense as* a *and* b.

We shall first prove this theorem for $a < b$ and integral powers of a and b. The statement is: If $0 < a < b$ and n is a positive integer, then $a^n < b^n$. Since $0 < a < b$, it follows from theorem 2 that $a/b < 1$ and, consequently, $(a/b)^n < 1$. Therefore, $(a/b)^n = a^n/b^n < 1$; hence, multiplying by the positive number b^n, we have $a^n < b^n$. The part of the theorem for $a > b$ follows in a similar manner.

We shall now prove that if $a < b$ then $\sqrt[n]{a} < \sqrt[n]{b}$. If $0 < a < b$, then $a/b < 1$ and we shall prove that $\sqrt[n]{a}/\sqrt[n]{b} < 1$ by the method of contradiction. Thus, we assume that $\sqrt[n]{a}/\sqrt[n]{b} = a^{1/n}/b^{1/n} \geq 1$ and show that this leads to a contradiction. If $a^{1/n}/b^{1/n} \geq 1$, then raising each member to the nth power as in the first part of this theorem, we have

$$\left(\frac{a^{1/n}}{b^{1/n}}\right)^n \geq 1^n.$$

Therefore, we are led to the statement that

$$\frac{a}{b} \geq 1.$$

This, however, contradicts the hypothesis that $a/b < 1$ and proves the theorem.

A similar argument can be used to prove that if $a > b > 0$ then $\sqrt[n]{a} > \sqrt[n]{b}$.

EXAMPLE

Show that $x + x^{-1} \geq 2$ for x positive.

SOLUTION

If we multiply each member of $x + x^{-1} \geq 2$ by the positive number x, we have

$$x^2 + 1 \geq 2x;$$

hence,

$$x^2 - 2x + 1 \geq 0, \text{ adding } -2x \text{ to each member.}$$

The left member is the square of the real number $x - 1$; therefore, it is nonnegative. The proof is complete since we began with the questionable inequality and arrived at a statement known to be true by means of reversible operations.

Exercise 14.1

Prove the following inequalities.

1. If $a > b$, then $a - c > b - c$.

2. If $a > b$, then $a + c > b + c$.

3. If $a > b$, then $am < bm$ for $m < 0$.

4. If $a > b$, then $ap > bp$ for $p > 0$.

5. If $a \neq b$ and $ab > 0$, then $a/b + b/a > 2$.

6. If $a \neq b$ and $b > 0$, then $(a^2 - b^2)/2b > a - b$.

7. If $b > 1$, then $b^2 > b$.

8. If $0 < a < 1$, then $a^2 < a$.

9. If $a > b > 0$, then $a^2 > b^2$.

10. If $a > b > 0$, then $a^2 - b^2 > (a - b)^2$.

11. If $a > b > .5$, then $a^2 - a > b^2 - b$.

12. If $a > 1 + 1/(b - 1)$ and $b > 1$, then $a + b < ab$.

13. If $a > b > 0$, then $a^3 > b^3$.

14. If $a > b > 0$, then $a^3 - b^3 > (a - b)^3$.

15. If $0 < a < b$, then $(a + 1)/(b + 1) > a/b$.

16. If $b > 0$ and $a > 1 + b$, then $a/b > (b + 1)/(a - 1)$.

17. If $a > 1$, then $(a - 1)(a + 1) < (a^2 - 1)(a^2 + 1)$.

18. If $a > b > 0$, then $1/a^2 + 1/b^2 > 2/ab$.

19. If $a > 1$, then $(a - 1)/(a + 1) < (a^2 - 1)/(a^2 + 1)$.

20. If a and b are positive integers and $a > b$, then $1/2a < (a - b)/(a + b)$.

14.2. Linear Inequalities

The solution of a linear inequality consists of the set of values of the independent variable for which the inequality is a true statement. This set of values can be obtained algebraically by use of the first three theorems of the last article. The set can be obtained graphically by changing the inequality to an equivalent one with zero as one member and $L(x)$ as the other, then sketching the graph of $y = L(x)$, and observing that values of the variable less than or greater than the value of x for which $y = L(x)$ crosses the X axis constitute the solution.

EXAMPLE

1. Solve $5x - 2 > 2x + 4$

SOLUTION

If we add $2 - 2x$ to each member as justified by the first theorem of Art. 14.1, we obtain

$$5x - 2 + 2 - 2x > 2x + 4 + 2 - 2x$$

$3x > 6$, collecting like terms,

$x > 2$, dividing each member by 3

GRAPHICAL SOLUTION

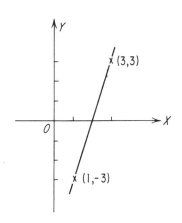

As a first step, we shall add $-2x - 4$ to each member of the given equation to obtain an inequality with one member zero. Thus we have $3x - 6 > 0$ and setting the left member equal to y gives $y = 3x - 6$. Since this is a linear equation, we need find only two points on the graph in order to draw it. If $x = 1$, then $y = -3$ and if $x = 3$ then $y = 3$; hence, $(1, -3)$ and $(3, 3)$ are on the graph. If we draw a line through these two points (Fig. 14-1), we see that it appears to cross the X axis at $x = 2$. Since we want values of x for which the function $3x - 6 > 0$, the solution is the set of values of x for which $x > 2$.

FIGURE 14-1

14.3. *Linear Inequalities That Involve Absolute Values*

The inequality $| ax + b | < c$ is read "the absolute value of $ax + b$ is less than c" and is satisfied by the set of values of x for which $ax + b < c$ and at the same time $ax + b > -c$. These last two inequalities can be combined into $-c < ax + b < c$.

EXAMPLES

1. Solve $|2x + 7| < 3$

SOLUTION

We must find the set of values of x for which $2x + 7 < 3$ and $2x + 7 > -3$ simultaneously. This can be re-phrased as the set of values of x for which $-3 < 2x + 7 < 3$. If we add -7 to each member of this inequality, we get $-10 < 2x <$

-4. Now dividing by the positive number 2, we obtain $-5 < x < -2$. Therefore the solution is the set of values of x such that $-5 < x < -2$. This is all values of x in the interval between -5 and -2.

The inequality $|ax + b| > c$ is read " the absolute value of $ax + b$ is greater than c", and it is satisfied by the set of values of x for which $ax + b > c$ and the set for which $ax + b < -c$, since in each case $|ax + b| > c$. Thus, the solution consists of two sets that are separated by the interval from $x = -c$ to $x = c$.

2. Solve $|3x + 5| > 8$

SOLUTION

This inequality is satisfied by the set of values of x for which $3x + 5 > 8$ and also by the set for which $3x + 5 < -8$. If we add -5 to each member of each of these linear inequalities, we get $3x > 3$ and $3x < -13$. Now, dividing each member by the positive number 3, we see that the solution consists of $x > 1$ and $x < -13/3$.

Exercise 14.2

Solve the following conditional inequalities algebraically and graphically.

1. $3x + 5 > x + 9$.   **2.** $5x - 1 > 3x + 5$.

3. $4x - 1 > -2x - 7$.   **4.** $7x + 4 > 2x - 6$.

5. $6x + 5 < 2x - 15$.   **6.** $4x + 7 < 2x + 1$.

7. $8x - 7 < 5x - 1$.   **8.** $2x - 1 < -x + 8$.

9. $2x - 1 > 4x + 1$.   **10.** $3x - 4 > 5x + 4$.

11. $-3x + 5 > 2x - 10$.   **12.** $6x - 11 > 8x - 11$.

13. $-2x + 1 < x - 5$.   **14.** $-5x + 4 < 3x - 12$

15. $-5x + 7 < -3x + 1$.   **16.** $-4x + 3 < -x + 9$.

17. $|2x + 3| < 5$.   **18.** $|3x - 1| < 8$.

19. $|5x + 4| < 14$.   **20.** $|4x - 3| < 5$.

21. $|3x - 2| > 4$.   **22.** $|2x + 7| > 1$.

23. $|6x + 5| > 13$.   **24.** $|7x - 8| > 6$.

14.4. Nonlinear Inequalities

A nonlinear inequality can be solved algebraically and graphically, just as a linear inequality. The first step in the algebraic solution is to get zero as one member of the inequality and then to factor the other member into linear factors. After this is done, we make use of the fact that the product of two numbers with like signs is positive and the product of two numbers with unlike signs is negative then solve the resulting pairs of linear inequalities simultaneously as illustrated in the examples.

EXAMPLES

1. Solve $6x^2 - 5x > 6$ algebraically.

SOLUTION

If we add -6 to each member of the given inequality, we get $6x^2 - 5x - 6 > 0$. If we factor the left member, we have $(2x - 3)(3x + 2) > 0$. Since the product is positive, the two factors are of the same sign. Hence, we must find the set of values for which

(1) $2x - 3 < 0$ and $3x + 2 < 0$ simultaneously,

and the set of values of x for which

(2) $2x - 3 > 0$ and $3x + 2 > 0$ simultaneously.

If we solve the pair of inequalities in (1), we find that $x < 3/2$ and $x < -2/3$. Consequently, the solution as furnished by inequalities (1) is $x < -2/3$ since x is automatically less than $3/2$ if it is less than $-2/3$. We shall now solve the pair given in (2). If that is done, we have $x > 3/2$ and $x > -2/3$; hence, the solution as furnished by (2) is $x > 3/2$. Therefore, the solution consists of the set $x < -2/3$ and the set $x > 3/2$.

In order to solve a quadratic inequality graphically, we change it to an equivalent one with zero as one member and $Q(x)$ as the other, then sketch the graph of $y = Q(x)$ and observe that the inequality is satisfied by all real values of x, by no real values, by the values between the roots of $Q(x) = 0$, or by the set of numbers larger than the larger root and the set that is smaller than the smaller root.

2. Solve $3x^2 - 8 > x$ graphically.

SOLUTION

As a first step we shall add $-x$ to each member to obtain an inequality with zero as one member. Thus, we have $3x^2 - x - 8 > 0$ and setting the left member equal to y gives $y = 3x^2 - x - 8$. We now locate several points on the curve by assigning values arbitrarily to x and calculating each corresponding value of y. If $x = 2$, then $y = 3(2^2) - 2 - 8 = 2$. This point $(2, 2)$ and several others are given in the table.

x	2	1	0	-1	-2
y	2	-5	-8	-4	6

If we draw a curve through these points and estimate its intersections with the X axis, we see that they are approximately $x = -1.5$ and $x = 1.8$ (Fig. 14-2). Consequently, the solutions of the inequality $3x^2 - 8 > x$ are $x < -1.5$ and $x > 1.8$.

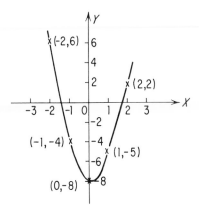

FIGURE 14-2

Exercise 14.3

Solve the following nonlinear inequalities algebraically and graphically.

1. $2x^2 - 3x - 5 > 0$.

2. $3x^2 - 2x - 8 > 0$.

3. $2x^2 + 11x + 12 > 0$.

4. $3x^2 - 14x + 15 > 0$.

5. $6x^2 - 11x - 7 < 0$.

6. $8x^2 - 22x + 15 < 0$.

7. $6x^2 + 17x + 5 < 0$.

8. $10x^2 - 39x + 14 < 0$.

9. $2x^2 - 3x + 2 > 0$.

10. $3x^2 - 5x + 4 > 0$.

11. $2x^2 + x + 1 < 0$.

12. $5x^2 - 4x + 1 < 0$.

13. $10x^2 - 21x - 8 > 0$.

14. $6x^2 + 7x - 4 > 0$.

15. $6x^2 + x - 11 > 0$.

16. $12x^2 - 13x - 12 > 0$.

17. $6x^2 + 7x - 25 < 0$.

18. $6x^2 - x - 17 < 0$.

19. $10x^2 - 31x - 16 < 0$.

20. $15x^2 + 19x + 3 < 0$.

15

COMPOUND INTEREST

15.1. Introduction

If money is loaned, the sum loaned is called the *principal* and the money paid by the borrower to the lender for the use of the principal is called *interest*. The interest is sometimes paid when it is due; at other times it is added to the principal and this sum used as a new principal. If the latter is done, we say that we are using *compound interest*. If the interest is added to the principal at equal intervals, this time interval is called the *interest period* or *conversion period*. The time interval between the date on which the money is invested and the one on which it is repaid is called the *term* of the investment. The sum to which the principal and interest on it grow at the end of the term is called the *maturity value* or *accumulated value*.

EXAMPLE

1. If $100 is borrowed at 5% per year compound interest and accumulates to $121.55 in four years, then the principal is $100, the interest rate is 5%, the conversion period is one year, the term is four years, and the accumulated value is $121.55.

15.2. The Accumulated and Present Values

We shall now develop a formula for the accumulated value at compound interest. In order to be specific, we shall find an expression for the value S to which a principal of P accumulates if it is invested for n periods at rate i per period.

If P is invested for one period at rate i per period, the interest is Pi and the accumulated value is $P + Pi = P(1 + i)$; hence, we see that the accumulated value at the end of one period is obtained by multiplying the principal for the period by $(1 + i)$. Since this factor must be used for each period, it follows that

$$(1) \qquad\qquad S = P(1 + i)^n$$

is the accumulated value if P is invested for n periods at i per period at compound interest.

Values of $(1 + i)^n$ for some combinations of i and n are given in Table V.

EXAMPLES

1. Find the maturity value if \$200 is invested for 6 years at 4% compound annually.

SOLUTION

Substituting $P = \$200$, $n = 6$, and $i = 4\%$ in (1) leads to

$$S = \$200(1.04)^6$$
$$= \$200(1.2653), \text{ using Table V,}$$
$$= \$253.06.$$

2. How much must one repay if he borrows \$250 for eight years at a quarterly rate of 1.5%.

SOLUTION

In this problem, $P = \$250$, $i = 1.5\%$ and $n = (4)(8) = 32$; hence, substituting in (1), we have

$$S = \$250(1.015)^{32}$$
$$= \$250(1.6103), \text{ using Table V,}$$
$$= \$402.58.$$

Equation (1) can be solved for P by multiplying each member by $(1 + i)^{-n}$. If this is done, we have

$$(2) \qquad\qquad P = S(1 + i)^{-n}.$$

By use of this equation, we can determine the sum P that must be invested at rate i per period in order to produce S in n periods. This sum is often called the *present value* but it might better be called the value at the beginning of the term since it is the value at that time. Values of $(1 + i)^{-n}$ for some values of n and i are given in Table VI.

EXAMPLE

3. What sum must be invested so as to accumulate to $640.05 in 5 years at a semi-annual rate of 2.5%.

SOLUTION

In this problem, $S = \$640.05$, $n = (5)(2) = 10$, and $i = 2.5\%$. Therefore, (2) becomes

$$P = \$640.05(1.025)^{-10}$$
$$= \$640.05(.7812) = \$500.01.$$

Exercise 15.1

Find either the present or accumulated value in Problems 1 through 16.

1. $P = \$300$, $i = 3\%$, $n = 7$.
2. $P = \$500$, $i = 5\%$, $n = 6$.

3. $P = \$1200$, $i = 2.5\%$, $n = 14$.
4. $P = \$800$, $i = 6\%$, $n = 4$.

5. $P = \$736$, $i = 4\%$, $n = 9$.
6. $P = \$548$, $i = 2\%$, $n = 42$.

7. $P = \$1974$, $i = 1.5\%$. $n = 46$.
8. $P = \$5969$, $i = 3\%$, $n = 38$.

9. $S = \$400$, $i = 6\%$, $n = 11$.
10. $S = \$900$, $i = 4\%$, $n = 13$.

11. $S = \$700$, $i = 2.5\%$. $n = 33$.
12. $S = \$800$, $i = 1.5\%$, $n = 50$.

13. $S = \$3241$, $i = 5\%$, $n = 25$.
14. $S = \$5976$, $i = 3\%$, $n = 37$.

15. $S = \$6984$, $i = 2\%$, $n = 48$.
16. $S = \$7088$, $i = 6\%$, $n = 18$.

17. To what sum will $750 accumulate in 4 years if money is worth 2% per six months?

18. Find the accumulated value if $5760 is invested for 13 quarters at 1.5% per quarter.

19. How much money must be invested now in order for it to accumulate to $2993 in 7 years if money is worth 5% per year?

20. What is the present value of $1492 due in 6 years provided money is worth 1.5% per quarter?

21. The United States paid $7,200,000 to Russia for Alaska in 1867. To what sum would this money have accumulated by 1967 if invested at 4%?
NOTE. Since Table V gives no values of n greater than 50, we must make use of the fact that $(1 + i)^m(1 + i)^n = (1 + i)^{m+n}$.

22. In 1803, the United States paid France $15,000,000 for the Louisiana Territory. If this money had been invested at 4% compounded annually, to what sum would it have accumulated by 1965?

23. To what sum would it have accumulated by 1966, if the $24 paid to the Indians for Manhattan Island in 1626 had been invested at 3% compounded annually?

24. How much must be invested for a newborn child at 5% compounded annually to accumulate to $17,000 on his sixty-fifth birthday?

25. Mr. Toliferra bought a farm. What was the cash price if he made a down payment of $8000 and signed a noninterest-bearing note for $12,000 due in 3 years. Assume money is worth 5% converted annually.

26. Mr. Schantz invested $700 for two years at 4% and then signed a noninterest-bearing note for $900 due in four years and was able to buy a lot in his home town. If money is worth 6% compounded annually, what was the cash price of the lot?

27. On May 17, 1899, Mr. Soloman borrowed $6000 and signed a note due in 4 years with interest at 3% per six months. He paid $1200 on May 17, 1901 and $3500 on November 17, 1902. How much did he still owe when the note was due?

28. How much was due on maturity date if Mr. Sandrat borrowed $4300 on June 22, 1955, signed an eight year note bearing 5% interest converted annually and made payments of $1100 on June 22 of 1957, 1959, and 1961?

15.3. Solution for n

At times we may want to know how long it takes for a specified principal to accumulate to a given amount at a known rate. This query can be answered, since the relation

(1) $$S = P(1 + i)^n$$

can be solved for n by use of logarithms or by interpolation. We shall illustrate each procedure.

EXAMPLE

1. How long will it take $2756 to accumulate to $3725 at 2% converted annually?

SOLUTION BY LOGARITHMS

(2) $$3725 = 2756(1.02)^n, \text{ substituting in (1)}$$

If we take the logarithm of each member of (2), we get

$$\log 3725 = \log 2756(1.02)^n$$
$$= \log 2756 + n \log 1.02$$

since $\log MN = \log M + \log N$ and $\log N^p = p \log N$. Therefore,

$$n \log 1.02 = \log 3725 - \log 2756$$
$$n(.0086) = 3.5711 - 3.4403, \text{ using Table III,}$$
$$= .1308,$$

$$n = \frac{.1308}{.0086}$$

We now have determined n as a fraction and can evaluate that fraction by use of logarithms or other means. If we use logarithms, we have

$$\log n = \log .1308 - \log .0086$$
$$\log .1308 = 9.1166 - 10$$

$$\frac{\log .0086 = 7.9345 - 10}{\log n = 1.1821,} \text{ subtracting,}$$

$$n = 15.21$$

Therefore the term is 15.21 years, and we shall express it in terms of years, months, and days.

$$\begin{aligned} \text{term} &= 15.21 \text{ yeras} \\ &= 15 \text{ years} + .21(12 \text{ months}) \\ &= 15 \text{ years } 2.52 \text{ months} \\ &= 15 \text{ years } 2 \text{ months} + .52(30 \text{ days}) \\ &= 15 \text{ years } 2 \text{ months } 16 \text{ days.} \end{aligned}$$

SOLUTION BY INTERPOLATION

At the beginning of the solution by logarithms, we substituted the given values in $S = P(1 + i)^n$ and got

(2) $$3725 = 2756(1.02)^n.$$

Solving for 1.02^n by dividing by 2756 gives

$$1.02^n = \frac{3725}{2756} = 1.3516.$$

We shall now look in the 2% column of values of $(1 + i)^n$ for 1.3516. It is not there; so, we resort to interpolation. Thus

$$1 \left(\begin{array}{l} 1.02^{15} = 1.3459 \\ 1.02^n = 1.3516 \\ 1.02^{16} = 1.3728 \end{array} \right. .0057 \right) .0269$$

$$\frac{.0057}{.0269}(1) = .21.$$

Consequently, the term is 15.21 years and can be reduced to 15 years 2 months, 16 days as in the solution by logarithms.

15.4. Solution for i

We can solve $S = P(1 + i)^n$ for i by use of logarithms and by interpolation. We shall illustrate each procedure.

EXAMPLE

1. At what rate converted annually will $500 accumulate to $1773 in 27 years?

SOLUTION BY LOGARITHMS

If we substitute the known values in $S = P(1 + i)^n$, we get

$$1773 = 500(1 + i)^{27};$$

hence, dividing by 500, we see that

$$(1 + i)^{27} = \frac{1773}{500} = 3.5460.$$

Therefore,

$$\log (1 + i)^{27} = \log 3.5460$$
$$27 \log (1 + i) = 0.5497$$
$$\log (1 + i) = 0.0204, \text{ dividing by } 27,$$

and $\qquad\qquad\qquad 1 + i = 1.048, \quad \text{interpolating.}$

Consequently, $\qquad\qquad\qquad i = 4.8\%.$

SOLUTION BY INTERPOLATION

If we substitute the given values in $S = P(1 + i)^n$, we have

$$1773 = 500(1 + i)^{27};$$

hence, dividing by 500, we get

$$(1 + i)^{27} = \frac{1773}{500}$$

$$= 3.5460.$$

We now look for 3.5460 across from 27 in the table of values of $(1 + i)^n$ and find that it is not there. Consequently, we resort to interpolation. Thus,

$$.01 \begin{cases} 1.04^{27} = 2.8834 \\ (1 + i)^{27} = 3.5460 \\ 1.05^{27} = 3.7335 \end{cases} \begin{matrix} .6626 \\ \\ \end{matrix} \Bigg) .8501$$

$$\frac{.6626}{.8501}(.01) = .008$$

Therefore, $\qquad\qquad\qquad i = .048 = 4.8\%.$

Exercise 15.2

Find the term in Problems 1 through 12 in years, months, and days as well as in thousandths of a year. Use logarithms or interpolation or both as directed by your instructor. The answers given were obtained by interpolation.

1. $P = \$550$, $S = \$1300$, $i = 2.5\%$.　　**2.** $P = \$740$, $S = \$1406$, $i = 1.5\%$.

3. $P = \$327$, $S = \$500$, $i = 4\%$.　　**4.** $P = \$628$, $S = \$1222$, $i = 5\%$.

5. $P = \$327$, $S = \$500$, $i = 2\%$.　　**6.** $P = \$628$, $S = \$1222$, $i = 1.5\%$.

7. $P = \$258$, $S = \$1057$, $i = 4\%$.　　**8.** $P = \$1352$, $S = \$4103$, $i = 3\%$.

9. $P = \$258$, $S = \$1057$, $i = 6\%$.　　**10.** $P = \$1352$, $S = \$4103$, $i = 5\%$.

11. $P = \$943$, $S = \$1902$, $i = 3\%$.　　**12.** $P = \$841$, $S = \$2500$, $i = 2.5\%$.

Find the periodic rate to the nearest hundredth of a per cent in each of Problems 13 through 24. Use logarithms or interpolation or both as directed by your instructor. The given answers were obtained by interpolation.

13. $P = \$327$, $S = \$571$, $n = 26$.　　**14.** $P = \$371$, $S = \$719$, $n = 43$.

15. $P = \$603$, $S = \$907$, $n = 21$.　　**16.** $P = \$506$, $S = \$738$, $n = 9$.

17. $P = \$603$, $S = \$907$, $n = 10$.　　**18.** $P = \$506$, $S = \$838$, $n = 26$.

19. $P = \$521$, $S = \$2101$, $n = 36$.　　**20.** $P = \$673$, $S = \$2144$, $n = 34$.

21. $P = \$521$, $S = \$2101$, $n = 24$.　　**22.** $P = \$673$, $S = \$2144$, $n = 22$.

23. $P = \$709$, $S = \$1417$, $n = 28$.　　**24.** $P = \$428$, $S = \$1279$, $n = 45$.

25. How long is required for an investment to double if it accumulates at 4% compounded annually?

26. How long is needed for \$75 to accumulate to \$100 if the rate is 3% converted annually?

27. If a savings bank pays 4% converted annually, how long is needed for an investment of \$140 to accumulate to \$200?

28. How long must \$328 be invested at 5% converted annually to accumulate to \$583?

29. If \$75 accumulates to \$100 in 9 years find the annual rate at which it was invested.

30. At what rate converted annually must money be invested to double in 13 years?

31. What rate converted annually was paid by U.S. savings bonds during the period in which it took 10 years for \$75 to accumulate to \$100?

32. How much sooner in months and days will money double if compounded at 2.5% per 6 months than if compounded at 5% per year?

15.5. Nominal and Effective Rates

If the conversion period in compound interest is different from a year, the stated annual rate is called the *nominal rate*. This term is used since an annual rate is named in most business transactions along with the number of times per year that it is to be compounded. We shall use j and m to represent the nominal rate and number of times per year that it is compounded respectively. The symbol $j_{(m)}$ is often used to represent a nominal rate of j compounded m times per year. The rate per period is j/m; consequently, if 1 is invested at $j_{(m)}$ for a year, it accumulates to $(1 + j/m)^m$.

   The annual rate at which an investment increases if invested at the rate j converted m times per year is called the *effective rate*, and we shall represent it by e. If 1 is invested at an effective rate of e for a year, it accumulates to $1 + e$. Consequently *the relation between the nominal rate j converted m times per year and the effective rate e is*

(1) $$1 + e = \left(1 + \frac{j}{m}\right)^m.$$

This equation can be solved readily for e in terms of j and m, and we then have the effective rate that is equivalent to the nominal rate j converted m times per year.

EXAMPLE

   1. What is the effective rate if the nominal rate is 6% compounded quarterly?

SOLUTION

   If we substitute $j = 6\% = .06$ and $m = 4$ in (1), we have

$$1 + e = 1.015^4, \quad .06 \div 4 = .015,$$
$$= 1.0614, \text{ using table V.}$$

Consequently, $e = 6.14\%.$

15.6 Accumulated and Present Value at a Nominal Rate

If we substitute the value of $1 + e$ from

$$1 + e = \left(1 + \frac{j}{m}\right)^m$$

in $$S = P(1 + e)^n,$$

we get
$$S = P\left[\left(1 + \frac{j}{m}\right)^m\right]^n;$$

hence, raising a power to a power, we find that

(1)
$$S = P\left(1 + \frac{j}{m}\right)^{mn}$$

is the sum to which **P** *will accumulate in* **n** *years at the nominal rate* **j** *converted* **m** *times per year.*

EXAMPLE

1. If $700 is invested for six years at 5% converted semiannually, what is the accumulated value?

SOLUTION

In this problem, $P = \$700$, $n = 6$, $j = 5\%$, and $m = 2$; hence, $j/m = 2.5\%$ and $mn = 12$. Putting these values in (1), we have

$$S = \$700(1.025)^{12} = \$700(1.3449) = \$941.43.$$

Equation (1) can be solved for P by multiplying each member by $(1 + j/m)^{-mn}$. If this is done, we see that

(2)
$$P = S\left(1 + \frac{j}{m}\right)^{-mn}$$

is the value at the beginning of the term of **S** *due in* **n** *years if money is worth the nominal rate* **j** *converted* **m** *times per year.* The value at the beginning of the term is often referred to as the *present value.*

EXAMPLE

2. If money is worth 6% converted quarterly, what is the present value of $900 due in 5 years?

SOLUTION

In this problem, $j = 6\%$, $m = 4$, $S = \$900$, and $n = 5$. If we substitute these values in (2), it becomes

$$P = \$900\left(1 + \frac{.06}{4}\right)^{-4(5)} = \$900(1.015)^{-20}$$

$$= \$900(.7425) = \$668.25.$$

Exercise 15.3

Find either P or S in Problems 1 through 20. In each, n is the term in years.

1. $P = \$250, j = 6\%, m = 2, n = 7.$　　**2.** $P = \$340, j = 5\%, m = 2, n = 8.$

3. $P = \$420, j = 8\%, m = 2, n = 3.$　　**4.** $P = \$530, j = 4\%, m = 2, n = 11.$

5. $P = \$290, j = 6\%, m = 4, n = 9.$　　**6.** $P = \$840, j = 8\%, m = 4, n = 10.$

7. $P = \$1170, j = 6\%, m = 3, n = 5.$　　**8.** $P = \$1260, j = 9\%, m = 3, n = 8.$

9. $S = \$630, j = 6\%, m = 4, n = 10.$　　**10.** $S = \$770, j = 8\%, m = 4, n = 5.$

11. $S = \$1060, j = 6\%, m = 3, n = 4.$　　**12.** $S = \$1720, j = 9\%, m = 3, n = 2.$

13. $S = \$490, j = 6\%, m = 2, n = 8.$　　**14.** $S = \$530, j = 5\%, m = 2, n = 7.$

15. $S = \$960, j = 4\%, m = 2, n = 17.$　　**16.** $S = \$280, j = 8\%, m = 2, n = 13.$

17. $S = \$550, j = 6\%, m = 2, n = 7.5.$　　**18.** $S = \$670, j = 5\%, m = 2, n = 11.5.$

19. $S = \$340, j = 6\%, m = 4, n = 6.25.$　　**20.** $S = \$880, j = 8\%, m = 4, n = 8.75.$

21. Mr. Olivier deposited $600 in a bank that paid 4% converted semiannually. How much was to his credit after 3 years and 6 months?

22. Mr. Sisti borrowed $400 and agreed to repay it and interest at 6% compounded quarterly at the end of 30 months. How much interest did he pay?

23. In order to accumulate money to use as a down payment on a house, Mr. Smythe deposited $600 at the beginning of each 6 months until 5 payments had been made. How much was then available if the fund drew interest at 5% converted semiannually?

24. A father deposited $1000 to his son's account on the day of his birth and similar amounts when the boy was 2, 5, and 9 years old. If these sums draw interest at 4% converted annually, how much was to his credit on his 18th birthday?

25. A farm sold for a promise of $17,000 in 4 years. What was the equivalent cash price if money was worth 5% converted annually?

26. A house was sold for a down payment of $4000 and promises of $5000 in 2 years and $6000 in 4 years. What cash price could have been accepted if money is worth 6% converted quarterly?

27. Mr. Rock put some money in a bank that pays 5% converted semiannually. He drew out $1100 after 18 months and a year later had $1400 to his credit. How much did he deposit?

28. If a bank pays 4% compounded annually, how much must one deposit in order to be able to withdraw $2000 at the end of one year and $3000 at the end of two years?

15.7. Equations of Value

There are times when it is desirable to replace one set of financial obligations by another. An equation which states that the value of one set of obligations is equal to the value of another set on a specified debt is called an *equation of value*. It is immaterial what date is chosen but essential that the same date be used for all of the obligations. The date chosen for use in an equation of value is called the *comparison date* or *focal date*. It is desirable to choose the comparison date so as to have as simple an equation as possible. If one person chooses one focal date and another person chooses a different one, the resulting equations differ only in that one can be obtained from the other by multiplying through by a power of $(1 + j/m)$.

EXAMPLE

1. What equal payments made 2 and 4 years from now will equitably repay a note of $600 due in 3 years with interest at 5% converted semiannually and $900 due in 5 years with interest at 6% compounded annually provided money is worth 4% converted semiannuallly?

SOLUTION

We shall choose five years from now as the comparison date and represent each of the equal payments by x. We can now fill in the table as shown for the reasons given below it.

Debts (value when due)	Debts (value on focal date)	Payments (value on focal date)
$600(1.025)^6$	$600(1.025)^6 (1.02)^4$	$x(1.02)^6$
$900(1.06)^5$	$900(1.06)^5$	$x(1.02)^2$

The $600 is due in 3 years with interest at $j = 5\%$, $m = 2$, and must be accumulated for another 2 years at $j = 4\%$, $m = 2$; the $900 is due in 5 years with interest at $r = 6\%$; the first x is due in 2 years and must be accumulated for 3 years at $j = 4\%$, $m = 2$; and the second x is due in 4 years and must be accumulated for a year at $j = 4\%$, $m = 2$. Hence, equating the values of the two sets of obligations on the focal date, we have

$$x(1.02)^6 + x(1.02)^2 = \$600(1.025)^6(1.02)^4 + \$900(1.06)^5$$
$$1.1262x + 1.0404x = \$600(1.1597)(1.0824) + \$900(1.3382)$$
$$2.1666x = \$600(1.2553) + \$1204.38$$
$$= \$753.18 + \$1204.38$$
$$= \$1957.56.$$
$$x = \frac{\$1957.56}{2.1666} = \$903.52.$$

Exercise 15.4

Determine x in Problems 1 through 16 so that the payments equitably discharge the debts.

1. Debt: $600 due in 3 years with interest at 5% converted semiannually. Payments: x in 5 years. Money worth 4%.

2. Debt: $800 due in 2 years with interest at 6%. Payments: x in 3 years. Money worth 4%.

3. Debt: $1700 due in 3 years with interest at 5% converted semiannually. Payments: x in 2 years. Money worth 4%.

4. Debt: $2300 due in 4.5 years with interest at 6% compounded quarterly. Payments: x in 3 years. Money worth 5% converted semiannually.

5. Debt: $600 due in 1.5 years with interest at $j = 5\%$, $m = 2$, and $800 due in 3 years with $i = 4\%$, Payments: x in 4 years. Money worth $j = 6\%$, $m = 4$.

6. Debt: $700 due in 4 years with interest at $j = 4\%$, $m = 2$ and $900 due in 2.5 years with interest at $j = 6\%$, $m = 4$. Payments: x in 3 years. Money worth $j = 5\%$, $m = 2$.

7. Debt: $1000 due in 3 years with interest at $j = 6\%$, $m = 2$, and $600 due in 1.5 years without interest. Payments: x in 2 years. Money worth $j = 5\%$, $m = 2$.

8. Debt: $1100 due in 6 years with interest at $j = 6\%$, $m = 4$, and $800 due in 3 years with interest at $j = 5\%$, $m = 2$. Payments: x in 2.5 years. Money worth $j = 6\%$, $m = 2$.

9. Debt: $3200 due in 3.5 years with interest at $j = 5\%$, $m = 2$. Payments: x in 2 years and $2x$ in 4 years. Money worth $j = 4\%$, $m = 2$.

10. Debt: $1700 due in 4 years with interest at $i = 5\%$. Payments: $2x$ in 3 years and x in 5 years. Money worth $j = 6\%$, $m = 2$.

11. Debt: $4700 due in 3 years with interest at $j = 5\%$, $m = 2$. Payments: x in 1.5 years and $3x$ in 2.5 years. Money worth $j = 6\%$, $m = 4$.

12. Debt: $3300 due in 6 months with interest at $j = 6\%$, $m = 4$. Payments: x in 1 year and $2x$ in 2 years. Money worth $j = 5\%$, $m = 2$.

13. Debt: $700 due in 2 years with interest at $i = 5\%$ and $1600 due in 3 years with interest $j = 6\%$, $m = 4$. Payments: x at the end of 1 year and also at the end of 2 years. Money worth $j = 5\%$, $m = 2$.

14. Debt: $1300 due in 4.5 years with interest at $j = 4\%$, $m = 2$ and $1900 due in 3 years and 9 months with interest at $j = 6\%$, $m = 4$. Payments: x in 2 years and 3 months and $3x$ in 4 years. Money worth $j = 6\%$, $m = 4$.

15. Debt: $9000 due in 5 years with interest at $i = 5\%$ and $7200 due in 4.25 years

with interest at $j = 6\%$, $m = 4$. Payments: $2x$ in 3.75 years and x in 4.5 years. Money worth $j = 6\%$, $m = 4$.

16. Debt: $7000 due in 2.5 years with interest at $j = 5\%$, $m = 2$ and $6200 due in 4 years with interest at $i = 4\%$. Payments: x in 2 years and $3x$ in 5 years. Money worth $j = 6\%$, $m = 2$.

17. Mr. Tompkins bought a business for $37,000 but could pay only $12,000 in cash. If he paid $10,000 at the end of 1 year and $8000 at the end of 2 years, what additional payment 3 years after the purchase will equitably finish paying off the debt provided money is worth 5%?

18. Mr. Senoj bought a farm for $19,200 and paid $5200 in cash. He also paid $5000 after 1 year and $6000 after 3 years. If money is worth 4%, how much did he still owe just after the time of the last payment?

19. Mr. H. Tims bought a house and still owed $15,000 after the down payment. What equal payments 1, 3, and 5 years from the date of purchase will finish paying for the house if money is worth $j = 6\%$, $m = 2$.

20. If money is worth $j = 5\%$, $m = 2$, what equal payments now, 1 year from now, 2 years from now, and 3 years from now will accumulate to $18,700 in 2 years after the last payment?

16

GEOMETRIC PROGRESSIONS, SIMPLE ANNUITIES

16.1. Introduction to Geometric Progressions

We frequently have occasion to deal with a set of quantities in mathematics. If a set is or can be arranged in a definite order, it is called a *sequence*. We shall now work with a particular type of sequence that is described in the following statement. *A sequence of numbers so related that each after the first can be obtained from the immediately preceding one by multiplying by a fixed number is called a **geometric progression**. The fixed number is called the **common ratio**.*

EXAMPLES

1. Since each number after the first in 1, 3, 9, 27, 81 can be obtained from the one just before it by multiplying by the fixed number 3, the numbers form a geometric progression.

2. If 2 is the first term of a geometric progression and if the common ratio is 5, then 2, $5.2 = 10$, $5.10 = 50$, and $5.50 = 250$ are terms in the progression.

We shall represent

the first term by a,
the common ratio by r,
the number of terms by n,
the nth or last term by l, and
the sum of the terms by S.

The first few terms in the series are, then, a, ar, ar^2, ar^3, and ar^4. If we now notice that the exponent of r is one less than the number of the term in each of these and recall that each term is obtained from the one just before it by multiplying by r, we see that

$$1 = ar^{n-1}$$

*is the **n**th, or last term in the progression.*

EXAMPLE

3. If 3, 6, and 12 are the first three terms of a geometric progression, find the seventh term.

SOLUTION

The common ratio is 2, since that is the number by which each term must be multiplied to give the next. We now know that $a = 3$, $r = 2$, and $n = 7$, and want l; hence, we substitute the known values in

$$l = ar^{n-1}$$

and have

$$l = 3(2^{7-1}) = 192.$$

16.2. The Sum of a Geometric Progression

We shall now develop a formula for the sum S of a geometric progression of n terms with first term a and common ratio r. The sum is

(1) $S = a + ar + ar^2 + \cdots + ar^{n-2} + ar^{n-1}$

and, multiplying by r, we see that

(2) $rS = ar + ar^2 + ar^3 + \cdots + ar^{n-1} + ar^n.$

Consequently,

(3) $S - rS = a - ar^n,$

as obtained by subtracting each member of (2) from the corresponding member of (1). Now, factoring the left number of (3), we have

$$S(1 - r) = a - ar^n$$

and, dividing by $1 - r$, we find that

(4)
$$S = \frac{a - ar^n}{1 - r}, \qquad r \neq 1,$$

is the sum of the geometric progression described above. If we make use of the fact that $l = ar^{n-1}$, (4) can be put in the form

(4′)
$$S = \frac{a - rl}{1 - r}, \qquad r \neq 1.$$

EXAMPLE

1. Find the sum of a geometric progression of 7 terms if the first term is 2 and the second term is 6.

SOLUTION

We are given that $n = 7$ and $a = 2$, and can find that $r = 3$ by dividing the first term 2 into the second term 6. Now putting these values in (4), we have

$$S = \frac{2 - 2(3)^7}{1 - 3} = 2186.$$

Exercise 16.1

Write out all terms of each geometric progression described in Problems 1 through 8 by use of the definition.

1. $a = 2, r = 3, n = 5.$ **2.** $a = 3, r = 2, n = 4.$

3. $a = 1, r = 2, n = 6.$ **4.** $a = 2, r = 2, n = 5.$

5. $a = 64, r = \frac{1}{2}, n = 5.$ **6.** $a = 243, r = \frac{1}{3}, n = 6.$

7. $a = \frac{1}{4}, r = -2, n = 4.$ **8.** $a = \frac{1}{256}, r = 8, n = -5,$

Find the last term of the geometric progression in each of Problems 9 through 16 by use of $l = ar^{n-1}$.

9. $a = 3, r = 2, n = 4.$ **10.** $a = 2, r = 3, n = 6.$

11. $a = 729, r = \frac{1}{3}, n = 5.$ **12.** $a = 512, r = \frac{1}{2}, n = 7.$

13. $a = 6$, second term 12, $n = 5.$ **14.** $a = 3$, second term 9, $n = 4.$

15. $a = 2$, third term 18, $n = 6.$ **16.** $a = 256$, second term 128, $n = 5.$

Find the sum of each geometric progression in Problems 17 through 32.

17. $a = 1, r = 3, n = 5.$ **18.** $a = 2, r = 2, n = 6.$

19. $a = 128, r = \frac{1}{2}, n = 6$.   **20.** $a = 729, r = \frac{1}{3}, n = 5$.

21. Second term 4, third term 8, $n = 6$.

22. Second term 6, third term 18, $n = 5$.

23. Third term 2, fourth term 4, $n = 7$.

24. Third term 9, fourth term 27, $n = 6$.

25. $a = 3, r = 2, l = 48$.   **26.** $a = 4, r = 2, l = 128$.

27. $a = 625, r = \frac{1}{5}, l = \frac{1}{5}$.   **28.** $a = 243, r = \frac{1}{3}, l = 1$.

29. $a = 2, r = 1.04, n = 5$.   **30.** $a = 3, r = 1.05, n = 6$.

31. $a = 5, r = 1.02, n = 4$.   **32.** $a = 10, r = 1.03, n = 5$.

16.3. Introduction to Annuities

As much as we would like to be able to pay cash for the things we buy, most of us are obliged to make a down payment and complete the purchase by a series of payments in the future. In many cases, the future payments are equal. This is the usual situation in buying a house.

A sequence of equal payments at equal intervals is called an *annuity*. If a payment is made at the end of each interval, the annuity is an *ordinary annuity*. The payments that most of us make for rent or on the purchase of a house form an annuity.

The length of time between consecutive payments is called the *payment period* and the sum paid periodically is known as the *periodic rent* or *payment*. The length of time between consecutive additions of interest to the principal is, as before, called the *conversion period* or *interest* period. The time from the beginning of the first payment period to the end of the last is called the *term* of the annuity.

EXAMPLE

1. If, after a down payment, one finishes paying for his house by making a payment of $70.67 at the end of each month for 8 years, these payments form an annuity; furthermore, the periodic payment is $70.67, the payment period is 1 month, and the term is 8 years.

16.4. The Accumulated Value

We shall now find the sum that is accumulated if 1 is invested at the end of each period for n periods and if money is worth i per period. The first payment is invested for $(n - 1)$ periods, the second for $(n - 2)$ periods, the third for $(n - 3)$ periods, $\cdots$, the next to last for one period, and the last for zero periods. Consequently, by use of the formula for the accumulated value at compound interest, we see that

the first payment accumulates to $(1 + i)^{n-1}$,
the second payment accumulates to $(1 + i)^{n-2}$,
the third payment accumulates to $(1 + i)^{n-3}$,

.

.

.

the next to last payment accumulates to $(1 + i)^{1}$
the last payment accumulates to 1.

If taken in reverse order, these powers of $1 + i$ for a geometric progression of n terms with first term 1 and common ratio $1 + i$. Hence, using (4) of Art. 16.2, we find that the sum of these amounts is

$$\frac{1 - 1(1 + i)^n}{1 - (1 + i)} = \frac{1 - (1 + i)^n}{-i} = \frac{(1 + i)^n - 1}{i}.$$

Consequently, if we use $s_{\overline{n}|i}$ to represent this sum, we see that *the accumulated value of an annuity of* **1** *at the end of each period for* **n** *periods at rate i per period is*

$$s_{\overline{n}|i} = \frac{(1 + i)^n - 1}{i}.$$

The symbol $s_{\overline{n}|i}$ is read

"**s** angle **n** at **i**."

If the periodic payment is R and if the accumulated value is represented by S_n, we have

(1) $$S_n = R s_{\overline{n}|i}.$$

There are tables which give values of $s_{\overline{n}|i}$ for certain combinations of values of n and i. Table VII in this book is one such table.

EXAMPLE

1. How much will one have to his credit if he deposits $100 at the end of each 3 months for 5 years in a fund that pays 5% converted quarterly?

SOLUTION

These payments form an ordinary annuity with $R = \$100$, $n = 4(5) = 20$, $j = 5\%$, and $m = 4$; hence, the rate per period is $\frac{1}{4}(5\%) = 1\frac{1}{4}\%$. Putting these values in (1), we have

$$S_{20} = \$100 s_{\overline{20}|\,1.25\%}.$$

The value of $s_{\overline{20}|1.25\%}$, is in Table VII across from 20 and under 1.25% and is 22.5630. Consequently,

$$S_{20} = \$100(22.5630) = \$2256.30$$

is the accumulated value of this annuity.

16.5. The Present Value

The value of an annuity at the beginning of its term is called the *present value*. The time may or may not be the present but it is the beginning of the term of the annuity. We shall now develop a formula for the present value of an annuity of 1 paid at the end of each period for n periods with money worth i per period. This can be done readily by discounting the accumulated value of the annuity for its term at its rate. If we represent the present value by $a_{\overline{n}|i}$, we have

$$a_{\overline{n}|i} = s_{\overline{n}|i}(1 + i)^{-n} = \frac{(1 + i)^n - 1}{i}(1 + i)^{-n};$$

hence, performing the indicated multiplications, we see that

$$a_{\overline{n}|i} = \frac{1 - (1 + i)^{-n}}{i}$$

*is the present value of an annuity of one at the end of each period for **n** periods with money worth **i** per period.* If the periodic rent or payment is R instead of 1 and if the present value is represented by A_n, then

(1) $$A_n = Ra_{\overline{n}|i}.$$

There are tables which give values of $a_{\overline{n}|i}$ for some combinations of n and i. Table VIII is one such table.

EXAMPLE

1. Find the present value of an ordinary annuity of $50 per month for 4 years if money is worth 6% converted monthly.

SOLUTION

The term of this annuity is 4 years $= 4(12)$ periods $= 48$ periods, the rate per period is $\frac{1}{12}(6\%) = \frac{1}{2}\%$, and the periodic rent is $50; hence, (1) becomes

$$A_{48} = \$50a_{\overline{48}|1/2\%}$$
$$= \$50(42.5803), \text{ using Table VIII,}$$
$$= \$2129.02.$$

Exercise 16.2

Find the present and accumulated value of each ordinary annuity described in Problems 1 through 20. In each of them, R represents the periodic payment, j is the nominal rate, m is the number of times per year that interest is converted, and t is the term in years.

1. $R = \$300, j = 3\%, m = 2, t = 7.$ **2.** $R = \$500, j = 3\%, m = 1, t = 19.$

3. $R = \$250, j = 4\%, m = 4, t = 12.$ **4.** $R = \$700, j = 4\%, m = 2, t = 23.$

5. $R = \$60, j = 5\%, m = 12, t = 4.$ **6.** $R = \$80, j = 5\%, m = 4, t = 9.$

7. $R = \$110, j = 6\%, m = 12, t = 3.$ **8.** $R = \$40, j = 6\%, m = 4, t = 11.$

9. $R = \$75, j = 6\%, m = 3, t = 15.$ **10.** $R = \$80, j = 6\%, m = 2, t = 25.$

11. $R = \$170, j = 3\%, m = 1, t = 50.$ **12.** $R = \$85, j = 3\%, m = 2, t = 25.$

13. $R = \$60, j = 4\%, m = 2, t = 17.$ **14.** $R = \$45, j = 4\%, m = 4, t = 12.$

15. $R = \$60, j = 5\%, m = 4, t = 9.$ **16.** $R = \$70, j = 5\%, m = 12, t = 3.$

17. $R = \$120, j = 6\%, m = 2, t = 24.$ **18.** $R = \$80, j = 6\%, m = 3, t = 16.$

19. $R = \$60, j = 6\%, m = 4, t = 12.$ **20.** $R = \$20, j = 6\%, m = 12, t = 4.$

21. A family sold a piece of property in order to be able to send their son to college. What was the cash price if money is worth 5% converted monthly and they received $125 at the end of each month for 4 years?

22. An insurance company sold an ordinary annuity certain of $4000 per year for 20 years. If money is worth 3%, what did they receive?

23. Mr. Tim bought some property for $70.60 at the end of each quarter for 12 years. What was the cash equivalent if money is worth 4% converted quarterly?

24. Mrs. Mott bought a used car for $60 at the end of each month for 2 years. What was the equivalent cash price if money is worth 6% converted monthly?

25. The Kingsleys deposited $45 at the end of each month for 4 years and then were able to buy a tractor for cash. What was its price if money is worth $j = 5\%, m = 12$?

26. The Outermeyers accumulated a fund for their child's education by making a deposit of $400 at the end of each year for 16 years. How much was available if money is worth 3%?

27. What is the value of the machinery in a plant if a fund for replacing it can be accumulated by depositing $600 at the end of each quarter for 12 years in a bank that pays interest at 4% compounded quarterly?

28. The Youngbloods accumulated a fund for a European trip by depositing $200 at the end of each 6 months in a fund that drew interest at 4% converted semi-annually. How much was in the fund after 3.5 years?

29. The owner of some property had two offers on it. One was $5000 and $800 at the end of each year for 9 years with interest at 3%. The other was $6100 cash and $700 at the end of each 6 months for 4 years with interest at 4% converted semiannually. Which offer should he take?

30. A man can buy a lot for $125 at the end of each month for 40 months or for $5500 in one payment after 40 months. Which should he choose if money is worth 6% converted monthly?

31. A man asked $20,000 for his house. He was offered $7000 cash and $1450 at the end of each 6 months for 5 years with interest at $j = 6\%$, $m = 2$. What is the difference between the price asked and the offer?

32. Mr. Soubsky has a car for sale for $1350. He is offered $250 cash and $50 at the end of each month for 2 years. What is the difference between the cash price and the offer if money is worth 5% converted monthly?

16.6. The Periodic Payment

If we know the periodic interest rate, the number of periods in the term, and either the accumulated value or the present value, we can find the periodic payment by solving (1) of Art. 16.4 or (1) of Art. 16.5 for R. These formulas are

$$S_{\overline{n}|} = Rs_{\overline{n}|i} \quad \text{and} \quad A_{\overline{n}|} = Ra_{\overline{n}|i}$$

and, if solved for R, give

$$R = \frac{S_{\overline{n}|}}{s_{\overline{n}|i}} \quad \text{and} \quad R = \frac{A_{\overline{n}|}}{a_{\overline{n}|i}}.$$

There are tables of values of $1/s_{\overline{n}|i}$ and of $1/a_{\overline{n}|i}$, but they are not given in this book; hence, we must find R by dividing S_n by $s_{\overline{n}|i}$ or A_n by $a_{\overline{n}|i}$.

EXAMPLES

1. What must the periodic payment be in order to accumulate $1200 by a series of 18 equal monthly payments if money is worth $j = 6\%$, $m = 12$, and each payment is made at the end of the month?

SOLUTION

Since we have the value at the end of the term we must use S_n. We know that

$$S_n = \$1200, \quad i = \frac{6\%}{12} = \frac{1}{2}\%, \quad \text{and} \quad n = 18;$$

hence, $S_n = Rs_{\overline{n}|i}$ becomes

$$\$1200 = Rs_{\overline{18}|1/2\%}.$$

Therefore,

$$R = \frac{\$1200}{s_{\overline{8}|1/12\%}}$$

$$= \frac{\$1200}{18.7858}, \text{ by use of Table VII,}$$

$$= \$63.88.$$

2. How much must be paid at the end of each month for 30 months to pay off the balance of $1900 still due on a car after the down payment? Assume money is worth 5% converted monthly.

SOLUTION

In this problem we have $n = 30$, $i = \frac{5}{12}$% and $A_n = \$1900$, and want to find R. Putting the known values in the formula for the present value of an ordinary annuity, we get

$$\$1900 = R a_{\overline{30}|5/12\%};$$

hence,

$$R = \frac{\$1900}{a_{\overline{30}|5/12\%}};$$

$$= \frac{\$1900}{28.1457}, \text{ using Table VIII,}$$

$$= \$67.51.$$

Exercise 16.3

Find the periodic payment for each ordinary annuity in Problems 1 through 16. The rate in each is j converted m times per year for the term of t years. Assume that the payment and interest periods coincide.

1. $A_n = \$6000$, $j = 6\%$, $m = 4$, $t = 12$.

2. $A_n = \$2800$, $j = 5\%$, $m = 12$, $t = 3$.

3. $A_n = \$4500$, $j = 4\%$, $m = 4$, $t = 11$.

4. $A_n = \$3000$, $j = 3\%$, $m = 1$, $t = 47$.

5. $A_n = \$17,000$, $j = 6\%$, $m = 12$, $t = 4$.

6. $A_n = \$22,000$, $j = 4\%$, $m = 2$, $t = 25$.

7. $A_n = \$16,000$, $j = 6\%$, $m = 2$, $t = 23$.

8. $A_n = \$19,000$, $j = 5\%$, $m = 4$, $t = 12$.

9. $S_n = \$1427$, $j = 5\%$, $m = 12$, $t = 3$.

10. $S_n = \$1873, j = 3\%, m = 1, t = 50.$

11. $S_n = \$3883, j = 4\%, m = 2, t = 17.$

12. $S_n = \$2998, j = 5\%, m = 4, t = 9.$

13. $S_n = \$9800, j = 6\%, m = 4, t = 8.$

14. $S_n = \$7200, j = 4\%, m = 4, t = 7.$

15. $S_n = \$4900, j = 6\%, m = 12, t = 3.$

16. $S_n = \$6300, j = 6\%, m = 2, t = 24.$

17. A farmer borrowed \$6800 to build a dryer for his rice. How much must be paid on the debt at the end of each year in order to have the debt paid off in 4 years if money is worth 3%?

18. Mr. Sampler borrowed \$4300 to renovate and air-condition his house and repaid it by monthly payments at the end of each month. If money is worth 5% converted monthly. how much must each payment be in order to repay the debt in 27 months?

19. Mr. S. A. Jones bought a car and owed \$3700 after a down payment. He agreed to pay off the debt by 36 equal monthly payments. How much was each if the first was made one month after the purchase, provided money is worth $j = 6\%, m = 12$?

20. Sammy Tiller bought a business for \$21,000 and agreed to pay for it by 10 equal quarterly payments. If the first payment was made 6 months after the purchase and if money was worth $j = 6\%, m = 4$, how much was each payment?

21. Mr. Smart decided to accumulate the necessary \$3100 to buy a car. If he made 17 equal monthly payments and if money is worth $j = 5\%, m = 12$, how much was each payment?

22. What sum must be deposited at the end of each month for 48 months if money is worth $j = 5\%, m = 12$ to accumulate \$5519.69?

23. How much must be deposited at the end of each 6 months in order to have \$3964 at the time of the 22nd payment if money is worth 4% compounded semiannually?

24. A business anticipates that it will need \$7800 at the end of 3 years and decides to accumulate it by means of 3 equal annual payments. How much is each if money is worth 3%?

25. Mr. Michelle had an endowment policy for \$10,000 payable at age 65. He chose to take 20 equal semiannual payments. If the first was made when he was 65 years and 6 months of age and if money is worth $j = 4\%, m = 2$, how much was each payment?

26. A piece of machinery cost \$39,438, will probably last 6 years, and will have a scrap value of \$4500. If money is worth 4% converted semiannually, what payment at the end of each 6 months will provide the money needed for a replacement?

27. Mr. Scholz sold his farm for $38,500 and got a down payment of $12,500 and a promise of 13 equal semiannual payments with the first due 6 months after the sale. How much was each payment if money is worth $j = 4\%$, $m = 2$?

28. A farmer borrowed $15,000 with the agreement that it was to be repaid in 4 equal annual payments. If money is worth 3% and if the first payment was made 4 years after the loan, how much was each payment?

16.7. Installment buying

There are many situations in which it is desirable to have an object or a piece of property before we have enough cash to pay for it. This can be arranged by giving the seller a certain amount of cash and promising him a sequence of equal payments at equal intervals for a specified time. This arrangement is called *installment buying* or buying on the installment plan. If done wisely, it can benefit the buyer and the seller; if attempted foolishly, it can break the buyer and make a second-hand dealer out of the seller.

If the sequence of payments that follow the cash or down payment are equal and are made at the ends of equal intervals, they form an ordinary annuity. If interest and payment periods coincide, then $Ra_{\overline{n}|i}$ is the value at the time of purchase of n payments of R each at rate i per period. Furthermore, if we represent the cash price by P and the cash payment by C, then $P - C$ is the sum still owed after the down payment. Consequently,

(1) $$P - C = Ra_{\overline{n}|i}.$$

We know the value of each symbol except i and can find it as shown below.

EXAMPLE

1. What interest rate is being charged if an article that can be bought for $375 cash is bought for a down payment of $35 and a payment of $40 at the end of each month for 9 months?

SOLUTION

In this problem, $P = \$375$, $C = \$35$, $R = \$40$, and $n = 9$; hence, (1) becomes

$$375 - 35 = 40a_{\overline{9}|i},$$

$$a_{\overline{9}|i} = \frac{340}{40} = 8.5.$$

This value of $a_{\overline{n}|i}$ is not in the table; hence, we shall resort to interpolation. Thus

$$1/4\% \left(\begin{array}{c} a_{\overline{9}|1\%} = 8.5660 \\ a_{\overline{9}|} = _i = 8.5 \\ a_{\overline{9}|1\ 1/4\%} = 8.4623 \end{array} \right) .0660 \right) .1037$$

$$\frac{.0660}{.1037} \frac{1}{4}\% = .159\%$$

Therefore $i = 1\% + .159\% = 1.159\%$; consequently, $j = 12(1.159\%) = 13.91\%$ to the nearest one-hundredth of one per cent.

The high rate of interest charged by firms that sell on the installment plan is often justified since it is their only way of getting paid for the expenses entailed in book-keeping, the loss of interest on the unpaid balance, and possible capital loss if the payments on the goods are not completed.

Exercise 16.4

Find the nominal interest rate to the nearest hundredth of a percent in each of Problems 1 through 12 under the assumption that the payment period and interest period coincide. In each problem, the notation of equation (1) is used.

1. $P = \$210$, $C = \$30$, $R = \$23$ per month for 8 months.

2. $P = \$415$, $C = \$33$, $R = \$70$ per month for 6 months.

3. $P = \$400$, $C = \$40$, $R = \$39$ per month for 10 months.

4. $P = \$225$, $C = \$40$, $R = \$25$ per month for 8 months.

5. $P = \$720$, $C = \$100$, $R = \$80$ per month for 8 months.

6. $P = \$360$, $C = \$40$, $R = \$34$ per month for 10 months.

7. $P = \$850$, $C = \$90$, $R = \$74$ per month for 12 months.

8. $P = \$1123$, $C = \$123$, $R = \$70$ per month for 15 months.

9. $P = \$2936$, $C = \$286$, $R = \$140$ per month for 20 months.

10. $P = \$3147$, $C = \$647$, $R = \$73$ per month for 43 months.

11. $P = \$1838$, $C = \$188$, $R = \$86$ per month for 22 months.

12. $P = \$3762$, $C = \$562$, $R = \$103$ per month for 40 months.

In each of Problems 13 through 20 the price P, down payment C, and the number of monthly payments n are given, and the problem is to find each periodic payment under the assumption that they form an ordinary annuity. In 13 through 16, use 6% as the nominal rate and in the others use 5%.

13. $P = \$276$, $C = \$29$, $n = 9$.

14. $P = \$508$, $C = \$58$, $n = 12$.

15. $P = \$317$, $C = \$32$, $n = 8$.

16. $P = \$486$, $C = \$50$, $n = 18$.

17. $P = \$2047$, $C = \$167$, $n = 16$.

18. $P = \$2388$, $C = \$418$, $n = 24$.

19. $P = \$3263$, $C = \$583$, $n = 36$.

20. $P = \$4176$, $C = \$419$, $n = 30$.

21. A room air conditioner can be bought for a down payment of $42.50 and $17.50 at the end of each month for a year. What rate of interest converted monthly is being charged if the cash price is $242.50?

22. A farmer can buy a combine for $6700 cash or for $700 down and $1550 at the end of each 6 months for 2 years. What rate of interest converted semi-annually is being charged?

23. A combination refrigerator and deepfreeze can be bought for a down payment of $84 and a payment of $21 at the end of each month for 19 months. If the same box can be bought for $444 cash, what rate of interest converted monthly is being charged?

24. Simpleson can buy a lot for $5300 or for a down payment of $1800 and $100 at the end of each month for 40 months. What rate of interest converted monthly is being charged?

25. A boat can be bought for $2931 or for $431 down and the remainder paid off in 24 equal monthly payments. How much should each payment be if money is worth 5% converted monthly?

26. How much should each payment be if a farm can be bought for $26,800 cash or for a down payment of $5800 and equal payments at the end of each year for 5 years. Assume that money is worth $j = 3\%$.

27. A car is priced at $3226 and can be bought for a down payment of $726 and equal payments at the end of each month for 30 months. How much is each payment if money is worth $j = 6\%$, $m = 12$?

28. Mr. Samot had a $5700 addition built on his house. He agreed to pay for it by 8 equal semiannual payments with the first due 6 months after the work was finished. How much should each be if money is worth $j = 6\%$, $m = 2$?

17

RATIO, PROPORTION, VARIATION

17.1. Ratio

The indicated division of a by b is called *the ratio of a to b*. The division may be indicated by $a \div b$, a/b, $\frac{a}{b}$, or $a:b$. The quantities a and b may be magnitudes of the same kind but they need not be. If they are of the same kind, they must be expressed in the same unit for $a \div b$ to have a meaning. Thus, in order to find the ratio of 8 inches to 3 feet, we reduce 3 feet to 36 inches; then the desired ratio is $\frac{8}{36} = \frac{2}{9}$. If a and b represent magnitudes of the same kind and are expressed in the same unit, then a/b gives the answer to the question "a is what multiple or what fractional part of b?"

If a and b do not represent magnitudes of the same kind, then the ratio $a:b$ represents the amount of a that corresponds to one unit of b. Thus if \$3.60 is the cost of five pounds of bacon, then \$3.60/5 lb., or \$.72 per pound, is the cost or price of bacon.

17.2. Proportion

If one person bought P pounds of apples for A cents and another bought p pounds of the same kind of apples for a cents at the same time and place, then

$$\frac{A}{P} = \frac{a}{p}$$

since each fraction is the cost in cents of one pound of apples. This illustrates the following definition: *a statement that two ratios are equal is called a* ***proportion***. If the ratios are $a \div b$ and $c \div d$, then

(1)
$$\frac{a}{b} = \frac{c}{d}$$

is a proportion formed from stating that they are equal. This is an equation, since it is a statement that two quantities are equal; furthermore, it is a fractional equation or an equation that involves fractions. Consequently it can be solved for the remaining one of the four quantities a, b, c, and d if three of them are known.

EXAMPLE

1. If a 510-lb bale of cotton cost $156.57, how much would a 490-lb bale of the same grade cost at the same time and place?

SOLUTION

We can set up a proportion because the cost divided by the weight in each case gives the price per pound. Consequently, if c is the cost of the 490-lb bale, then

$$\frac{c}{490} = \frac{156.57}{510}$$

Now, multiplying through by 490, we have

$$c = \frac{490(156.57)}{510} = 150.43;$$

hence, the cost is $150.43.

If, in (1), $c = b$, we call their common value a *mean proportional* between (or to) a and d; furthermore, d is called the *third proportional* to a and b.

EXAMPLES

2. Find a mean proportional between 4 and 9.

SOLUTION

If we let x represent the mean proportional, then

$$\frac{4}{x} = \frac{x}{9}$$

$36 = x^2$, multiplying through by $9x$;

consequently, $x = \pm 6.$

3. Find the third proportional to 27 and 18.

SOLUTION

If we represent the third proportional by x, then

$$\frac{27}{18} = \frac{18}{x}.$$

Therefore, multiplying through by $18x$, we have

$$27x = (18)(18) = 324$$

and $x = 12.$

Exercise 17.1

Express the ratio in each of Problems 1 through 16 as a fraction, and remove any factors that occur in both the numerator and denominator of a fraction.

1. 3 inches to 1 foot.

2. 5 yards to 2 feet.

3. 2 miles to 660 yards.

4. 2 yards to 10 inches.

5. 3 hours to 45 minutes.

6. 3 weeks to 6 days.

7. 1 week to 56 hours.

8. 2.5 hours to 18 minutes.

9. 88 miles to 2 hours.

10. $16 to 40 pounds.

11. $7620 to 12 months.

12. 204 miles to 12 gallons.

13. 84 students to 3 classes.

14. 102 women to 100 men.

15. 2205 persons to 7 square miles.

16. 32 apples to 8 boys.

Find the value or values of x in each of the following proportions.

17. $\dfrac{x}{3} = \dfrac{2}{6}.$

18. $\dfrac{x}{6} = \dfrac{2}{3}.$

19. $\dfrac{x}{4} = \dfrac{15}{20}$.

20. $\dfrac{x}{7} = \dfrac{18}{14}$.

21. $\dfrac{2}{x} = \dfrac{6}{9}$.

22. $\dfrac{3}{x} = \dfrac{7}{5}$.

23. $\dfrac{7}{x} = \dfrac{21}{12}$.

24. $\dfrac{4}{x} = \dfrac{6}{7}$.

25. $\dfrac{x}{8} = \dfrac{2}{x}$.

26. $\dfrac{x}{4} = \dfrac{9}{x}$.

27. $\dfrac{x}{25} = \dfrac{1}{x}$.

28. $\dfrac{x}{3} = \dfrac{27}{x}$.

Find the third proportional to each pair of members in Problems 29 through 32.

29. 2, 6.   **30.** 8, 4.   **31.** 2, 5.   **32.** 3, 9.

Find the mean proportion between the pair of numbers given in each of Problems 33 through 36.

33. 27, 3.   **34.** 16, 4.   **35.** 81, 1.   **36.** 4, 9.

37. The density of a body is defined as the ratio of its mass to its volume. If the mass of 60 cubic centimeters of aluminum is 112 grams, find its density.

38. The pitch of a roof is the distance the roof rises per unit of horizontal distance covered. Find the pitch of a roof if a rafter 13 feet long extends 12 feet horizontally.

39. The grade of a highway is the distance it rises divided by the distance along the road. Find the grade of a road that rises 396 feet in a mile.

40. The specific gravity of a body is the ratio of its weight to the weight of an equal body of water. If a cubic foot of water weighs 62.5 pounds and a cubic foot of mercury weighs 850 pounds, what is the specific gravity of mercury?

41. Divide 28 into two parts that are in the ratio 5 to 2.

42. A boy 4 feet tall casts a shadow that is 14 feet long. His older brother is standing by him and is 5 feet tall. How long is the shadow of the older brother?

43. If 8 pounds of bacon cost $2.96, how much should 5 pounds of the same grade cost at the same time?

44. The ratio of the heights of a mother and father is the same as the ratio of the heights of their daughter and son. How tall is the father if the mother is 5 feet 6 inches, the daughter is 4 feet 7 inches and the son is 5 feet?

17.3. Variation

There are many situations in which each of two or more quantities vary in such a way that their product or quotient is a constant. For example, at a given temperature, the pressure times the volume of a confined gas is a constant; furthermore, the weight of any body of water divided by its volume gives the weight of a unit of volume, and that is a constant. These two examples illustrate the following pair of definitions.

One quantity *varies directly** as a second if it is a constant times the second.

EXAMPLE

1. The statement that y varies directly as x can be written as an equation in the form $y = kx$.

One quantity *varies inversely* as a second if it is a constant times the reciprocal of the second.

EXAMPLE

2. The statement that y varies inversely as x can be written as an equation in the form $y = k\left(\dfrac{1}{x}\right)$.

There are other types of variation and one of them is described in the following definition. One quantity *varies jointly* as two or more others if it is a constant times their product.

EXAMPLE

3. The statement that w varies jointly as x and y can be written as an equation in the form $w = kxy$.

The constant in each of the above definitions is called the *constant of variation*.

A problem can make use of a combination of the types of variation described above.

EXAMPLE

4. The statement that w varies jointly as x and y and inversely as t can be put in equation form as

$$w = \frac{kxy}{t}.$$

* At times "directly" is omitted.

17.4. Problems Solved by Use of Variation

Any statement of variation includes a constant and at least two variables. If we know a set of corresponding values of the variables, we can evaluate the constant by substituting them in the equation that expresses the variation and then solving for the constant.

EXAMPLE

1. If y varies as x and is 63 for $x = 18$, find the constant of variation.

SOLUTION

The equation which expresses the variation can be obtained by use of the definition of direct variation and is

(1) $$y = kx.$$

If we substitute the set of corresponding values of the variables in (1), we have

$$63 = k\,18;$$

hence, dividing by 18, we see that

$$k = \frac{63}{18} = 3.5.$$

Consequently, putting this in (1), it becomes

$$y = 3.5x.$$

After the constant of variation has been determined and substituted in the equation which expresses the variation, we can evaluate any one of the variables provided a set of corresponding values for all others is known.

EXAMPLE

2. If y varies as x and is 63 for $x = 18$, find y for $x = 10$.

SOLUTION

We found the constant of variation for this situation in Example 1, used it in the equation which expresses the variation, and got

(2) $$y = 3.5x.$$

All that we need to do now is substitute 10 for x in (2) and find y. Thus,

$$y = (3.5)(10) = 35$$

is the value of y which corresponds to $x = 10$.

It is often possible to compare two quantities by use of variation without finding the value of either.

E X A M P L E

3. The lateral surface of a right circular cylinder varies jointly as the radius of the base and the altitude of the cylinder. Compare the lateral surfaces of two cylinders if the one has a radius of 4 inches and an altitude of 6 inches and the other has 6 inches for its radius and 8 inches for its altitude.

SOLUTION

If we represent the surfaces by L_1 and L_2 and make use of the definition of the types of variation, we have

$$L_1 = krh = k4(6) = 24k$$

and

$$L_2 = krh = k6(8) = 48k.$$

Now we see that

$$\frac{L_1}{L_2} = \frac{24k}{48k} = \frac{1}{2};$$

consequently,

$$L_1 = \frac{1}{2}L_2.$$

Exercise 17.2

1. Express each of the following statements as an equation: s varies directly as t; p varies inversely as q; b varies jointly as c and d; h varies directly as j and inversely as m.

2. If a varies directly as b and is 14 for $b = 7$, find the value of a if $b = 3$.

3. If s varies inversely as t and $s = 12$, $t = 3$ are corresponding values, find the value of s for $t = 9$.

4. If w varies jointly as x and y and is 24 for $x = 3$ and $y = 4$, find the value of w for $x = 8$ and $y = 1.75$.

5. If y varies jointly as x and w and is 30 for $x = 5$ and $w = 3$, find y for $x = 6$ and $w = 4$.

6. If s varies inversely as t and is 3 for $t = 4$, find s for $t = 6$.

7. If y varies as x and is 18 for $x = 3$, find y for $x = 6$.

8. If y varies as x and inversely as w, find y for $x = 8$ and $w = 2$ provided $y = 1$ for $x = 1$ and $w = 4$.

9. The circumference of a circle varies as the radius. Find the circumference of a circle of radius 5 if one of radius 3 has a circumference of 18.84.

10. The distance travelled by a freely falling body varies directly as the square of the time in flight. How far will a body fall in 7 seconds if it falls 144 feet in 3 seconds?

11. The power required to propel a ship varies as the cube of the speed. If 5184 horsepower is required to drive the ship at 12 knots, what horsepower is required for a speed of 18 knots?

12. The period of a complete vibration of a pendulum varies as the square root of the length. If a pendulum 16 inches long has a period of 1 second, what is the period of one 36 inches long?

13. The weight of a wire used for a clothesline varies directly as the length. If 30 feet of the wire weighs 1.8 pounds, what is the weight of 50 feet?

14. The exposure time necessary to obtain a good negative varies as the square of the *f*-number on the camera shutter. If 1/100 of a second is required when the shutter is set on *f*/8, what exposure time is required under the same light conditions if the shutter is set at *f*/16.

15. At any given time, the length of a shadow varies directly as the height of the object casting it. If a pole 9 feet tall casts a shadow 13.5 feet long, how tall is a wall that casts a shadow 27 feet long?

16. The amount of hydrogen produced by adding sodium to water varies as the weight of the sodium. If 24 ounces of sodium produces 1 ounce of hydrogen, how much hydrogen will be produced by 108 ounces of sodium?

17. One of Kepler's laws states that the square of the time needed by a planet to make one revolution about the sun varies directly as the cube of the average distance of the planet from the sun. If Mars requires 670 days for a revolution and is on the average 1.5 times as far from the sun as is the earth, how long does the earth require for a revolution?

18. The time needed to make an enlargement from a photographic negative varies directly as the area. If 7 seconds is required to make a 5 by 7 enlargement, how long is needed to make an 8 by 10 enlargement?

19. The volume of a cube varies directly as the cube of an edge. If a 4 inch cube has a volume of 64 cubic inches, what is the volume of a cube that is 7 inches on a side?

20. On the ocean, the square of the distance in miles to the horizon varies as the height that the observer is above the surface of the water. If a 6 foot man on a surfboard can see 3 miles, how far can a person see if he is in a balloon 600 feet above the water?

21. The intensity of light varies inversely as the square of the distance from the source. If an object 4 feet from a source is illuminated by 27 candlepower what is the illumination on a similar object that is 6 feet from the source?

22. The gravitational attraction of the earth for an object varies inversely as the square of the distance of the object from the center of the earth. If a man weighs 180 pounds on the surface of the earth, how much will he weigh if 200 miles above the surface? Assume the radius of the earth is 4000 miles.

23. The resistance of a wire of given length varies inversely as the cross-sectional area. If a wire of diameter .15 inch has a resistance of 60 ohms, what is the resistance of a wire of the same length and same material with a diameter of .06 inch?

24. The time required to build the walk of a building varies inversely as the number of men on the job. If 24 men take 18 days to erect the walk for a building, how long would be required by 36 men?

25. The volume of a right circular cylinder varies jointly as the height and the square of the radius. If the volume of right circular cylinder of radius 5 inches and height 8 inches is 628 cubic inches, what is the volume of another of radius 8 inches and height 5 inches?

26. The amount of fuel used by a ship travelling at a uniform speed varies jointly as the distance travelled and the square of the velocity. If a ship used 33 tons of fuel on a trip at 20 knots, how much would it use on a trip twice as long at 15 knots?

27. The simple interest earned by a given principal varies jointly as the rate and the time. If a certain principal yields $48 in 3 years at 4 per cent, how much would it yield in 2 years at 6%?

28. The wind force on a flat vertical surface varies jointly as the area of the surface and the square of the wind velocity. If the force on 1 square foot is 1 pound when the wind velocity is 12 miles per hour, what is the pressure on a 6 by 10 foot signboard when the wind velocity is 48 miles per hour?

29. The weight of a green log varies jointly as the length and the square of the radius. If a log 4 inches in diameter and 7 feet long weighs 42 pounds, find the weight of a 10 foot log of radius 3 inches.

30. The volume of a box varies jointly as the length, width, and thickness. Find the volume of a box that is 3 by 4 by 5 feet, if one 2 by 3 by 4 feet has a volume of 24 cubic feet.

31. The maximum safe load for a beam of given length varies jointly as the breadth and the square of its depth. If a beam of given length is 6 inches deep and 4 inches wide and can safely bear a load of 1200 pounds, find the largest safe load that a beam of the same length can bear if it is 4 inches deep and 6 inches wide.

32. The amount of paint required to paint a wall varies jointly as the area of the wall and the thickness of the coat of paint. If one-half of a gallon is required for a coat .01 inch thick on a certain wall, how much is used for a coat .015 inch thick on a wall that is $\frac{2}{3}$ as long and twice as high as the first wall?

33. If the kinetic energy of a body varies as the square of its velocity, compare the kinetic energy of a car going 10 miles per hour with that of the same car travelling 40 miles per hour.

34. The air resistance to a moving object is approximately proportional to the square of the speed of the object. Compare the air resistance of an automobile travelling 30 miles per hour to that of the same automobile travelling 60 miles per hour.

35. The weight of a dam of a certain design varies as the cube of its height. Compare the weights of two such dams if the height of the first is 1.7 times that of the second.

36. The landing speed of a plane varies as the square root of its gross load. Compare the landing speeds of two planes if the first has a gross load of 7569 pounds and the second a gross load of 13,456 pounds.

37. The amount of electric energy converted into heat in a given time varies jointly as the resistance of the conductor and the square of the current. Compare the heat produced by a certain resistance and a known current with that produced by half that much current and twice that resistance.

38. The electrical resistance of a uniform wire varies directly as the length and inversely as the area of the cross section. Compare the resistance of 900 feet of wire with a diameter of $\frac{1}{32}$ inch with that of 500 feet of diameter $\frac{1}{16}$ inch.

39. The force of attraction between two spheres varies directly as the product of their masses and inversely as the square of the distance between their centers. Compare the attraction between two bodies of masses m_1 and m_2 with that between two of masses $2m_1$ and $3m_2$ if the centers of the last two are twice as far apart as the first two.

40. The acceleration produced in an object by a force applied to it varies directly as the force and inversely as the mass of the object. Compare the acceleration produced by a force of 250 dynes acting on a mass of 50 grams to that of a force of 300 dynes acting on 75 grams.

18

INTRODUCTORY CONCEPTS
OF ANALYTIC GEOMETRY

18.1. The Field and Method of Analytic Geometry

Plane analytic geometry includes the study of points, lines, curves, angles, and areas in a plane; solid analytic geometry is made up of the study of points, lines, planes, curves, and surfaces in three dimensional space. The details of each investigation are carried on by establishing a correspondence between equations and geometric configurations. Part of the work is in learning to construct a curve that corresponds to a given equation and the remainder in forming the equation when the curve, or sufficient conditions to determine it, are given.

Analytic geometry differs in procedure from the geometry studied in high school in that the former makes use of a coordinate system and the latter does not.

18.2. The Distance Between Two Points

We shall need to be able to find the distance between two points on many occasions in our study of analytic geometry and shall now develop a formula for use at such times. We shall designate the two points by $P_1(x_1, y_1)$ and $P_2(x_2, y_2)$, as shown in Fig. 18-1. We then draw a parallel to the X axis through P_1 and one to the Y axis through P_2 and designate their intersection by P_3. It has the indicated coordinates since it is the same distance and direction from the Y axis as P_2 and the same distance and direction from the X axis as P_1. If we knew the length of $P_1 P_3$ and of $P_3 P_2$, we could find the distance between P_1 and P_2 by use of the Pythagorean theorem. Those lengths are:

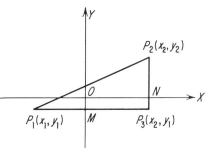

FIGURE 18-1

$$P_1P_3 = MP_3 + P_1M = MP_3 - MP_1 = x_2 - x_1 \quad \text{and}$$

$$P_3P_2 = NP_2 + P_3N = NP_2 - NP_3 = y_2 - y_1$$

Consequently,

$$(P_1P_2)^2 = (P_1P_3)^2 + (P_3P_2)^2$$

$$= (x_2 - x_1)^2 + (y_2 - y_1)^2;$$

hence,

$$P_1P_2 = \sqrt{(x_2 - x_1)^2 + (y_2 - y_1)^2}$$

If P_1 had been chosen to the right of P_2, then we would have had $P_1P_3 = x_1 - x_2$; furthermore, if P_1 had been chosen above P_2, then we would have had $P_3P_2 = y_1 - y_2$. The choice of relative positions of P_1 and P_2 is immaterial since the square of a number and of the negative of the number are equal. Therefore, we are able to say that *the length d of the line segment that joins $P_1(x_1, y_1)$ and $P_2(x_2, y_2)$ is*

$$d = \sqrt{(x_2 - x_1)^2 + (y_2 - y_1)^2}$$

EXAMPLES

1. Find the length of the line segment between $P_1(2, 7)$ and $P_2(-1, 3)$.

SOLUTION

If we substitute the given values of x_1, x_2, y_1 and y_2 in the distance formula, we see that

$$d = \sqrt{(-1-2)^2 + (3-7)^2}$$
$$= \sqrt{(-3)^2 + 4^2}$$
$$= \sqrt{9+16} = 5$$

is the distance between the given points.

2. Find the point on the X axis that is equidistant from $P_1(5, 3)$ and $P_2(1, -7)$.

SOLUTION

We shall designate the point by $P(x, 0)$ since it is on the X axis; hence, we must determine x so that $PP_1 = PP_2$. If we now use the distance formula and then square each member of the resulting equation, we have

$$(x-5)^2 + (0-3)^2 = (x-1)^2 + (0+7)^2$$
$$x^2 - 10x + 25 + 9 = x^2 - 2x + 1 + 49, \text{ expanding,}$$
$$-8x = 16, \text{ collecting like terms,}$$
$$x = -2, \text{ dividing by } -8$$

Therefore, $(-2, 0)$ is the point on the X axis that is equidistant from $P_1(5, 3)$ and $P_2(1, -7)$.

Exercise 18.1

By use of the distance formula prove that each of the following triples of points lie on a line

1. $(3, -1)$, $(5, 5)$, $(-1, -13)$.　　　　**2.** $(-5, 0)$, $(-2, 6)$, $(4, 18)$.

3. $(-8, -6)$, $(-6, -2)$, $(0, 10)$.　　　　**4.** $(-3, 1)$, $(-1, 4)$, $(3, 10)$.

Prove that each of the following triples of points can be used as the vertices of an isosceles but not an equilateral triangle.

5. $(1, -2)$, $(5, 4)$, $(6, -1)$.　　　　**6.** $(7, -3)$, $(1, 5)$, $(8, 4)$.

7. $(8, 13)$, $(0, 3)$, $(19, -4)$.　　　　**8.** $(-9, 2)$, $(3, -7)$, $(3, 8)$.

Prove that each of the following triples of points can be used as the vertices of a right triangle.

9. $(-1, 5)$, $(7, 1)$, $(5, -3)$.　　　　**10.** $(3, 14)$, $(13, -10)$, $(8, -11)$.

11. $(0, 0)$, $(-8, 6)$, $(-1, 7)$.　　　　**12.** $(-5, -1)$, $(19, -11)$, $(12, 6)$.

In each of Problems 13 through 16, determine x or y so that $PP_1 = PP_2$.

13. $P(x,1)$, $P_1(4, 3)$, $P_2(-2, -5)$.

14. $P(x, 4)$, $P_1(3, 0)$, $P_2(7, -2)$.

15. $P(0, y)$, $P_1(5, -1)$, $P_2(1, 3)$.

16. $P(-3, y)$, $P_1(6, 5)$, $P_2(-2, -3)$.

17. By means of an equation, state that (x, y) is 3 units from $(3, 4)$.

18. State that (x, y) is 5 units from $(2, -3)$ by means of an equation.

19. By means of an equation give the condition under which (x, y) is equidistant from $(3, 4)$ and $(2, -1)$.

20. State that (x, y) is equidistant from $(6, -2)$ and $(-3, 1)$ by means of an equation.

18.3. Division of a Line Segment in a Given Ratio

We shall consider the segment that connects $P_1(x_1, y_1)$ and $P_2(x_2, y_2)$ and determine the coordinates of the point $P(x, y)$ which is on the segment P_1P_2 and which divides that segment in two pieces such that

$$\frac{P_1P}{PP_2} = \frac{r_1}{r_2},$$

where r_1 and r_2 are any two nonzero real numbers. As an aid in doing this, we shall now construct Fig. 18-2 by dropping perpendiculars from P_1, P,

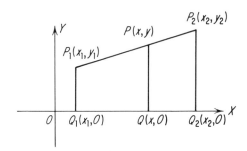

FIGURE 18-2

and P_2 to the X axis and call their feet Q_1, Q, and Q_2, respectively. The coordinates of these points are as given in the figure since each Q is the same distance and direction from the Y axis as the corresponding P. We now have the three parallels P_1Q_1, PQ, and P_2Q_2 cut by two transversals; hence, the segments into which the transversals are divided are proportional. Therefore

(1)
$$\frac{r_1}{r_2} = \frac{P_1 P}{P P_2} = \frac{Q_1 Q}{Q Q_2}$$

We now notice from the figure that $Q_1 Q = x - x_1$ and $Q Q_2 = x_2 - x$; hence,

$$\frac{Q_1 Q}{Q Q_2} = \frac{x - x_1}{x_2 - x}$$

and by use of (1)

$$\frac{r_1}{r_2} = \frac{x - x_1}{x_2 - x}$$

Solving this for x, we find that *the x coordinate of the point that divides the segment from P_1 to P_2 in the ratio r_1 to r_2 is*

(2)
$$x = \frac{r_2 x_1 + r_1 x_2}{r_1 + r_2}, \qquad r_1 + r_2 \neq 0$$

We can show similarly that

(3)
$$y = \frac{r_2 y_1 + r_1 y_2}{r_1 + r_2}, \qquad r_1 + r_2 \neq 0$$

These formulas for x and y are called *the point of division formulas.* The special case in which $r_2 = r_1$ is known as the *midpoint formula.* We have

$$x = \frac{r_2 x_1 + r_1 x_2}{r_1 + r_2}$$

$$= \frac{r_1 x_1 + r_1 x_2}{r_1 + r_1}, \qquad \text{if } r_2 = r_1$$

$$= \frac{x_1 + x_2}{2}$$

by dividing the numerator and denominator by r_1. Similarly,

$$y = \frac{y_1 + y_2}{2}$$

Consequently, *the coordinates of the midpoint of $P_1 P_2$ are*

$$x = \frac{x_1 + x_2}{2} \quad and \quad y = \frac{y_1 + y_2}{2}$$

Figure 18-2 was drawn with P between P_1 and P_2, but the same formulas for x and y would have been obtained had P_1 or P_2 been between P and the other. If P is between P_1 and P_2, then P_1P and PP_2 are of the same sign since they are measured in the same direction. If P is not between P_1 and P_2 then P_1P and PP_2 are measured in opposite directions and r_1 and r_2 are of opposite signs.

EXAMPLES

1. Find the coordinates of the point that divides the segment from $P_1(-4, 6)$ to $P_2(1, -9)$ into the ratio 3 to 2.

SOLUTION

In this problem, we have $x_1 = -4$, $y_1 = 6$, $x_2 = 1$, $y_2 = -9$, $r_1 = 3$, and $r_2 = 2$. By substituting these values in the point of division formulas, we see that

$$x = \frac{(2)(-4) + (3)(1)}{3 + 2} = \frac{-8 + 3}{5} = -1$$

and

$$y = \frac{(2)(6) + (3)(-9)}{3 + 2} = \frac{12 - 27}{5} = -3.$$

Consequently, the desired point is $P(-1, -3)$. The point is between P_1 and P_2 since r_1 and r_2 are of the same sign. The reader should notice that P is $\frac{3}{5}$ of the way from P_1 to P_2.

2. Determine the point that divides the segment from $P_1(4, -3)$ to $P_2(10, 6)$ into the ratio 5 to -2.

SOLUTION

Since r_1 and r_2 are of opposite signs, the point of division is external to the segment. By use of (2) and (3), we see that

$$x = \frac{(-2)(4) + (5)(10)}{-2 + 5} = \frac{-8 + 50}{3} = 14$$

and

$$y = \frac{(-2)(-3) + (5)(6)}{-2 + 5} = \frac{6 + 30}{3} = 12$$

Consequently, $\dfrac{P_1P}{PP_2} = \dfrac{5}{-2}$ if $P(x, y)$ is $P(14, 12)$.

Exercise 18.2

Find the point that divides the segment from P_1 to P_2 in the given ratio in Problems 1 through 12.

1. $P_1(3, -2)$, $P_2(9, 1)$, 2 to 1.

2. $P_1(-6, 1)$, $P_2(15, 15)$, 3 to 4.

3. $P_1(-5, 3)$, $P_2(5, 18)$, 2 to 3.

4. $P_1(2, -6)$, $P_2(9, 8)$, 2 to 5.

5. $P_1(4, 2)$, $P_2(2, -2)$, 3 to -2.

6. $P_1(-3, -5)$, $P_2(6, 4)$, 2 to -5.

7. $P_1(-1, 7)$, $P_2(5, -2)$, 4 to -1.

8. $P_1(0, -3)$, $P_2(6, 5)$, 5 to -3.

9. $P_1(2, a)$, $P_2(7, a - 5)$, 3 to 2.

10. $P_1(a, -1)$, $P_2(-6a, 6 + 7a)$, 2 to 5.

11. $P_1(2, -3)$, $P_2(3 + a, -1 + 2a)$, a to 1.

12. $P_1(0, 1)$, $P_2(a + b, 1 + 2a + 2b)$, a to b.

13. Find the midpoint of the segment that connects $P_1(3, -2)$ and $P_2(7, 4)$.

14. What are the coordinates of the point that divides the segment between P_1 $(4, -3)$ and $P_2(-2, 5)$ into the ratio 1 to 1?

15. What are the coordinates of the center of a circle if $P_1(7, -1)$ and $P_2(1, -7)$ are opposite ends of a diameter?

16. If $P_1(2, 3)$ is one end of a diameter of a circle with center at $(-1, 5)$, find the coordinates of the other end of the diameter.

17. Find the point of intersection of the medians of a triangle with vertices at $P_1(6, 0)$, $P_2(-3, -2)$, and $P_3(3, -4)$.

18. If the vertices of a triangle are at $P_1(2, -5)$, $P_2(6, -3)$, and $P_3(13, -7)$, find the point of intersection of the medians.

19. Show that the diagonals of the quadrilateral with vertices at $(6, 4)$, $(11, 6)$, $(9, 2)$, and $(4, 0)$ bisect one another.

20. Do the diagonals of the quadrilateral with vertices at $(4, 1)$, $(7, 3)$, $(5, -2)$, and $(8, -3)$ bisect one another?

21. Find the vertices of a triangle if the midpoints of its sides are at $(2, 3)$, $(5, -3)$, and $(-1, 4)$.

22. The midpoints of the sides of a triangle are at $(3, 1)$, $(1, 5)$, and $(0, 0)$. Find the vertices.

23. If two vertices of a triangle are $(6, 3)$ and $(8, -5)$, find the third vertex provided the medians meet at $(5, -3)$.

24. The medians of a triangle meet at $(3, -1)$ and two vertices are at $(8, 7)$ and $(-2, -1)$. Find the third vertex.

18.4. Slope of a Line

In order to be able to find the slope of a line, we shall need the following definition. The *angle of inclination* or *inclination* of an undirected line is the smallest positive angle from the positive X axis to the line, provided the line is not parallel to the X axis. The angle of inclination is indicated by θ in Figs. 18-3(a) and 18-3(b). If the line is parallel to the X axis, the inclination is defined as zero.

The *slope* of a line is the tangent of the angle of inclination and is often designated by m; hence, we write

$$m = \tan \theta$$

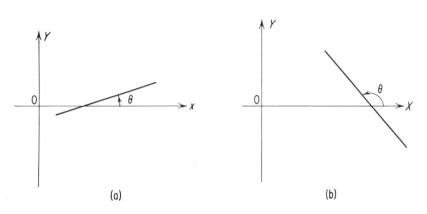

(a)   (b)

FIGURE 18-3

as a symbolic form of the slope of a line. Thus, if a line makes an angle of $60°$ with the X axis, its slope is $m = \tan 60° = \sqrt{3}$.

Since there is no number to represent $\tan 90°$, we must exclude lines with an inclination of $90°$ in all discussions which involve slope.

We shall now develop a formula for the slope of a line through the points $P_1(x_1, y_1)$ and $P_2(x_2, y_2)$. These points and a line through them are shown in Fig. 18-4. The point M is also shown. It is the intersection of a parallel to the X axis through P_1 and a parallel to the Y axis through P_2; hence, its coordinates are as given in the figure. The angle MP_1P_2 is equal to θ since P_1M is parallel to the X axis. Therefore,

$$\tan \theta = \tan MP_1P_2 = \frac{MP_2}{P_1M} = \frac{y_2 - y_1}{x_2 - y_1}.$$

Furthermore, it can be shown that $\tan \theta$ has this same value if θ is greater than 90°. Consequently, we know that *the slope of a line through $P_1(x_1,y_1)$ and $P_2(x_2,y_2)$ is*

$$m = \frac{y_2 - y_1}{x_2 - x_1}$$

This equation is known as the *slope formula*.

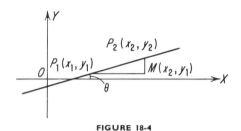

FIGURE 18-4

EXAMPLE

1. Find the slope of the line through $(3, -2)$ and $(6, 8)$.

SOLUTION

If we use the slope formula, we find that the slope of the line is $m = [8 - (-2)]/(6 - 3) = 10/3$. The angle of inclination can now be determined by use of a table of values of trigonometric functions and is $73°20'$ to the nearest ten minutes.

18.5. Parallel and Perpendicular Lines

We shall consider two lines, L_1 and L_2 (Fig. 18-5), neither of which is parallel to the Y axis. If they are parallel, then their angles of inclination, θ_1 and θ_2, are equal; hence, $\tan \theta_1 = \tan \theta_2$ and the slopes of the lines are equal. Conversely, if the slopes of the lines are equal, i.e., if $\tan \theta_1 = \tan \theta_2$, then $\theta_1 = \theta_2$, since the slope of a line is never as large as 180°. Therefore the lines are parallel. Thus we see that *two lines are parallel if their slopes are equal*; *the converse is also true*.

We shall now consider two perpendicular lines, L_1 and L_2 (Fig. 18-6), and draw the coordinate axes with the origin at their intersection so that their angles of inclination are angles in standard position. There is no loss of generality in doing this since it can be done for any pair of intersecting lines. If P and R are on L_1 and L_2 at a distance r from the origin, and if we drop a perpendicular PQ from P to the X axis and one from R to the Y axis at S, then the triangles OQP and OSR are congruent. Therefore, corresponding sides RS and QP

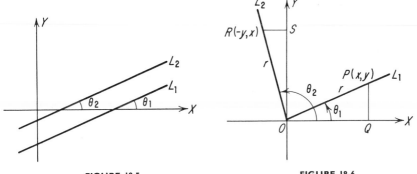

FIGURE 18-5   FIGURE 18-6

are equal, as are OS and OQ; however, the abscissa of R is $SR = -RS$. Consequently, if the coordinates of P are (x, y), it follows that those of R are $(-y, x)$. Now,

$$\tan \theta_2 = \tan (\theta_1 + 90°)$$

$$= \frac{x}{-y} = -1 \div \frac{y}{x}$$

$$= \frac{-1}{\tan \theta_1}$$

and $m_2 = -1/m_1$ or $m_1 m_2 = -1$, since $\tan \theta_2 = m_2$ and $\tan \theta_1 = m_1$. Conversely, if $m_2 = -1/m_1$, then $\tan \theta_2 = -1/\tan \theta_1$ and $\theta_2 = \theta_1 + 90°$. Consequently, the lines are perpendicular. Thus, we have shown that *two lines are perpendicular if the product of their slopes is* -1; *the converse is also true.*

EXAMPLE

1. Show that the line L_1 through $(2, -1)$ and $(6, 7)$ is parallel to the line L_2 through $(-1, 2)$ and $(1, 6)$, perpendicular to L_3 through $(3, 5)$ and $(1, 6)$, and neither parallel nor perpendicular to L_4 through $(5, -2)$ and $(6, -4)$.

SOLUTION

The slope of L_1 is $m_1 = [7 - (-1)]/(6 - 2) = 2$ and the slopes of the other lines are $m_2 = (6 - 2)/[1 - (-1)] = 2$, $m_3 = (6 - 5)/(1 - 3) = -1/2$, and $m_4 = [-4 - (-2)]/(6 - 5) = -2$. Since $m_2 = m_1$, it follows that L_2 and L_1 are parallel; furthermore, $m_3 m_1 = -1$ and, as a result, L_3 and L_1 are perpendicular. Finally, $m_4 \neq m_1$ and $m_4 m_1 \neq -1$; consequently, L_4 is neither parallel to nor perpendicular to L_1.

18.6. The Area of a Triangle

We shall now derive a formula for use in finding the area of a triangle in terms of the coordinates of the vertices of the triangle. We shall consider any triangle with vertices at $P_1(x_1, y_1)$, $P_2(x_2, y_2)$, and $P_3(x_3, y_3)$ as in Fig. 18-7 and embed it in a rectangle by drawing parallels to the X axis through the highest and lowest vertices and parallels to the Y axis through the vertices

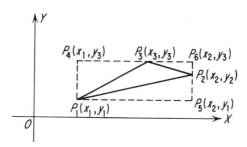

FIGURE 18-7

that are farthest to the left and to the right. Thus, we have the rectangle $P_1P_5P_6P_4$. Now we make use of the fact that the area of a rectangle is the product of its base and altitude, and the area of a triangle is one-half the product of its base and altitude. We designate the area of the triangle with vertices P_1, P_2, and P_3 by $A(P_1P_2P_3)$ and from Fig. 18-7 have

$$A(P_1P_2P_3) = A(P_1P_5P_6P_4) - A(P_1P_5P_2) - A(P_2P_6P_3) - A(P_3P_4P_1)$$
$$= (x_2 - x_1)(y_3 - y_1) - \tfrac{1}{2}[(x_2 - x_1)(y_2 - y_1) +$$
$$+ (x_2 - x_3)(y_3 - y_2) + (x_3 - x_1)(y_3 - y_1)].$$

Expanding and collecting like terms, we have

$$A(P_1P_2P_3) = \tfrac{1}{2}[x_1y_2 + x_2y_3 + x_3y_1 - x_1y_3 - x_2y_1 - x_3y_2].$$

This expression can be more readily recalled if it is written as a determinant. If this is done, we see that

$$A(P_1P_2P_3) = \frac{1}{2} \begin{vmatrix} x_1 & y_1 & 1 \\ x_2 & y_2 & 1 \\ x_3 & y_3 & 1 \end{vmatrix}$$

is the area of the triangle with vertices at $P_1(x_1, y_1)$, $P_2(x_2, y_2,)$ and $P_3(x_3, y_3)$ or the negative of it according as the determinant is positive or negative.

The proof is beyond the scope of this book but it can be shown that the determinant is positive if the vertices are numbered in a counterclockwise direction and negative if numbered in a clockwise direction.

EXAMPLE

1. Find the area of the triangle with vertices at $P_1(3, 2)$, $P_2(-1, 6)$, and $P_3(0, -5)$.

SOLUTION

Since the vertices are numbered in a counterclockwise order, the determinant will be positive. The area is

$$\frac{1}{2}\begin{vmatrix} 3 & 2 & 1 \\ -1 & 6 & 1 \\ 0 & -5 & 1 \end{vmatrix} = \frac{1}{2}\begin{vmatrix} 3 & 2 & 1 \\ -1 & 6 & 1 \\ 0 & -5 & 1 \end{vmatrix}\begin{matrix} 3 & 2 \\ -1 & 6 \\ 0 & -5 \end{matrix}$$

$$= \tfrac{1}{2}[(3)(6)(1) + (2)(1)(0) + (1)(-1)(-5) - (1)(6)(0) - (3)(1)(-5) - (2)(-1)(1)]$$
$$= \tfrac{1}{2}[18 + 0 + 5 - 0 + 15 + 2]$$
$$= 20.$$

Exercise 18.3

Find the slope of the line through the pair of points in each of Problems 1 through 8.

1. (3, 2) and (5, 4).

2. (8, 1) and (5, 3).

3. (4, 7) and (9, 8).

4. (2, 3,) and (7, 5).

5. $(-1, -3)$ and $(-4, 2)$.

6. $(7, -2)$ and $(-4, 3)$.

7. $(9, -5)$ and (4, 1).

8. $(-5, 3)$ and $(3, -5)$.

Clarify the pair of lines in each of Problems 9 through 20 as parallel, perpendicular, or neither.

9. Through (2, 5) and (1, 7), through (2, 3) and $(4, -1)$.

10. Through (1, 0) and (3, 2), through (4, 4) and (3, 5).

11. Through $(3, -2)$ and $(-2, 3)$, through (7, 3) and $(3, -7)$.

12. Through (6, 5) and $(4, -1)$, through (4, 5) and (7, 4).

13. Through $(5, -2)$ and (7, 2), through (0, 3) and (4, 5).

14. Through (11, 8) and (2, 5), through $(-8, -1)$ and (1, 2).

15. Through $(7, -2)$ and (4, 3), through $(-4, 6)$ and $(2, -4)$.

16. Through $(0, -5)$ and (7, 1), through (13, 4) and (7, 11).

17. Through (9, 4) and $(4, -5)$, through $(8, -2)$ and (13, 7).

18. Through (6, 0) and (0, 6), through $(-3, 4)$ and $(4, -3)$.

19. Through (5, 5) and (7, 3), through (6, 3) and (11, 8).

20. Through $(3, -7)$ and $(5, -10)$, through $(6, 4)$ and $(0, 8)$.

21. Determine x so that the slope of the line through $(x, 3)$ and $(4, -1)$ is 2.

22. Find the value of x for which the slope of the line through $(2, -5)$ and $(x, 9)$ is 7.

23. For what value of y is the slope of the line through $(4, y)$ and $(6, 8)$ equal to 3?

24. If the slope of the line through $(2, -3)$ and $(3, y)$ is 5, what is the value of y?

Find the area of the triangle with vertices as given in each of Problems 25 through 32.

25. $(4, 1)$, $(-2, 3)$, $(6, 0)$.   **26.** $(-3, 1)$, $(-2, -5)$, $(4, 0)$.

27. $(7, -3)$, $(4, 1)$, $(-2, 5)$.   **28.** $(0, 2)$, $(-1, 4)$, $(5, -3)$.

29. $(5, 6,)$ $(3, -1)$, $(-2, 0)$.   **30.** $(-1, -5)$, $(4, 3)$, $(0, 2)$.

31. $(7, -6)$, $(-1, -2)$, $(3, 4)$.   **32.** $(-4, 7)$, $(3, 2)$, $(5, -1)$.

33. Show in two ways that $(-5, 2)$, $(-3, 3)$, and $(1, 5)$ lie on a line.

34. Show in two ways that $(-4, 9)$, $(-3, 6)$, and $(-1, 0)$ lie on a line.

35. Determine y so that $(3, 1)$, $(-2, 4)$, and $(8, y)$ lie on a line.

36. Determine x so that $(6, -3)$, $(3, -1)$, and $(x, 5)$ lie on a line.

37. Show that $(6, 10)$, $(-4, -14)$, and $(-11, 3)$ can be used as the vertices of an isosceles right triangle.

38. Prove that $(-2, 5)$, $(6, 3)$, $(5, 0)$ and $(-3, 2)$ can be used as the vertices of a parallelogram.

39. Show that $(2, 7)$, $(8, -12)$, $(6, 6)$, and $(3, -2)$ can be used as the vertices of a trapezoid.

40. Show that the quadrilateral with vertices at $(-3, 1)$, $(6, 4)$, $(4, 6)$, and $(-1, 4)$ is not a trapezoid and does not contain a right angle.

18.7. Analytic Proofs

Many theorems, including some from plane geometry, can be demonstrated quite readily by making use of a coordinate system and the concepts and formulas of this chapter. A proof is said to be analytic if a coordinate system is used in the demonstration.

We shall now give an analytic proof and at the same time point out some general principles to be used in all analytic proofs.

EXAMPLE

1. The line segment which joins the midpoints of the nonparallel sides of a trapezoid is parallel to the other two sides and half as long as the two together.

SOLUTION

We must be sure that the proof is for a general rather than for a specialized trapezoid. We should also make use of the fact that the work will be somewhat simpler if we make a wise choice of the position of the coordinate axes. Consequently, we shall draw a nonspecialized trapezoid *and then* put in the coordinate axes as in Fig. 18-8. The work is often simplified if we put a coordinate axis along

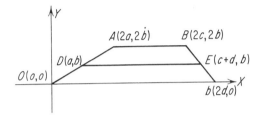

FIGURE 18-8

one side and a vertex at the origin; hence, we shall do that. In anticipation of using the midpoints of two sides, we shall use $(2a, 2b)$ as the coordinates of one vertex so as to avoid the later use of fractions. Now, since one side is along the X axis, the y coordinate of the other vertex of the side that contains A must be $2b$; hence, we shall use $2c$ and $2b$ as the coordinates of B. Finally, $2d$ and 0 will be used as the coordinates of the fourth vertex, since it is on the X axis. We now use the midpoint formula and see that the midpoints of the nonparallel sides are $D(a, b)$ and $E(c + d, b)$. Consequently, the slope of the segment between them is $(b - b)/(c + d - a) = 0$ and it is parallel to the parallel sides of the trapezoid. Finally, the sum of the lengths of OC and AB is

$$OC + AB = 2d + (2c - 2a)$$
$$= 2(d + c - a)$$

and $$DE = c + d - a;$$

consequently, the length of DE is half the sum of the lengths of the parallel sides.

Exercise 18.4

Prove each of the following theorems analytically.

1. The diagonals of a rectangle are equal in length.

2. The diagonals of a rhombus are perpendicular.

3. The diagonals of an isosceles trapezoid are of equal length.

4. Two medians of an isosceles triangle are of equal length.

5. The midpoint of the hypotenuse of a right triangle is equidistant from the three vertices.

6. The line segments that join the midpoints of opposite sides of a quadrilateral bisect each other.

7. The lines which join the midpoints of the sides of a triangle divide it into four equal triangles.

8. If a quadrilateral has two opposite sides equal and parallel, it is a parallelogram.

9. If taken in order, the segments that connect the midpoints of the sides of a quadrilateral form a parallelogram.

10. If taken in order, the segments that connect the midpoints of the sides of a rectangle form a rhombus.

11. The segments that connect the midpoints of opposite sides of a quadrilateral bisect each other.

12. The segments which connect the midpoints of the successive sides of a square form a square if taken in order.

13. In any triangle, the sum of the squares of two sides is equal to half the square of the third side plus twice the square of the median drawn to that side.

14. In any right triangle, the segment that joins the vertex of the right angle and the midpoint of the hypotenuse is half as long as the hypotenuse.

15. In any triangle, four times the sum of the squares of the medians is equal to three times the sum of the squares of the sides.

16. The segments that join the midpoints of consecutive sides of a right triangle form another right triangle.

17. If the diagonals of a trapezoid are equal, the figure is an isosceles trapezoid.

18. If the diagonals of a parallelogram are equal, the figure is a rectangle.

19. If the diagonals of a parallelogram are perpendicular, the figure is a rhombus.

20. If the diagonals of a rectangle are perpendicular, the figure is a square.

19

STATISTICS

19.1. Introduction

The study of statistics deals with collecting, exhibiting, studying, and drawing conclusions from data as well as checking the reliability of the conclusions. Before, during, or after the collection of the data, we must decide on the relevancy of the available data. The next step is to decide on a way of exhibiting the data. We then study the array, draw conclusions from it, and test the validity of these conclusions.

Modern statistics requires the use of advanced mathematics; hence, we shall be content with studying some of the traditional and elementary aspects of the subject.

Some rearrangement of data after they have been collected is almost always necessary in order to make facts stand out. One such rearrangement is called a *frequency distribution*, and it consists of separating the data into classes and showing the number of items in each class instead of considering each item separately.

EXAMPLE

1. An instructor gave a quiz to a class of 40 students. The grades, in the order in which the papers were corrected, were

75, 91, 86, 49, 58, 78, 94, 74, 89, 93, 54, 68, 73, 53, 67, 97, 74, 84, 72, 59,
76, 67, 85, 52, 88, 71, 87, 71, 66, 58, 71, 78, 46, 78, 79, 70, 80, 67, 71, 79.

The lowest grade was 46, and the highest was 97; hence, the lowest class must begin at 46 or less, and the highest must end at 97 or more. We can arbitrarily choose the number of classes and their size so long as the entire data range is included. We shall choose 9 classes of 6 each beginning at 45.5. We can now make a table which shows the class, the tally, and the frequency in each class.

Class	Tally	Frequency
45.5 to 51.5	||	2
51.5 to 57.5	|||	3
57.5 to 63.5	|||	3
63.5 to 69.5	++++	5
69.5 to 75.5	++++ ++++	10
75.5 to 81.5	++++ ||	7
81.5 to 87.5	||||	4
87.5 to 93.5	||||	4
93.5 to 99.5	||	2
Total:		40

The largest and smallest values in a class are called the *class boundaries* and their difference is called the *class width*. Half of the sum of the class boundaries is called the *class mark*.

At times it is not practicable to use all items of data. The term *population* is used to indicate the totality of all items. If it is decided not to use the entire population, there are methods for obtaining a sample. One is called a *random sample*; it is obtained so that each member of the population has an equal chance of being selected. The other is called a *selective sample* and it is obtained in the hope that all characteristics of the population are represented in proper proportion. This is the method used by various institutes of public opinion.

EXAMPLE

2. The class boundaries of the first class in the above table are 45.5 and 51.5, the class width is $51.5 - 45.5 = 6$, and the class mark is $\frac{1}{2}(45.5 + 51.5) = 48.5$.

19.2. Central Tendencies

Any quantity that locates or gives a central tendency or central value of a frequency or distribution is called an *average* or *measure of central tendency*.

There are several such quantities. There are advantages and limitations of each. The average that is the most desirable depends on the use to which the average is to be put or the reason for choosing one.

One average of a group of data or items is called the *arithmetic mean*; it is the sum of the items divided by the number of items. If a frequency distribution is used, the class mark is used as the value of each item.

EXAMPLE

1. If the weights of the linemen on a football team are 248, 201, 201, 220, 197, 208, and 202, find their arithmetic mean.

SOLUTION

We shall find the sum of these weights and then divide it by 7, since that is the number of items. The sum is

$$248 + 201 + 201 + 220 + 197 + 208 + 202 = 1477.$$

Consequently, the arithmetic mean is $1477/7 = 211$.

Another of the averages is called the *median*; it is the middle item if there is an odd number of items. It is the arithmetic mean of the two middle items if there is an even number. Hence, 202 is the median of the weights given in the example, since there are three items larger than 202 and the same number smaller than 202. Furthermore, the median of 3, 7, 9, and 14 is $\frac{1}{2}(7 + 9) = 8$.

If we want the median of data that have been put in classes, we use the class mark as the value of the item and count it the number of times indicated by the frequency. Consequently, the median of the grades collected in classes in the table given in Art. 19.1 is 72.5, since that is the class mark of the class that contains the middle items.

The third measure of central tendency is called the *mode*; it is the item which occurs most frequently. The mode of the data given in the example of this article is 201, since that item occurs more frequently than any other.

Exercise 19.1

Find the arithmetic mean, the median, and the mode of the numbers in each of Problems 1 through 4.

1. 23, 23, 29, 34, 36, 38, 41.

2. 64, 68, 71, 71, 71, 75, 77.

3. 574, 365, 688, 580, 1280, 2840, 2891, 1630.

4. 35, 42, 48, 50, 60, 63, 69, 74, 81.

5. The salaries of the employees of a business firm were $3300, $3700, $3700, $4100, $5300, $5900, $8500, and $13,000. Find the arithmetic mean, the median and the mode. Why are the mean and mode not representative averages in this problem?

6. The weights in ounces of the fish caught by two boys were 6, 6, 6, 8, 8, 11, 17, and 42. Find the mean, the median and the mode. Why are the mean and mode not representative averages?

7. The contributions of the members of a college department for the United Givers were $1, $15, $18, $28, $30, $35, $35, $35, and $100. Find the mean, the median, and the mode.

8. The incomes of 8 lawyers were $3800, $4300, $4300, $7000, $8600, $10,000, $14,000, and $72,000. Find the mean, the median and the mode. Why are the mode and the mean not representative averages?

Arrange the numbers in each of Problems 9 through 12 in frequency distribution as suggested, and then find the mean, the median, and the mode.

9. Use class widths of 5 beginning with 17: 17, 18, 21, 22, 22, 26, 27, 28, 29, 29, 30, 31, 32, 33, 33, 34, 38, 41, 42, 43, 43, 44.

10. Use class widths of 15 beginning with 18: 18, 18, 18, 19, 19, 23, 27, 29, 31, 32, 36, 36, 39, 42, 47, 48, 54, 63, 65, 73, 84.

11. Use class widths of $80 with the following salaries beginning with $510: $510 $525, $550, $592, $600, $607, $620, $700, $740, $744, $775, $775, $800, $830, $844.

12. Use class widths of 3 inches with the following heights; begin with 53 inches: 53, 54, 54, 56, 56, 57, 59, 59, 61, 61, 62, 63, 64, 65, 66, 66.

19.3. Measures of Dispersion

In the last article we discussed the arithmetic mean, the median, and the mode. They are measures of central tendency. It would be desirable to have a statistical measure which gives an indication of the extent to which the items are distributed about the average. There are several measures that do this by taking into consideration the amount each item deviates from the arithmetic mean. One of them is called the *standard deviation*. In order to give a formula for it, we shall let x_1, x_2, $\cdots$, x_n represent the items and M their arithmetic mean; then

$$\sigma = \sqrt{\frac{(x_1 - M)^2 + (x_2 - M)^2 + \cdots + (x_n - M)^2}{n}}$$

is the standard deviation of the items.

EXAMPLE

1. Find the standard deviation of the following grades made on a mathematics test: 20, 35, 43, 50, 56, 63, 67, 71, 73, 74, 74, 76, 78, 81, 83, 86, 89, 91, 94, 96.

SOLUTION

In order to be able to find the standard deviation, we must have the arithmetic mean. It is readily seen to be 70 by adding the items and dividing by the number of them, 20. We shall arrange the remainder of the work in the following table.

Item	*Item — Mean = Deviation*	*Square of Deviations*
20	—50	2500
35	—35	1225
43	—27	729
50	—20	400
56	—14	196
63	—7	49
67	—3	9
71	1	1
73	3	9
74	4	16
74	4	16
76	6	36
78	8	64
81	11	121
83	13	169
86	16	256
89	19	361
91	21	441
94	24	576
96	26	676
Sum: 1400	0	7850

Since there are 20 items and the sum of the squares of the deviations from the mean is 7850, the standard deviation is

$$\sigma = \sqrt{\frac{7850}{20}} = 19.8.$$

If the items are arranged in a frequency distribution, the class mark is used as the value of the item and the item is used as many times as it occurs in the frequency distribution.

EXAMPLE

2. Arrange the items that occur in Example 1 in class widths of 11 and then find the standard deviation.

SOLUTION

We must first make the frequency distribution and then compute the arithmetic mean as a preliminary to finding the standard deviation.

Class	20–30	31–41	42–52	53–63	64–74	75–85	86–96
Tally	l	l	ll	ll	⊞	llll	⊞
Frequency ..	1	1	2	2	5	4	5

We must now multiply the frequency of each class by the class mark, find the sum of these products, and divide by the number 20 of items in order to find the arithmetic mean. Thus the arithmetic mean M is

$$M = \frac{1(25) + 1(36) + 2(47) + 2(58) + 5(69) + 4(80) + 5(91)}{20}$$

$$= 69.55.$$

The remainder of the work of computing the standard deviation will be arranged in a table.

Class Mark	Frequency	Deviation	(Freq.) (Dev.)²	
25	1	−44.55	1(1984.7025)	= 1984.7025
36	1	−33.55	1(1125.6025)	= 1125.6025
47	2	−22.55	2(508.5025)	= 1017.0050
58	2	−11.55	2(133.4025)	= 266.8050
69	5	−0.55	5(0.3025)	= 1.5125
80	4	10.45	4(109.2025)	= 436.8100
91	5	21.45	5(460.1025)	= 2300.5125
			Sum	= 7132.9500

Consequently, $\sigma = \sqrt{\dfrac{7132.95}{20}} = 18.9.$

19.4 The Normal Curve

The graph of

$$y = Ce^{(-1/2)(x/\sigma)^2},$$

where e is approximately 2.718, x is the deviation of an item from the arithmetic mean, C is the value of y for $x = 0$, and σ is the standard deviation from the mean, is called the *normal frequency curve* and is shown in Fig. 19-1. The frequency distribution associated with the normal frequency curve is

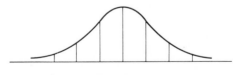

FIGURE 19-1
The normal frequency curve.

called a *normal distribution*. It is beyond the scope of this book to do so, but it can be shown that for a normal distribution with arithmetic mean M and standard deviation σ, about $\frac{2}{3}$ of the items come between $M - \sigma$ and $M + \sigma$, about 95% of the items come between $M - 2\sigma$ and $M + 2\sigma$, and about 99% of the items come between $M - 3\sigma$ and $M + 3\sigma$.

EXAMPLE

1. If the data of Example 1 of Art. 19.3 formed a normal distribution, then about $\frac{2}{3}$ of the items would be between $70 - 19.8 = 50.2$ and $70 + 19.8 = 89.8$, since $M = 70$ and $\sigma = 19.8$. Actually 13 of the 20 items are in that range. Furthermore, 19 items would be between $70 - 2(19.8) = 30.4$ and $70 + 2(19.8) = 109.6$, since 95% of 20 is 19. There are 19 in this range.

Note. In order for items or scores to be distributed according to a normal frequency curve, there should be a large number of items. There is no reason to expect that the grades of a single class will or should form a normal distribution.

Exercise 19.2

1. The average low temperatures of New York City for each month of 1961 beginning in January were 22°, 30°, 34°, 42°, 51°, 63°, 70°, 69°, 65°, 52°, 42°, and 30°. Find the standard deviation.

2. To the nearest $10, the per capita incomes in the 13 western states in 1961 were $1890, $1990, $1810, $2070, $1960, $1810, $2270, $2420, $1990, $2380, $2270, $3000, and $2780. Find the standard deviation.

3. The intelligence quotients of the members of an advanced mathematics class were 118, 121, 126, 128, 130, 132, 136, 142, 144, and 163. Find the standard deviation.

4. The incomes of a businessman on one of his projects for each year of a 10 year period were $3100, $3580, $3610, $3400, $3790, $3030, $3000, $3750, $3990 and $3650. Find the standard deviation.

5. The daily maximum Fahrenheit temperatures in New York City in January 1961 were:

43, 36, 35, 32, 34, 45, 53, 45, 29, 34, 44, 43, 47, 50, 50,
35, 45, 41, 26, 19, 21, 21, 22, 26, 19, 21, 20, 21, 28, 25, 29.

Divide the data into class widths of 7 beginning with 19, and find the standard deviation.

6. Put the following grades in class widths of 11, and find the standard deviation:

44, 53, 54, 58, 61, 63, 67, 70, 71, 73,
75, 78, 81, 83, 85, 88, 90, 92, 95, 98.

7. The salaries of the members of the mathematics department in a large university in 1963 in hundreds of dollars were:

62, 63, 67, 78, 81, 85, 93, 95, 99, 110, 113, 115,
120, 121, 125, 130, 145, 150, 180, and 180.

Find the standard deviation after arranging in class widths of 17.

8. The weights of the masculine members of a calculus class were:

123, 135, 141, 146, 153, 157, 162, 163, 169, 173,
175, 184, 187, 191, 198, 201, 214, 217, 234, 240.

Arrange in class widths of 17 beginning with 123 and find the standard deviation.

9. The arithmetic mean of the weights of the 9000 men registered in a university was 157 pounds and the standard deviation was 13. Describe the distribution by telling approximately the number whose weight is between 144 and 170, between 131 and 183, and between 118 and 196.

10. The arithmetic mean of the grades of the 3000 students on a mathematics placement test was 34 and the standard deviation was 11. Find the approximate number of grades between 23 and 45, between 12 and 56, and between 1 and 67.

11. The arithmetic mean of the weights of 2400 women students was 119 pounds with a standard deviation of 9 pounds. Approximately how many weighed between 110 and 128, between 101 and 137, and more than 146 or less then 92?

12. The arithmetic mean of the ages of the 1500 graduate students was 28 with a standard deviation of 3. About how many were between 25 and 31, over 34 or under 22, and between 19 and 37?

19.5. Simple Correlation

We shall now discuss and give a formula for a quantity which is called the *coefficient of correlation*. It is a measure of the relation between two sets of quantities or of the relation of both to a third quantity. A high degree of correlation between the grades in mathematics and French made by a boy might indicate that both sets of grades were closely related to the degree of intelligence and industry of the boy; however, a high degree of correlation between the grades made in trigonometry and in analytical geometry probably would indicate that there is a relationship between one's ability in the two subjects.

We shall consider two sets of items or quantities and represent them by $x_1, x_2, \cdots, x_n$ and $y_1, y_2 \cdots, y_n$. If the arithmetic means are M_x and M_y and the standard deviations are σ_x and σ_y, then

$$\sigma_x^2 = \frac{(x_1 - M_x)^2 + (x_2 - M_x)^2 + \cdots + (x_n - M_x)^2}{n}$$

and

$$\sigma_y^2 = \frac{(y_1 - M_y)^2 + (y_2 - M_y)^2 + \cdots + (y_n - M_y)^2}{n}.$$

These formulas can be expressed more compactly if we let

$$x_1 - M_x = X_1, \cdots, \qquad x_n - M_x = X_n,$$
$$y_1 - M_y = Y_1, \cdots, \qquad y_n - M_y = Y_n.$$

Thus
$$\sigma_x^2 = X_1^2 + X_2^2 + \cdots + X_n^2 = \frac{\sum X^2}{n}$$

and
$$\sigma_y^2 = \frac{Y_1^2 + Y_2^2 + \cdots + Y_n^2}{n} = \frac{\sum Y^2}{n}.$$

It is now possible for us to state that

$$r = \frac{X_1 Y_1 + X_2 Y_2 + \cdots + X_n Y_n}{n\sigma_x \sigma_y} = \frac{\sum XY}{n\sigma_x \sigma_y}$$

*is called the **coefficient of correlation** of the items,* x_1, x_2, $\cdots$, x_n *and the items* y_1. y_2, $\cdots$, y_n. Its value may be anywhere between $+1$ and -1. The correlation is said to be perfect if it is $+1$ or -1. If r is between $+1$ and -1, then the correlation between the two sets of items decreases as the absolute value of r decreases.

EXAMPLE

Find the correlation between the following heights and corresponding weights; the heights are given first in inches and each is followed by the corresponding weight in pounds:

64, 120; 65, 124; 66, 126; 67, 131; 68, 150;
69, 142; 70, 158; 71, 160; 72, 180; 73, 174.

Height	Weight	$X = x - M_x$	$Y = y - M_y$	XY	X^2	Y^2
64	120	-5	-30	150	25	900
65	124	-4	-26	104	16	676
66	126	-3	-24	72	9	576
67	131	-2	-19	38	4	361
68	150	-1	0	0	1	0
69	142	0	-8	0	0	64
70	158	1	8	8	1	64
71	160	2	10	20	4	100
72	180	3	30	90	9	900
73	174	4	24	96	16	576
74	185	5	35	175	25	1225
11 | 759	11 | 1650		Totals:	753	110	5442

$$M_x = 69 \qquad M_y = 150$$

$$\sigma_x{}^2 = \frac{110}{11} = 10, \quad \sigma_y{}^2 = \frac{5442}{11} = 494.73.$$

Now substituting these values in

$$r = \frac{\sum XY}{n\sigma_x \sigma_y}$$

gives

$$r = \frac{753}{11\sqrt{10}\sqrt{494.73}}$$

$$= \frac{753\sqrt{10}\sqrt{494.73}}{11(10)(494.73)}, \text{ rationalizing,}$$

$$= .97.$$

Consequently the given data indicate a high degree of correlation between height and weight.

19.6. Correlation with Data in Classes

The items are sufficiently numerous at times that it is desirable to put them in classes before finding the coefficient of correlation. If this is done each item is used as though its value were the class mark and each class mark is used as many times as it occurs. The procedure will be illustrated by an example.

EXAMPLE

Arrange the following data on corresponding ages and blood pressures into seven classes and then find the coefficient of correlation. The age is given first in each case

20, 112; 20, 125; 21, 120; 22, 127; 22, 138; 23, 126; 24, 127; 25, 130; 26, 130; 27, 132; 28, 131; 29, 145; 30, 132; 31, 138; 32, 137; 35, 134; 36, 138; 37, 169; 39, 140; 41, 140; 42, 141; 44, 140; 46, 140; 47, 107; 48, 144; 50, 162; 51, 110; 54, 142; 57, 143; 60, 144; 63, 122; 66, 150.

SOLUTION

Since we are to use seven classes, and the range in ages is from 20 to 66 inclusive, each class width should be about $\frac{1}{7}(66 - 20)$ but large enough to cover the entire range; hence, we shall use each as 7, beginning with 20. Furthermore, each class width for the blood pressures should be approximately $\frac{1}{7}(169 - 107)$ and we shall use 9, beginning with 107. The data will be arranged with the ages horizontal and blood pressures vertical and the tallies shown.

	20–26	27–33	34–40	41–47	48–54	55–61	62–68
107–115	I			I	I		
116–124	I						I
125–133	⫫I	III					
134–142	I	II	III	IIII	I		
143–151		I			I	II	I
152–160							
161–169		I			I		

This table will now be replaced by one that uses the class marks and shows each frequency. The row and the column headed by f_y and f_x, respectively, show the frequencies of the classes. For example, the 9 below the age 23 indicates there were 9 persons in the group with class mark 23 and the 6 in this column shows that 6 of them had a blood pressure with class mark 129. The arithmetic mean of the ages, to the nearest integer, is

	23	30	37	44	51	58	65	f_y	Y	$f_y Y^2$	YX
111	1			1	1			3	-24	1728	-96
120	1						1	2	-15	450	-180
129	6	3						9	-6	324	684
138	1	2	3	4	1			11	3	99	9
147		1			1	2	1	5	12	720	864
156								0	21	0	0
165			1		1			2	30	1800	360
f_x	9	6	4	5	4	2	2	32		5121	1641
X	-15	-8	-1	6	13	20	27				
$f_x X^2$	2025	384	4	180	676	800	1458	5527			
XY	1080	0	-39	-72	273	480	-81	1641			

$$M_x = \frac{9(23) + 6(30) + 4(37) + 5(44) + 4(51) + 2(58) + 2(65)}{32}$$

$$= 38.$$

That of the blood pressures is

$$M_y = \frac{3(111) + 2(120) + 9(129) + 11(138) + 5(147) + 2(165)}{32}$$

$$= 135.$$

The row headed by X gives the value of each $x - M_x$. Thus, the first entry $-15 = 23 - 38$. The column headed by Y gives the value of each $y - M_y$. The column and the row headed by $f_y Y^2$ and $f_x X^2$ give the indicated products. Each entry in the column headed by XY takes into account the value of Y, its frequency, the distribution of this frequency among the X's, and the values of those X's. Thus the fourth entry in the YX column is

$$9 = 3[1(-15) + 2(-8) + 3(-1) + 4(6) + 1(13)].$$

The entries in the row headed by XY are obtained in a similar manner. For example, the second entry in this row is

$$0 = -8[3(-6) + 2(3) + 1(12)].$$

The value of XY is computed as a row and as a column in order to have a check on the work.

After the table is completed, the computation of the coefficient of correlation is not difficult but does require that we find σ_x and σ_y. They come from

$$\sigma_x^2 = \frac{\Sigma f_x X^2}{n} = \frac{5527}{32} = 172.72$$

and
$$\sigma_y^2 = \frac{\Sigma f_y Y^2}{n} = \frac{5121}{32} = 160.03;$$

hence,
$$\sigma_x = 13.1 \quad \text{and} \quad \sigma_y = 12.7.$$

Therefore,

$$r = \frac{\Sigma XY}{n\sigma_x\sigma_y} = \frac{1641}{32(13.1)12.7} = .31, \text{ to two significant figures.}$$

Consequently, the given items do not indicate any appreciable correlation between age and blood pressure.

Exercise 19.3

1. Find the coefficient of correlation between the heights and weights of the masculine members of a mathematics department as given:

66, 125; 66, 170; 67, 160; 68, 140; 68, 160; 69, 185; 70, 170; 70, 160; 70, 145; 70, 140; 70, 155; 70, 160; 71, 160; 71, 175; 71, 150; 71, 155; 71, 145; 72, 180; 73, 170; 74, 190; 75, 210; 75, 175; 75, 185; 76, 160; 76, 225.

2. Find the correlation between the ages and blood pressures listed:

40, 140; 39, 137; 34, 148; 29, 125; 31, 136; 40, 144; 39, 152; 32, 141; 29, 136; 28, 130; 35, 138; 58, 147; 32, 137; 37, 143; 38, 131; 42, 141; 56, 141; 34, 147; 61, 127; 59, 115; 63, 142; 45, 123; 39, 133; 51, 139; 36, 132.

3. Find the correlation between the heights and weights of women as listed:

66, 140; 63, 153; 67, 160; 61, 105; 69, 170; 65, 133; 62, 116; 68, 146; 68, 131; 63, 139; 66, 162; 64, 136; 64, 127; 70, 152; 65, 119; 64, 133; 65, 121; 68, 146; 68, 130; 69, 163; 66, 132; 66, 119; 69, 165; 62, 97; 70, 180.

4. Find the correlation between the incomes in hundreds of dollars and shoe sizes of men as given:

30, 9; 32, 13; 40, 7; 43, 13; 49, 10; 50, 7; 53, 13; 58, 9; 63, 15; 67, 8; 71, 11; 79, 7; 84, 13; 87, 9; 92, 11; 98, 15; 99, 6; 103, 12; 106, 7; 110, 11; 110, 6; 115, 10; 123, 9; 129, 10; 134, 9.

5. The IQ of each of 30 students and the number of quality credits earned by each during his freshman year is listed; find the correlation between the two sets of items:

105, 19; 106, 20; 108, 15; 112, 30; 114, 27; 116, 35; 118, 30; 118, 36; 119, 23; 121, 34; 122, 50; 126, 28; 123, 46; 123, 60; 124, 48; 125, 37; 125, 64; 126, 39; 127, 63; 128, 70; 130, 67; 130, 43; 131, 73; 134, 50; 135, 78; 137, 95; 140, 80; 143, 67; 150, 83; 164, 90.

6. The first number in each pair below is the annual savings in tens of dollars of the man whose salary is the number of hundred dollars given by the second number; find the correlation between the two sets of numbers:

> 4, 20; 5, 22; 0, 27; 6, 30; 10, 35; 70, 36; 20, 44; 8, 41; 12, 63; 60,
> 48; 340, 55; 50, 60; 0, 65; 70, 70; 60, 80; 110, 82; 40, 83; 20,
> 87; 0, 90; 60, 95; 200, 100; 120, 123; 60, 120; 40, 120;
> 89, 125; 100, 130; 0, 132; 40, 136; 320, 140; 205, 142.

7. The following are the grades of a class of 30 students on the final examinations in calculus and in French; find the correlation between them:

> 49, 67; 56, 50; 58, 85; 58, 61; 59, 63; 62, 53; 62, 71; 64,
> 63; 64, 52; 65, 73; 67, 67; 68, 64; 68, 83; 69, 62; 70, 72;
> 78, 88; 75, 69; 76, 83, 77, 82; 78, 72; 79, 47; 86, 93; 84,
> 87; 86, 83; 87, 98; 90, 97; 93, 84; 95, 99; 99, 95; 99, 87.

8. The following are the number of hours spent in the student union building during a semester and the grade points earned during the semester by 30 students; find the correlation between the pair of sets of data:

> 2, 40; 7, 15; 10, 55; 14, 53; 16, 38; 21, 43; 24, 44; 25, 42;
> 17, 56; 32, 41; 39, 39; 42, 19; 44, 37; 50, 32; 53, 31; 57,
> 40; 60, 27; 63, 31; 66, 24; 70, 22; 76, 27; 80, 53; 85, 25;
> 87, 19; 90, 20; 93, 36; 103, 30; 117, 21; 127, 18; 140, 12.

Arrange the data in each of Problems 9 through 16 in class marks as suggested and find the coefficient of correlation. Use each mean to the nearest integer.

9. Use class marks of 3 inches for the heights beginning with 66 and class marks of 25 pounds for the weights beginning with 125 for the data in Problem 1.

10. Use the data of Problem 2 with class marks of 9 for the ages and 11 for the blood pressures beginning with 28 and 115, respectively.

11. Use the data of Problem 3 with class marks of 3 inches and 25 pounds beginning with the smallest entry of each.

12. Use class marks of 21 and 2 with the data of Problem 4 beginning with the smallest entry in each set.

13. Begin with the smallest entry for the IQ and for the number of grade points in Problem 5 and use class marks of 17 and 21.

14. Use class marks of 71 for savings and 25 for salary for the data of Problem 6 beginning with the smallest entry for each.

15. Use class marks of 11 for each set of grades given in Problem 7 beginning with the smallest entry of each.

16. Begin with the smallest entry for each, use class marks of 29 for the time in the student union building and 9 for the grade points with the data in Problem 8.

20

PERMUTATIONS, COMBINATIONS, PROBABILITY

20.1. Introduction

We shall deal with arrangements of a set of objects or events. Each object or event is called an *element*; furthermore, each collection of elements is called a *combination* regardless of the order in which they are arranged and each arrangement of the elements in a combination is called a *permutation*. Thus, *ab* and *ba* are two permutations of the same combination. Some of the elements may be alike, and we may use all or only a part of the elements in an arrangement. Much of our work will be based on the following principle:

If a first event can happen in h_1 ways and if, after it has occurred, a second can happen in h_2 ways, then the two events can happen in $h_1 h_2$ ways in the indicated order.

EXAMPLE

In how many ways can 2 people be picked from a group of 7?

SOLUTION

The first one may be picked in 7 ways, since there are 7 people. After one is selected, 6 remain, and the second may be chosen in 6 ways. Consequently, 2 people can be selected from 7 in $(7 \times 6) = 42$ ways.

The principle stated above can be extended to more than two events as follows: If after the first two events have happened, a third can happen in h_3 ways then the three can happen in $(h_1 h_2)h_3 = h_1 h_2 h_3$ ways. If then a fourth can happen in h_4 ways, the four can happen in $(h_1 h_2 h_3)h_4 = h_1 h_2 h_3 h_4$ ways. This can be extended event by event until we see that *n events can happen in* $h_1 h_2 h_3 \cdots h_n$ *ways in the order indicated by the subscripts.*

20.2. Factorial Notation

Many of the formulas and equations of this chapter will contain the product of all integers from some specified one down to the number 1. There is a concise notation used for such a product. It is written as $n!$, is equal to

$$n(n-1) \cdots (2)(1)$$

and is called *n factorial*. This description does not account for $0!$ or $1!$, and we define $0! = 1! = 1$.

EXAMPLES

1. $4! = (4)(3)(2)(1) = 24$.

2. $7! = (7)(6)(5)(4)(3)(2)(1) = 5040$.

If the factorial notation is used, we can write the product of several consecutive integers as a quotient. If we multiply and divide by $(n - r)!$, we see that

$$n(n-1) \cdots (n-r+1) = n(n-1) \cdots (n-r+1) \frac{(n-r)!}{(n-r)!}$$

$$= \frac{n!}{(n-r)!}$$

EXAMPLE

3. $$(9)(8)(7)(6) = (9)(8)(7)(6)\frac{5!}{5!} = \frac{9!}{5!}.$$

20.3. Permutations of n Different Elements Taken
r at a Time

We shall now develop a formula for the number of permutations of n different elements taken r at a time. The first place in the arrangement can be filled in n ways, and there are then $(n - 1)$ elements unused. The second place can be filled in $(n - 1)$ ways with $(n - 2)$ elements unused. Continuing this procedure, the rth place can be filled in $n - (r - 1) = (n - r + 1)$ ways. Now, using $P(n,r)$ as a symbol for the number of permutations of the n different elements taken r at a time, and applying the extension of the general principle, we see that

$$P(n, r) = n(n - 1)(n - 2) \cdots (n - r + 1) \frac{(n - r)!}{(n - r)!} = \frac{n!}{(n - r)!}$$

Consequently *the number of permutations of **n** different elements taken **r** at a time is*

(1)
$$\boldsymbol{P(n, r) = \frac{n!}{(n - r)!}}.$$

If we put n in place of r in (1), we see that

$$P(n, n) = \frac{n!}{(n - n)!} = \frac{n!}{0!} = n!;$$

hence, *the number of permutations of **n** different elements taken **n** at a time is*

(2)
$$\boldsymbol{P(n, n) = n!}.$$

E XAMPLES

1. How many baseball teams can a coach field from 13 men if each can play any position on the team?

SOLUTION

The number of teams is equal to the number of permutations of 13 different things taken 9 at a time, since there are 13 men and a team consists of 9 men. Therefore the number of teams is

$$P(13, 9) = \frac{13!}{4!}.$$

This can be expressed as a single number by dividing 4! into 13!. If this is done, we have

$$P(13, 9) = 259{,}459{,}200.$$

2. How many different batting orders can a manager have after he has selected the 9 players for a baseball game?

SOLUTION

The number of batting orders is equal to the number of permutations of 9 different elements taken 9 at a time; hence, it is

$$P(9, 9) = 9! = 362,880.$$

20.4. Permutations of n Elements That Are Not All Different

We saw in (2) of Art. 20.3 that $n!$ is the number of permutations of n different elements taken n at a time. If n_1 of the elements are alike, then each permutation will be repeated $P(n_1, n_1) = n_1!$ times, since for a given arrangement of the elements the n_1 like elements can be arranged (permuted) in $P(n_1, n_1)$ ways without changing the arrangement. Furthermore, if other sets of $n_2, n_3 \cdots, n_k$ are alike then each permutation will be repeated $n_2!, n_3!, \cdots, n_k!$ times. Now we are able to say that *the number of permutations of* n *elements taken* n *at a time, if* n_1 *are alike,* n_2 *others are alike,* $\cdots$, n_k *others are alike is*

$$\boldsymbol{P} = \frac{\boldsymbol{n}!}{\boldsymbol{n_1}!\boldsymbol{n_2}! \cdots \boldsymbol{n_k}!}.$$

EXAMPLE

1. How many permutations can be made from the letters of the word Mississippi taken all at a time?

SOLUTION

The letter m occurs once, i occurs 4 times, s occurs 4 times, and p is used twice for a total of $n = 11$ letters. Hence,

$$P = \frac{11!}{1!4!4!2!} = \frac{11 \cdot 10 \cdot 9 \cdot 8 \cdot 7 \cdot 6 \cdot 5 \cdot 4!}{1 \cdot 4 \cdot 3 \cdot 2 \cdot 1 \cdot 4! \cdot 2 \cdot 1} = 34,650.$$

Exercise 20.1

Find the value in each of Problems 1 through 4.

1. 6!   **2.** 8!   **3.** 11!   **4.** 1!/0!

5. How many 3-digit numbers can be formed from the integers 2, 5, 7 if no repetitions are permitted?

6. In how many ways can a penny, a dime, a quarter and a dollar be distributed among 4 boys if each boy gets 1 coin?

7. How many football teams can be formed from 20 boys if every boy can play each of the 11 positions?

8. How many licence plates can be formed if the inscription on each consists of a letter of the alphabet followed by a 3-digit number?

9. How many committees consisting of a plumber, a carpenter, and a plasterer can be selected from 15 plumbers, 18 carpenters, and 6 plasterers?

10. How many sets of 4 types of dogs can be selected from 7 wolfhounds, 10 poodles, 21 Mexican hairless, and 14 Scotch terriers?

11. How many 4-digit numbers greater than 7000 can be formed from 4, 5, 6, 7, 8, and 9 if repetitions are permitted?

12. How many committees consisting of a redhead, a blond, and a brunette can be selected from 7 redheads, 10 blonds, and 6 brunettes?

Evaluate the symbol in each of Problems 13 through 16.

13. $P(6, 4)$.   **14.** $P(8, 3)$.   **15.** $P(8, 5)$.   **16.** $P(9, 2)$.

17. Prove that $P(n, n-1) = 2P(n, n-2)$.

18. Prove that $P(n, n) = r! \, P(n, n-r)$,

19. Show that $P(n, n-r) \, P(r, r-1) = P(n, n)$.

20. Show that $P(n, r) \, P(n-r, n-r) = P(n, n)$.

21. Five people enter a room that contains 9 chairs. In how many ways can they be seated?

22. How many 3-digit numbers can be formed from 3, 5, 7, and 9 if no digit is repeated in any number?

23. In how many orders can six people be seated about a round table?

24. How many sets of allies of 3 nations each can be formed from 8 nations?

25. In how many ways can 6 people be seated on a bench?

26. How many radio call signals of 4 different letters can be formed from our alphabet?

27. In how many ways can a red king, a black jack, a ten, a nine, a four, and a three be selected from a deck of cards?

28. Six people enter a room that contains 4 chairs. In how many ways can the chairs be occupied?

29. There are 15 boys out for a baseball team. How many teams can be selected if 3 can pitch, 2 can catch, 5 can play any infield position, and 5 can play any outfield position?

30. A school has a vacancy in each of the English, physics and mathematics departments. In how many ways can the vacancies be filled if 5 persons are available for the English position, 3 for the physics opening, and 5 for the mathematics place?

31. In how many ways can 6 different algebra books, 4 different geometry books and 3 different trigonometry books be arranged on a shelf so that all books on a subject are together?

32. In how many ways can the letters in "Tennessee" be arranged? Those in "Illinois"?

33. In how many ways can a committee of 3 be selected from 5 juniors, 6 seniors, and 4 sophomores if it must contain one of each?

34. How many sums of 25 cents each can be formed from 4 dimes and 4 nickels?

35. How many basketball teams can be formed from 4 guards, 3 centers, and 5 forwards?

36. A restaurant lists 7 salads, 4 meats, 6 vegetables, and 5 desserts. In how many ways can a patron order a meal consisting of a salad, a meat, 2 vegetables and a dessert?

20.5. Combinations

We pointed out in Art. 20.1 that a set of elements is called a combination regardless of the arrangement of the elements in the set. Therefore, the six permutations $abc, acb, bac, bca, cab,$ and cba are only one combination since the same elements are used in each. We shall use $C(n, r)$ as a symbol for the number of combinations of n elements taken r at a time and develop a formula for it. The number of permutations of a combination of r elements taken r at a time is $r!$; hence,

$$P(n, r) = r! \, C(n, r).$$

Now dividing through by $r!$, we see that *the number of combinations of n distinct elements taken r at a time is*

(1)
$$C(n, r) = \frac{P(n, r)}{r!}.$$

The form of this equation can be changed by using the expression for $P(n\,r)$, as given by (1) of Art. 20.3. Thus,

$$C(n, r) = \frac{n!}{(n - r)!\,r!}.$$

EXAMPLE

1. How many committees of 5 members can be selected from a class of 31 people?

SOLUTION

A committee is the same regardless of the order in which the members are selected, hence, the number of committees is equal to the number combinations of 31 things taken 5 at a time. This is

$$C(31,5) = \frac{31!}{26!5!} = 169,911.$$

Exercise 20.2

1. Find the value of $C(7, 4)$ and of $C(12, 5)$.

2. Show that $C(10, 3) = C(10, 7)$.

3. Show that $2C(9, 5) = 3C(9, 3)$.

4. Show that $5\,C(11, 6) = 7C(11, 7)$.

5. In how many ways can 5 marbles be selected from 20?

6. In how many ways can a group of 20 soldiers be divided into details of 4?

7. A tennis squad is made up of 8 men. In how many orders can each play every other one?

8. In how many ways can a party of 12 fishermen be assigned to 3 boats if each boat holds 4 men?

9. In how many ways can 100 bags of flour be distributed among 96 needy families?

10. How many different bridge hands of 13 cards can be formed from a deck of 52 cards? Leave in factorial form.

11. In how many ways can a bridge hand that consists of 5 clubs, 3 diamonds, 3 hearts, and 2 spades be dealt from a deck of 52 cards?

12. Fifteen apples are to be distributed equally among 15 boys. In how many ways can the distribution be made?

13. In how many ways can 4 men stay in 3 rooms if no room can have more than 2 occupants?

14. In how many ways can 6 men stay in 4 hotels if no hotel can care for more than 2 men?

15. How many straight walks are needed to connect each of 10 buildings with each other one if no 3 are in line?

16. In how many ways can a combination of 4 cars be made from a Stanley Steamer, a Chrysler, a Ford, a Dodge, a Hupmobile, and a Maxwell?

17. In how many ways can 4 animals of the same type be selected from 6 aardvarks and 5 armadillos?

18. In how many ways can a group of 6 faculty members and 2 deans be selected from 20 faculty members and 5 deans?

19. A pack of dogs is made up of 3 setters and 7 others. How many hunting teams of 4 each can be formed from them if each contains two or less setters?

20. In how many ways can a group of 6 Irishmen and 5 Englishmen be divided into groups so that each group is made up of 3 Irishmen and 2 Englishmen?

21. A drawer contains 12 quarters, 10 half dollars, and 5 silver dollars. In how many ways can $1.50 be withdrawn?

22. In how many ways can a girl pay her friend 25 cents if she has 10 dimes and 20 nickels?

23. In how many ways can $1.50 be selected from a box that contains 6 silver dollars, 11 half dollars, and 13 quarters?

24. How many sums of money can be formed from a dollar, a half dollar, a quarter, a dime, a nickel, and a penny?

25. Nine pigs are to be distributed between boys, A, B, and C who can take 4, 3, and 2 pigs, respectively. In how many ways can the distribution be made?

26. A box contains 20 grapefruit, 30 oranges, and 40 apples. In how many ways can a boy be given 3 of each?

27. A kennel has 69 dogs as lodgers. They include 2 spitz, 4 hounds, and 3 pointers. In how many ways can a group made up of 1 spitz, 2 hounds, 2 pointers, and 5 other dogs be selected?

28. How many committees made up of 5 juniors and 4 seniors can be selected from 52 juniors and 46 seniors?

29. In now many ways can 7 men be seated in 3 rooms if there are 3 chairs in each room and if the order of seating in the separate rooms is not considered?

30. In how many ways can 9 people be put in 3 rooms if each can take exactly 3 people?

31. In how many ways can 3 red and 4 black balls be withdrawn from a bag of 5 red and 6 black balls by withdrawing 7 balls without returning any?

32. There are 5 white and 6 red balls in a bag. If 5 are withdrawn simultaneously, in how many ways can they be 3 white and 2 red balls?

20.6. Mathematical Probability

The mathematical probability that an event will happen is a ratio that gives an arithmetic value of the chances in favor of it happening. We shall put this in a more precise form in the following definition.

*If an event can happen in **h** ways and fail to happen in f ways, then the probability that it will occur is*

$$(1) \qquad\qquad p = \frac{h}{h + f},$$

and the probability that it will fail to happen is

$$(2) \qquad\qquad q = \frac{f}{h + f}$$

*provided each of the **h** and f ways is equally likely.*

Since neither h nor f can be negative, it follows that the largest value of p and of q is 1; furthermore, adding corresponding members of (1) and (2), we see that

$$p + q = \frac{h}{h + f} + \frac{f}{h + f} = \frac{h + f}{h + f} = 1.$$

EXAMPLES

1. If 6 balls are drawn from a bag that contains 7 black and 5 white balls, what is the probability that 4 will be black and 2 white?

SOLUTION

There are $C(12, 6)$ ways in which 6 balls can be drawn from a bag that contains 12; furthermore, 4 black balls can be drawn from 7 in $C(7, 4)$ ways and 2 white balls can be drawn from 5 in $C(5, 2)$ ways. Therefore, the probability of drawing the stated combination is

$$p = \frac{C(7, 4)C(5, 2)}{C(12, 6)} = \frac{\dfrac{7!}{3!4!} \times \dfrac{5!}{3!2!}}{\dfrac{12!}{6!6!}} = \frac{25}{66}.$$

2. Two cards are withdrawn from a deck of 52 playing cards. If the first is not replaced before the second is drawn, what is the probability that both will be kings?

SOLUTION

Since there are 4 kings among the 52 cards, there are 4 ways of drawing a king from the complete deck. After the first drawing, there are 3 kings among the remaining 51 cards. Hence, there are $(4)(3) = 12$ ways of drawing 2 kings in 2 draws.

Consequently, the desired probability is $(4)(3)/(52)(51) = 1/221$, since there are $(52)(51)$ ways of drawing two cards from a deck if the first is not replaced before the second is drawn.

This problem could have been worked by observing that 2 cards can be selected from the deck of 52 in $C(52, 2) = 52!/50!2!$ ways and 2 kings can be selected from the 4 kings in $C(4, 2) = 4!/2!2!$ ways. Consequently, the probability called for is

$$p = \frac{\dfrac{4!}{2!2!}}{\dfrac{52!}{50!2!}} = \frac{(4)(3)}{(52)(51)} = \frac{1}{221}$$

Exercise 20.3

1. What is the probability of a head in 1 toss of a coin?

2. If 2 coins are tossed simultaneously, find the probability that both will fall tails; one will be heads and the other tails.

3. If 3 coins are tossed, what is the probability that all will be heads? all tails? two tails and one heads?

4. If 4 coins are tossed, what is the probability that all will be heads? three heads and one tails? two heads and two tails?

5. One dice is thrown one time. Find the probability that a two will show; that an even number will show.

6. If 2 dice are thrown, what is the probability that a total of 11 will show? a total of 7? a total of 3?

7. If 2 dice are thrown, what is the probability that two sixes will show? a five and a four? a total of five?

8. Find the probability of getting a two in one toss of 2 dice; an eight; a four.

9. If 1 card is drawn from a deck of 52, find the probability that it will be a queen; a club; a red card.

10. If 3 cards are drawn simultaneously from a deck of 52, what is the probability that all will be tens?

11. Three cards are drawn one at a time from a deck of 52 and each is replaced before the next is drawn. What is the probability that all will be hearts? deuces? of the same suit?

12. A card is drawn from a deck of 52 and not replaced before another is drawn. Find the probability that both will be hearts, 1 a heart and the other a club, both red.

13. A box contains 3 red, 5 white, and 4 green balls, If 1 is drawn at random, find the probability it is red; white; green; not white.

14. A box contains 5 red and 7 green balls. If 2 balls are drawn simultaneously at random, find the probability that both will be red; both green.

15. A box contains 7 green, 5 pink, and 4 black balls. Two balls were drawn simultaneously at random, find the probability that both were pink; 1 pink and 1 black.

16. If 4 red and 4 green balls are in a bag and are drawn one at a time and placed in a line, what is the probability that the colors will alternate?

17. A box contains 12 tickets numbered from 1 to 12. If 2 are drawn at random and simultaneously, find the probability that their sum is even.

18. A committee of 4 persons is to be picked at random from 5 men and 3 women. What is the probability that no women will be on it? one man and three women?

19. Two cards are numbered 1, 2 are numbered 2, 2 are numbered 3, and 2 are numbered 4, and then all placed in a bag. The cards are drawn out by 4 men and their wives. Find the probability that the cards drawn by each man and by his wife bear the same number.

20. The numbers 1 through 26 are written on similar cards and 2 are withdrawn after all are placed in a bag. What is the probability that both are even numbers? one even and the other odd?

21. If 4 cards are drawn simultaneously at random from a deck of 52, what is the probability that all will be clubs? all will be deuces? all will be black?

22. If 13 cards are drawn simultaneously at random from a deck of 52, what is the probability that all are hearts?

23. What is the probability that a 3-digit number made up of digits chosen from 4, 5, 6, and 7 will be divisible by 3?

24. Numbers from 1 through 300 are written on small pieces of paper, put in a box, shaken up, and one is withdrawn at random. Find the probability that the sum of the digits in the number on it is 3; that the number on it contains all three of the digits 1, 2, and 3.

25. If 3 balls are drawn simultaneously at random from a box that contains 4 yellow balls, 5 green balls, and 6 orange ones, what is the probability that 2 or more will be orange?

26. A drawer contains six $1 bills, seven $5 bills, and five $10 bills. If 2 bills are drawn simultaneously at random, find the probability that the sum of the two will be $2; $15.

27. If 3 bills are drawn simultaneously at random from the drawer of Problem 26, find the probability that their sum is $3; $20.

28. If 3 balls are drawn simultaneously at random from a box that contains 4 red balls, 5 black ones and 6 green ones, find the probability that no two will be the same color.

20.7. Mutually Exclusive Events

A set of events are said to be *mutually exclusive* if the occurrence of one of them excludes the occurrence of another of the set on the same trial. If a ball is drawn from a bag that contains red, green, and black balls, it cannot be red or green if it is black; hence, the drawing of a red, a green, or a black ball are mutually exclusive events. We shall now state and prove a theorem concerning mutually exclusive events.

THEOREM. *The probability that some one of a set of mutually exclusive events will occur in a single trial is the sum of the probabilities that the separate events will occur.*

Proof. If the events are E_1, E_2, $\cdots$, E_r and can occur in h_1, h_2, $\cdots$, h_r ways, respectively, and if one trial can occur in n ways, then the probabilities of the various events occurring is

$$p_1 = \frac{h_1}{n}, \quad p_2 = \frac{h_2}{n}, \quad \cdots, \quad p_r = \frac{h_r}{n}.$$

Furthermore, the sum of the ways in which the r events can happen is $(h_1 + h_2 + \cdots + h_r)$. Now if p is the probability that some one of the r events will happen, we have

$$p = \frac{h_1 + h_2 + \cdots + h_r}{n}$$

$$= \frac{h_1}{n} + \frac{h_2}{n} + \cdots + \frac{h_r}{n}$$

$$= p_1 + p_1 + \cdots + p_r.$$

EXAMPLE

1. If a bag contains 3 white, 4 green, and 5 red balls, the probability of getting a white ball if one ball is drawn is $\frac{3}{12}$ and the probability of getting a red one is $\frac{5}{12}$; hence, the probability of getting a white one or red one is

$$\frac{3}{12} + \frac{5}{12} = \frac{8}{12} = \frac{2}{3}.$$

Furthermore, the probability of getting a green one is $\frac{4}{12} = \frac{1}{3}$. Therefore, the probability of getting a white one, a red one or a green one is $\frac{2}{3} + \frac{1}{3} = 1$. This is as it should be since 1 is the symbol for certainty in probability.

20.8. Independent Events

A set of events are said to be *independent* if the occurrence of one of them has no effect on the probability of the occurrence of any other one. If a bag contains red and green balls, and if one is drawn and replaced before another is drawn, the drawing of a red ball in the first trial and a green ball in the second trial are independent events. The following theorem will enable us to work with independent events.

THEOREM. *The probability that all of a set of independent events will occur is the product of their separate probabilities.*

Proof. If there are two events E_1 and E_2, and if they can occur in h_1 and h_2 ways, respectively, and fail to happen in f_1 and f_2 ways, respectively, then the two events can occur in $h_1 h_2$ ways, and occur and fail to happen in $(h_1 + f_1)(h_2 + f_2)$ ways. Consequently, the probability p_{12} that both will occur is

$$p_{12} = \frac{h_1 h_2}{(h_1 + f_1)(h_2 + f_2)} = \frac{h_1}{h_1 + f_1} \times \frac{h_2}{h_2 + f_2}$$
$$= p_1 p_2.$$

We may now consider this as a single event, add another, and calculate the probability of all three occurring. In fact, we can add one event at a time until we have a total of r events and calculate the probability of all occurring after each has been added. If this is done and if $p_1, p_2, \cdots, p_r$ are the probabilities of the separate events, we will find that

$$p_{123 \cdots r} = p_1 p_2 p_3 \cdots p_r$$

as stated in the theorem.

EXAMPLE

1. If the probability that Jones will get a new car is .42, that Smith will plant tomatoes is .75, and that Burbank will get married is $\frac{2}{3}$; then the probability that all three events will take place is

$$p_{123} = (.42)(.75)\left(\frac{2}{3}\right) = .21.$$

20.9. Dependent Events

A set of events are said to be *dependent* if the occurrence of one of them affects the probability of the occurrence of the next. If a bag contains red and green balls and if one is drawn and not replaced, the probability of getting a ball of a specified color on the second drawing is affected by the first drawing. The following theorem will be useful in working with dependent events.

THEOREM. *If the probability that event* E_1 *will happen is* p_1; *if after it has happened, the probability that an event* E_2 *will happen is* p_2; *if, after they have occurred the probability that an event* E_3 *will come about is* p_3; $\cdots$; *if, after events* E_1 *to* E_{r-1} *have happened, the probability that an event* E_r *will occur is* p_r; *then the probability that all events will occur in the order indicated by the subscripts is the product* $p_1p_2p_3 \cdots p_r$.

Proof. If event E_1 can happen in h_1 ways and fail to occur in f_1 ways and if, after E_1 has come about, E_2 can occur in h_2 ways and fail to happen in f_2 ways, then the two events can occur in the order E_1E_2 in h_1h_2 ways, and they can occur or fail to happen in $(h_1 + f_1)(h_2 + f_2)$ ways. Consequently, the probability that they will occur in the order E_1E_2 is

$$p_{12} = \frac{h_1h_2}{(h_1 + f_1)(h_2 + f_2)}$$

$$= \frac{h_1}{h_1 + f_1}\frac{h_2}{h_2 + f_2} = p_1p_2.$$

The same argument used in connection with independent events can be used here to extend the proof to r events and give the statement made in the theorem.

EXAMPLE

1. A bag contains 3 white, 4 black, and 5 red balls. If 3 balls are withdrawn and none replaced, find the probability that the first 2 will be black and the third red.

SOLUTION

Since there are 4 black balls and a total of 12, the probability of getting a black ball on the first draw is $\frac{4}{12}$. After a black ball is drawn, there are 11 balls in the bag, and 3 are black; hence, the probability of getting a black ball on the second draw is $\frac{3}{11}$. After this, there are 10 balls in the bag. and 5 are red; hence, the probability of drawing a red ball is $\frac{5}{10}$. Consequently, the probability of drawing a black ball, a black ball, and a red ball in that order is

$$\frac{4}{12} \times \frac{3}{11} \times \frac{5}{10} = \frac{1}{22}.$$

Exercise 20.4

1. The probability that Tom will marry Sue is $\frac{1}{5}$, and the probability that he will marry Norma is $\frac{1}{4}$. What is the probability that he will marry one of them?

2. Determine the probability that a 7 or a 6 will be the sum of the dots showing if a pair of dice are tossed.

3. The probability that Sam's pig will win first place at a certain show is $\frac{2}{7}$ and the probability that George's pig will win first place at the same show is $\frac{3}{11}$. What is the probability that one of the two will win?

4. The probability that Alta will have her hair dyed red is $\frac{1}{8}$ and the probability that she will have it dyed blue is $\frac{1}{11}$. What is the probability that she will have it dyed red or blue?

5. The probabilities that two candidates for governor will be elected are $\frac{1}{3}$ and $\frac{1}{6}$. Find the probability that one of them will be elected.

6. Find the probability that neither a five nor a six will be thrown in one toss of a pair of dice.

7. The junior class in a small high school consists of 40 students. Of these, five are named Guidry and three are named Sanchez. What is the probability that the representative on the student council is named Guidry or Sanchez if he is chosen by lot?

8. If the probabilities that Tulane, Auburn, and Vanderbilt will win the Southeastern Conference championship are $\frac{1}{30}$, $\frac{1}{6}$, and $\frac{1}{35}$, find the probability that one of them will win it.

9. The probability that Abramowitz will be elected president of the Kiwanis Club in Detroit is $\frac{2}{3}$ and the probability that it will rain in Hoboken the day of the election is $\frac{5}{11}$. Find the probability that both will take place.

10. The probability that Mrs. Cassius will be chosen as president of her bridge club is $\frac{2}{3}$ and the probability that Mr. Cassius will be made chairman of the board of his organization is $\frac{1}{2}$. What is the probability that both will be chosen?

11. Sauer, Simpkins, and Soileau shoot at a target in the order listed above. What is the probability that each will be the first to hit the bull's-eye if the probability of each one's hitting it in one shot is $\frac{3}{5}$, $\frac{1}{3}$, and $\frac{3}{4}$, respectively?

12. One box contains 5 black and 3 white balls and another contains 7 black and 5 white balls. If one ball is drawn from each box, find the probability that both will be black; white; the same color.

13. What is the probability of throwing a total of 7 for 3 consecutive throws of 2 dice?

14. Two balls are drawn successively from a box that contains 4 white and 6 red balls. If the first ball is replaced before the second is drawn, find the probability that both balls will be red.

15. A die and a coin are tossed. What is the probability of obtaining a five and a tail?

16. A bag contains 5 red and 7 black balls. If a ball is drawn and replaced, then another drawing is made, find the probability that both will be black.

17. The probability of Broussard living to age 47 is .50 and the probability that he will be a grandfather if he does live to age 47 is .68. What is the probability that he will live to age 47 and be a grandfather?

18. The probability that Gremillion will be in the runoff for the Democratic nomination for governor is $\frac{3}{5}$, and the probability that he will win if he is in it is $\frac{2}{5}$. What is the probability that he will be elected governor?

19. A card is drawn from a deck of 52 and not replaced until after a second is drawn. What is the probability that both will be kings?

20. Three college friends agree that if all three or any two of them are alive in 20 years they will meet for a reunion. Find the probability that the meeting will take place if the probabilities of the friends living 20 years are $\frac{4}{5}$, $\frac{7}{10}$, and $\frac{5}{7}$.

21. A bag contains 6 white and 3 black balls. What is the probability that the first one will be black and the second white if 2 balls are drawn from the bag? Assume that the first is not replaced before the second is drawn.

22. The probability that Gustafson will be appointed chairman of a certain department is $\frac{5}{11}$. The probability that he will be select Guillot as his assistant if appointed is $\frac{2}{5}$. What is the probability that Guillot will be assistant to the chairman?

23. If 3 balls are drawn simultaneously from a bag that contains 3 pink, 5 yellow, and 8 green balls, find the probability that all are pink. Two are pink and one yellow. No two are the same color.

24. What is the probability of drawing 13 cards of one suit if 13 are drawn from a deck of 52.

20.10. Repeated Trials

If we know the probability that an event will occur in any one trial, we can find the probability that it will occur exactly r times in n independent trials by use of the following theorem.

THEOREM. *If p is the probability that an event will occur in any one trial, then $C(n, r)p^r(1 - p)^{n-r}$ is the probability that it will occur exactly r times in n independent trials.*

Proof. The probability that an event will happen in each of r trials and fail in each of $n - r$ trials is $p^r(1 - p)^{n-r}$, since occurring in one trial and failing to occur in another are independent events; furthermore, the r trials may be selected from the n trials in $C(n, r)$ ways. Consequently we have the statement given in the theorem.

EXAMPLE

1. Find the probability that exactly 3 fours will come up in 8 tosses of a die.

SOLUTION

Since the probability that a four will show in one toss is $p = \frac{1}{6}$, it follows that $1 - p = \frac{5}{6}$, furthermore $n = 8$ and $r = 3$. Consequently, the desired probability is

$$C(n, r)p^r(1 - p)^{n-r} = C(8, 3)\left(\frac{1}{6}\right)^3\left(\frac{5}{6}\right)^5$$

$$= \frac{8!}{5!3!} \times \frac{5^5}{6^8} = \frac{21,875}{209.952}.$$

We can find the probability of an event's occurring at least r times in n independent trials if we know the probability of its occurring in a single trial by a repeated use of the theorem given above. The probabilities that it will happen exactly r times, exactly $(r + 1)$ times, exactly $(r + 2)$ times, $\cdots$, exactly $(n - 1)$ times, and exactly n times in n trials are probabilities of mutually exclusive situations; hence, the theorem given above enables us to say that *the probability that an event will happen at least **r** times in **n** independent trials is*

$$C(n, r)p^r(1 - p)^{n-r} + C(n, r + 1)p^{r+1}(1 - p)^{n-r-1} +$$

$$\cdots + C(n, n - 1)p^{n-1}(1 - p)^1 + C(n, n)p^n,$$

*provided **p** is the probability that it will occur in any one trial.*

EXAMPLE

2. Find the probability of throwing at least 3 heads in 5 tosses of a coin.

SOLUTION

The desired probability is the sum of the probabilities of throwing exactly 3, exactly 4, and exactly 5 heads. In this problem, $p = 1 - p = \frac{1}{2}$; hence, the desired probability is

$$C(5, 3)\left(\frac{1}{2}\right)^3\left(\frac{1}{2}\right)^2 + C(5, 4)\left(\frac{1}{2}\right)^4\left(\frac{1}{2}\right)^1 + C(5, 5)\left(\frac{1}{2}\right)^5$$

$$= \frac{5!}{3!2!} \times \frac{1}{2^5} + \frac{5!}{4!1!} \times \frac{1}{2^5} + \frac{5!}{5!0!} \times \frac{1}{2^5} = \frac{1}{2}.$$

Exercise 20.5

1. If a coin is tossed 5 times, find the probability of throwing exactly 3 tails; at least 3 heads.

2. If 4 coins are tossed, what is the probability of getting 2 heads? at least two tails?

3. If 5 coins are tossed, what is the probability of obtaining at least 4 tails?

4. What is the probability of throwing exactly 4 tails in 6 tosses of a coin? at least 4 heads?

5. If a die is cast 7 times, find the probability of obtaining a six 3 or fewer times.

6. Find the probability of throwing exactly 2 fives if a die is cast 3 times; at least 2 fives.

7. If 2 dice are cast 4 times, what is the probability of getting exactly 2 tens? at least 2 tens?

8. If 2 dice are cast 5 times, what is the probability of getting exactly 2 sevens? at most 2 sevens?

9. If 7 cards are drawn from a deck of 52, and each is replaced before the next is drawn, find the probability of drawing exactly 5 clubs; at least 5 clubs.

10. If 5 cards are drawn from a deck of 52, and each is replaced before the next is drawn, what is the probability of drawing exactly 4 spades? at least 4 spades?

11. If 4 cards are drawn from a deck of 52, and each is replaced before the next is drawn, find the probability of drawing exactly 3 aces; at least 3 aces.

12. If 4 cards are drawn from a deck of 52 and each is replaced before the next is drawn, what is the probability of getting exactly 2 red cards? At least 2 black ones?

13. A bag contains 12 white, 8 black, and 4 red balls. Balls are drawn one at a time and each is replaced before the next is drawn. What is the probability of drawing 3 white balls in 5 drawings? three or more?

14. What is the probability of drawing exactly 3 red balls in 5 withdrawals with the conditions and data of Problem 13?

15. A bag contains 3 green, 4 purple, and 5 yellow balls. A ball is withdrawn, its color determined, then replaced. This is done five times. Find the probability that all five will be purple.

16. In Problem 15, find the probability that exactly 2 will be purple.

17. The probability that the champion archer in a girls' physical education class will hit the bull's eye in one shot is $\frac{3}{4}$. What is the probability that she will hit it exactly 3 times in 4 shots.

18. The probability that a gambler will win if he plays poker is $\frac{2}{5}$. What is the probability that he will win 3 times if he plays 5 times?

19. The probability that a football team will win on any Saturday is $\frac{2}{3}$. Find the probability that it will win 4 in a row and then lose the fifth.

20. If the probability of a person of age 17 living to age 48 is $\frac{3}{4}$, find the probability that exactly 4 out of 10 people aged 17 will live for 31 years.

21. If the probability that you will make an A in this course is $\frac{1}{7}$, find the probability that you and exactly one of two of your friends will all make A. Assume that each of them has the same chance for an A that you have.

22. The probability of John having a date with Hazel if he asks her is $\frac{2}{3}$. What is the probability that he will get exactly 3 dates with her if he asks her 4 times?

23. Seven students of equal ability work separately on a problem. If the probability of each solving it is $\frac{4}{5}$, what is the probability that exactly 4 of them will solve it?

24. The probability that Oscar will make a grade point average of 2.6 or better any semester is $\frac{3}{4}$. What is the probability of his making a 2.6 or better average 3 semesters in succession?

21

A GLIMPSE AT THE CALCULUS

21.1. Introduction

We studied constants, variables, and functions in Chapter 2 and shall now continue to develop the concepts treated there. In particular, we shall be interested in what happens to one variable as a related one behaves in a specified manner and in the comparative rates of change of two variables. If a variable gets closer and closer to a constant, we say that the variable approaches that constant. This statement is not dependent on whether the variable is ever equal to the constant. The only requirement is that the variable shall get as close as anyone wishes to the constant and, once having gotten within a specified distance, does not back off. For example, the variable $S(n) = 1/2 + 1/4 + \ldots + 1/2^n$ can be made as near 1 as anyone wishes by taking n sufficiently large. In fact $S(n)$ differs from 1 by $1/2^n$ for all positive integral values of n.

21.2. The derivative

If (x, y) is any point on a curve $y = f(x)$ and if $(x + \Delta x, y + \Delta y)$ is any other point on it, then Δx* is a change in x and $\Delta y = f(x + \Delta x) - f(x)$ is the corresponding change in y. The change Δx in x may be positive or negative and is often referred to as the increment of x. The increment Δy of y may also be of either sign. The ratio $\Delta y / \Delta x$ is the average rate of change of y relative to or with respect to x. There are conditions under which this ratio approaches a definite value as Δx approaches zero but we shall not investigate them. We indicate the value, if any, that $\Delta y / \Delta x$ approaches as Δx approaches zero by writing

$$\lim_{\Delta x \to 0} \frac{\Delta y}{\Delta x}.$$

If $y = x^2$, then $y + \Delta y = (x + \Delta x)^2$ and $y = (x + \Delta x)^2 - x^2 = 2x\Delta x + \overline{\Delta x}^2$

Consequently,

$$\frac{\Delta y}{\Delta x} = \frac{2x\Delta x + \overline{\Delta x}^2}{\Delta x} = 2x + \Delta x.$$

Therefore,

$$\lim_{\Delta x \to 0} \frac{\Delta y}{\Delta x} = \lim_{\Delta x \to 0} (2x + \Delta x) = 2x.$$

In arriving at this value, we assumed that *the limit of a sum is the sum of the limits*. We shall make further use of this assumption.

We are now in position to give an exceptionally important definition. It is:

If

$$\lim_{\Delta x \to 0} \frac{\Delta y}{\Delta x} = \lim_{\Delta x \to 0} \frac{f(x + \Delta x) - f(x)}{\Delta x}$$

*exists, it is called the **derivative of y $= f(x)$ with respect to x**.* The notations dy/dx, $D_x y$, y' and $f'(x)$ are used to indicate the derivate of $y = f(x)$ with respect to x.

21.3. Geometric Interpretation

In order to find a geometric significance for the derivative, we shall begin by sketching the graph of $y = f(x)$. We shall let $P(x, y)$ be any point on the curve and $Q(x + \Delta x, \ y + \Delta y)$ any other point on it. Then the slope of the

* This symbol is read delta x and is not a product.

secant PQ is $\Delta y/\Delta x$, since the slope of a line is the tangent of its angle of inclination. If we now recall from plane geometry that the limiting position, if any, of a secant is the tangent, we see that *the derivative can be interpreted geometrically as the slope of the tangent to the curve.*

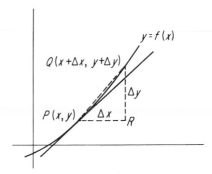

FIGURE 21-1

21.4. Finding the Derivative by the Δ-Process

We shall now outline a procedure for finding the derivative of $y = f(x)$ by use of the Δ-process. In so doing, we must keep in mind that we are trying to evaluate

$$\frac{dy}{dx} = \lim_{\Delta x \to 0} \frac{\Delta y}{\Delta x} = \lim_{\Delta x \to 0} \frac{f(x + \Delta x) - f(x)}{\Delta x}.$$

The procedure is to

1. *Write $y = f(x)$ for the function under consideration.*

2. *Find $y + \Delta y = f(x + \Delta x)$ by replacing x by $x + \Delta x$ each time it occurs in $f(x)$.*

3. *Subtract y from $y + \Delta y$ and thus obtain $\Delta y = f(x + \Delta x) - f(x)$.*

4. *Divide each member of the equation in 3 by Δx and thus obtain the fraction whose limit we want.*

5. *Find the limit as Δx approaches zero of the quotient obtained in 4.*

EXAMPLE

1. Find the derivative of $y = f(x) = x^3 + 2x$ and evaluate it for $x = 4$.

SOLUTION

In finding the derivative, we shall follow the steps outlined above. Thus, we have
1. $y = f(x) = x^3 + 2x$, the given function,
2. $y + \Delta y = f(x + \Delta x) = (x + \Delta x)^3 + 2(x + \Delta x)$, replacing x by $x + \Delta x$,
    $= x^3 + 3x^2(\Delta x) + 3x(\Delta x)^2 + (\Delta x)^3 + 2x + 2\Delta x$, expanding
3. $\Delta y = f(x + \Delta x) - f(x)$
    $= x^3 + 3x^2(\Delta x) + 3x(\Delta x)^2 + (\Delta x)^3 + 2x + 2\Delta x - x^3 + 2x$
    $= 3x^2(\Delta x) + 3x(\Delta x)^2 + (\Delta x)^3 + 2\Delta x$

4. $\dfrac{\Delta y}{\Delta x} = \dfrac{3x^2(\Delta x) + 3x(\Delta x)^2 + (\Delta x)^3 + 2\Delta x}{\Delta x}$, dividing by Δx,

$= 3x^2 + 3x(\Delta x) + (\Delta x)^2 + 2$, removing the common factor.

5. $\displaystyle\lim_{\Delta x \to 0} \dfrac{\Delta x}{\Delta y} = \lim_{\Delta x \to 0} \ [3x^2 + 3x(\Delta x) + (\Delta x)^2 + 2]$, indicating the limit,

$= 3x^2 + 2$, taking the limit

Consequently, we have found that

$$f'(x) = D_x(x^3 + 2x) = 3x^2 + 2.$$

In order to evaluate this derivative for $x = 4$, we need only replace x by 4 in $f'(x)$. Thus, we get $f'(4) = 3(4)^2 + 2 = 50$.

EXAMPLE

2. Find the slope of the curve represented by $y = f(x) = 3x^2 - 2$ at $(2, 10)$.

SOLUTION

We shall follow the steps outlined above.
1. $y = f(x) = 3x^2 - 2$, the given equation,
2. $y + \Delta y = f(x + \Delta x) = 3(x + \Delta x)^2 - 2$, replacing x by $x + \Delta x$,
3. $\Delta y = f(x + \Delta x) - f(x)$
    $= 3(x + \Delta x)^2 - 2 - [3x^2 - 2]$
    $= 6x(\Delta x) + (\Delta x)^2$

4. $\dfrac{\Delta y}{\Delta x} = \dfrac{6x(\Delta x) + (\Delta x)^2}{\Delta x}$, dividing by Δx,

$= 6x + \Delta x$, removing the common factor.

5. $\displaystyle\lim_{\Delta x \to 0} \dfrac{\Delta y}{\Delta x} = \lim_{\Delta x \to 0} \ (6x + \Delta x)$, indicating the limit,

$= 6x$, taking the limit.

Therefore, the formula for the slope is $6x$ and its value for $x = 2$ is $6(2) = 12$.

EXAMPLE

3. Find the value of the derivative of $y = f(x) = \sqrt{2x - 1}$ for $x = 5$.

SOLUTION

If we follow the usual procedure, we have
1. $y = f(x) = \sqrt{2x - 1}$, the given equation,
2. $y + \Delta y = f(x + \Delta x) = \sqrt{2(x + \Delta x) - 1}$, replacing x by $x + \Delta x$,

3. $\Delta y = \sqrt{2(x + \Delta x) - 1} - \sqrt{2x - 1}$

$$= \frac{\sqrt{2(x + \Delta x) - 1} - \sqrt{2x - 1}}{1} \cdot \frac{\sqrt{2(x + \Delta x) - 1} + \sqrt{2x - 1}}{\sqrt{2(x + \Delta x) - 1} + \sqrt{2x - 1}},$$

rationalizing the numerator,

$$= \frac{2(x + \Delta x) - 1 - (2x - 1)}{\sqrt{2(x + \Delta x) - 1} + \sqrt{2x - 1}}, \text{ multiplying,}$$

$$= \frac{2\Delta x}{\sqrt{2(x + \Delta x) - 1} + \sqrt{2x - 1}}, \text{ collecting terms.}$$

Now combining steps 4 and 5, we have

$$\lim_{\Delta x \to 0} \frac{\Delta y}{\Delta x} = \lim_{\Delta x \to 0} \frac{2\Delta x}{[\sqrt{2(x + \Delta x) - 1} + \sqrt{2x - 1}]\Delta x}$$

$$= \lim_{\Delta x \to 0} \frac{2}{\sqrt{2(x + \Delta x) - 1} + \sqrt{2x - 1}}, \text{ removing the common}$$

factor,

$$= \frac{1}{\sqrt{2x - 1}}$$

Therefore, $f'(x) = 1/\sqrt{2x - 1}$ and its value for $x = 5$ is $f'(5) = 1/\sqrt{10 - 1} = \frac{1}{3}$.

21.5. Velocity

If y in $y = f(x)$ represents distance and x represents time, then Δy represents the distance traveled in the time Δx; hence, $\Delta y/\Delta x$ is the average velocity during the time Δx. Consequently, taking the limit as the increment Δx of time approaches zero, we see that dy/dx *represents instantaneous velocity if y is distance and x is time.* Quite often, distance is represented by s, time by t, and velocity by v.

EXAMPLE

1. If distance in feet is given by $s = 16t^2$ and time in seconds by t, find the velocity at the end of 3 seconds.

SOLUTION

1. $s = 16t^2$, the given equation,
2. $s + \Delta s = 16(t + \Delta t)^2$, replace t by $t + \Delta t$,
3. $\Delta s = 16(t + \Delta t)^2 - 16t^2$, $f(t + \Delta t) - f(t)$,
    $= 32t(\Delta t) + 16(\Delta t)^2$

4. $\dfrac{\Delta s}{\Delta t} = \dfrac{32t(\Delta t) + 16(\Delta t)^2}{\Delta t}$, dividing by Δt,

$\qquad = 32t + 16(\Delta t)$, removing the common factor,

5. $\displaystyle \lim_{\Delta t \to 0} \dfrac{\Delta s}{\Delta t} = \lim_{\Delta t \to 0} [32t + 16(\Delta t)]$, indicating the limit,

$\qquad = 32t$, taking the limit.

Hence, the velocity is $v = 32t$ in terms of t. Consequently, the velocity at the end of 3 seconds is $v = (32)(3) = 96$.

Exercise 21.1

Find the derivative with respect to x of the function given in each of Problems 1 through 8.

1. $y = x^2 + 3$.

2. $y = x^2 - 5$.

3. $y = 2x^2 + x$.

4. $y = 3x^2 - x$.

5. $y = x^3 - 3x^2$.

6. $y = x^3 + 2x$.

7. $y = x^3 - 2x - 1$.

8. $y = x^3 - 3x + 5$.

Evaluate dy/dx for the indicated value of x in each of Problems 9 through 12.

9. $y = x^2 - 3x + 1$, $x = 2$.

10. $y = x^2 + 5x - 4$, $x = 3$.

11. $y = x^3 + 2x^2 - 6$, $x = -1$.

12. $y = x^3 - 3x^2 + 9$, $x = -2$.

Find the slope of the curve in each of Problems 13 through 16 for the given value of x.

13. $y = x^3 - 5x^2 + 4$, $x = 3$.

14. $y = x^3 + 2x^2 - x$, $x = -1$.

15. $y = 2x^2 - 3x + 5$, $x = -2$.

16. $y = 3x^2 - 5x + 1$, $x = 3$.

In each of Problems 17 through 20, find the velocity for the given value of t if distance is given by s.

17. $s = 3t + 5$, $t = 4$.

18. $s = 16t^2 - 5t$, $t = 3$.

19. $s = 2t^3 - 5t^2$, $t = 2$.

20. $s = t^3 - 4t^2 - 8$, $t = 2$.

Find $f'(x)$ in each of Problems 21 through 28.

21. $y = \sqrt{x}$.

22. $y = \sqrt{1 - x}$.

23. $y = \sqrt{x^2 + 1}$.

24. $y = \sqrt{x^2 - 4x}$.

25. $y = \dfrac{1}{x}$.

26. $y = \dfrac{1}{x^2}$.

27. $y = \dfrac{2}{x-1}$.

28. $y = \dfrac{3}{(2x+1)}$

21.6. Some Differentiation Formulas

Heretofore, we have obtained derivatives by use of the Δ-process. We shall now use that fundamental procedure to obtain some formulas for use in differentiation. We shall first find that the derivative of a constant is zero. If $y = c$, c a constant, then $y + \Delta y = c$. Consequently, $\Delta y = c - c = 0$, and $\Delta y / \Delta x = 0$. Therefore, $\lim\limits_{\Delta x \to 0} \dfrac{\Delta y}{\Delta x} = \lim\limits_{\Delta x \to 0} 0 = 0$ and we know that

(1) $$\frac{dc}{dx} = 0, \qquad c \text{ a constant.}$$

This is as one might expect since a constant does not change and the derivative is the rate of change of one quantity relative to another.

We shall now find a formula for the derivative of x^n, n a positive integer, by use of the Δ-process. Thus, using the usual steps,

(1) $\qquad y = x^n$, the given equation,

(2) $y + \Delta y = (x + \Delta x)^n$, replacing x by $x + \Delta x$,
$\qquad\quad = x^n + nx^{n-1}(\Delta x) + n(n-1)x^{n-2}(\Delta x)^2 + \cdots + (\Delta x)^n$, expanding

by the binomial theorem,*

(3) $\qquad \Delta y = nx^{n-1}(\Delta x) + n(n-1)x^{n-2}(\Delta x)^2 + \cdots + (\Delta x)^n$

(4) and (5) $\quad \lim\limits_{\Delta x \to 0} \dfrac{\Delta y}{\Delta x} = \lim\limits_{\Delta x \to 0} \left[nx^{n-1}\dfrac{\Delta x}{\Delta x} + n(n-1)x^{n-2}\dfrac{(\Delta x)^2}{\Delta x} + \right.$

$$\left. \cdots + \frac{(\Delta x)^n}{\Delta x} \right],$$

dividing by Δx and indicating the limit as Δx approaches zero,

$$= \lim\limits_{\Delta x \to 0} [nx^{n-1} + n(n-1)x^{n-2}(\Delta x) + \cdots + (\Delta x^{n-1})],$$

removing the common factor Δx from numerator and denominator,

$$= \lim\limits_{\Delta x \to 0} nx^{n-1} + \lim\limits_{\Delta x \to 0} [n(n-1)x^{n-2}(\Delta x) + \cdots + (\Delta x)^{n-1}],$$

using the limit of a sum equal to the sum of the limits,
$= nx^{n-1}$, each term in brackets contains a zero factor.

---

* If not familiar with the binomial theorem, see page 297 of F. W. Sparks and P. K. Rees, *College Algebra* (New York: McGraw-Hill Book Company, 1961.)

We now have shown that

(2)
$$\frac{dx^n}{dx} = nx^{n-1}, \qquad \textbf{\textit{n} an integer.}$$

This formula is true for n any constant and will be so used even though it is beyond the scope of this chapter to prove the statement.

EXAMPLE

1. Find the derivative with respect to x of x^6.

SOLUTION

If we apply the differentiation formula just derived we see that $D_x x^6 = 6x^5$.

We shall now find the derivative of a constant times a differentiable function of x. If we employ the usual steps of the Δ-process, and represent the function by u, we have

(1) $y = cu$, the given function,

(2) $y + \Delta y = c(u + \Delta u)$, giving each variable an increment,

(3) $\Delta y = c\Delta u$

(4) and (5) $\displaystyle\lim_{\Delta x \to 0} \frac{\Delta y}{\Delta x} = \lim_{\Delta x \to 0} \frac{c\Delta u}{\Delta x}$, dividing by Δx and indicating the limit

$\displaystyle = \lim_{\Delta x \to 0} c \lim_{\Delta x \to 0} \frac{\Delta u}{\Delta x}$, *assuming* that the limit of a product is the product of the limits,

$$= cD u_x u$$

Consequently, we know that *the derivative of a constant times a differentiable function is the constant times the derivative of the function* and write

(3)
$$\frac{dcu}{dx} = c\,\frac{du}{dx}$$

EXAMPLE

2. Find the derivative of $7x^4$.

SOLUTION

If we apply the formula just derived, we have

$$\begin{aligned}
D_x 7x^4 &= 7D_x x^4 \\
&= (7)4x^3, \; D_x x^n = nx^{n-1}, \\
&= 28x^3
\end{aligned}$$

We shall continue our derivation of differentiation formulas by finding the derivative of a sum. If we represent the functions by u and v and use the Δ-process, we have

(1) $y = u + v$, the given function,

(2) $y + \Delta y = u + \Delta u + v + \Delta v$, giving each function an increment,

(3) $\Delta y = \Delta u + \Delta v$

(4) and (5) $\lim\limits_{\Delta x \to 0} \dfrac{\Delta y}{\Delta x} = \lim\limits_{\Delta x \to 0} \left(\dfrac{\Delta u}{\Delta x} + \dfrac{\Delta v}{\Delta x} \right)$

$\qquad = \lim\limits_{\Delta x \to 0} \dfrac{\Delta u}{\Delta x} + \lim\limits_{\Delta x \to 0} \dfrac{\Delta v}{\Delta x}$, assuming that the limit of a sum

$\qquad$ is the sum of the limits,

$\qquad = D_x u + D_x v$

Therefore, we know that *the derivative of a sum is the sum of the derivatives* and write

(4) $$\frac{d(u + v)}{dx} = \frac{du}{dx} + \frac{dv}{dx}$$

If we make use of the four differentiation formulas obtained in this article, we can find the derivative of any polynomial since such a function is made up of the sum of terms of the form cx^n.

EXAMPLES

3. Find the derivative of $f(x) = 3x^4 - 5x^3 - 7x^2 + 8x - 13$, and evaluate it for $x = 2$.

SOLUTION

$\qquad f'(x) = D_x(3x^4 - 5x^3 - 7x^2 + 8x - 13)$
$\qquad\qquad = D_x 3x^4 - D_x 5x^3 - D_x 7x^2 + D_x 8x - D_x 13$, by 4
$\qquad\qquad = 3 D_x x^4 - 5 D_x x^3 - 7 D_x x^2 + 8 D_x x - 0$, by 3 and 1,
$\qquad\qquad = (3)4x^3 - (5)3x^2 - (7)2x + 8$, by 2,
$\qquad\qquad = 12x^3 - 15x^2 - 14x + 8$

Therefore, $f'(2) = 12(2^3) - 15(2^2) - 14(2) + 8 = 96 - 60 - 28 + 8 = 16$.

4. Find the value of x for which the derivative of $f(x) = x^2 - x - 2$ is zero and the intervals for which it is negative and positive.

SOLUTION

First we shall find the derivative. It is

$$f'(x) = 2x - 1$$

Consequently,

$$f'(x) = 0 \text{ for } x = \tfrac{1}{2},$$
$$f'(x) < 0 \text{ for } x < \tfrac{1}{2}, \text{ and}$$
$$f'(x) > 0 \text{ for } x > \tfrac{1}{2}.$$

Exercise 21.2

Find the derivative of the function in each of Problems 1 through 8.

1. $y = x^3 + 2x^2 + 3.$
2. $y = x^4 - 7x^2 + 2x.$

3. $y = 2x^5 - 3x^4 + 7x + 11.$
4. $y = 3x^7 - 7x^3 - 19x + 4.$

5. $y = x^2 + \dfrac{3}{x}.$
6. $y = x + \dfrac{1}{x} - \dfrac{2}{x^2}.$

7. $y = 2x^3 - \sqrt{x}.$
8. $y = \sqrt[3]{x} + \dfrac{2}{\sqrt{x}}.$

Find the value of the derivative for the specified value of x in each of Problems 9 through 12.

9. $y = x^2 - 3x + 8\sqrt{x}, x = 4.$
10. $y = x^{-1} + x^2 + 6\sqrt[3]{x}, x = 1.$

11. $y = 5x - \dfrac{16}{\sqrt{x}}, x = 4.$
12. $y = 24\sqrt[3]{x} - \dfrac{96}{\sqrt[3]{x}}, x = 8.$

Find the slope of the curve at the given point in each of Problems 13 through 16.

13. $y = x^3 - 6\sqrt[3]{x}, (1, -5).$
14. $y = 3x^2 - 20\sqrt{x}, (4, 8).$

15. $y = x - \dfrac{8}{x}, (2, -2).$
16. $y = x^3 - 27x^{-2}, (3, 24).$

Find the velocity in each of Problems 17 through 20 for the sepcified value of t if the distance is as given by s.

17. $s = t^2 - 2t + 6\sqrt{t}, t = 9.$
18. $s = t^3 - \dfrac{4}{t} + \dfrac{4}{t^2}, t = 2.$

19. $s = 12\sqrt[3]{t} + \dfrac{128}{t} + 7, t = 8.$
20. $s = t^{-2} + \dfrac{6}{\sqrt{t}}, t = 1.$

Find the value or values of x for which the derivative is zero in each of Problems 25 through 32. Find the range on x for which the slope is negative and for which it is positive.

21. $y = x^2 - 2x + 3.$
22. $y = x^2 + 4x + 5.$

23. $y = x^2 - 3x + 1.$
24. $y = 2x^2 + 5x + 6.$

25. $y = 2x^3 - 9x^2 + 12x + 1.$ **26.** $y = x^3 - 3x^2 - 9x - 4.$

27. $y = 4x^3 + 9x^2 - 12x + 7.$ **28.** $y = 2x^3 - x^2 - 4x + 3.$

21.7. Maxima and Minima of Functions

We shall now be interested in determining points that are higher than any other points in the vicinity and on a curve. Such points are called *relative maxima*. We can find the point readily if we know its abscissa. The point A (Fig. 21-2) is a relative maximum since it is higher than any other point on the curve and in the vicinity of A. If the curve is rising on the left of and sufficiently near A, has zero for a derivative at A and is falling to the right of and sufficiently near A, then A is higher than any other point on the curve and sufficiently near A. Consequently, A is a relative maximum point. If we now make use of the interpretation of a derivative as the slope of the curve along with the slope being the tangent of the angle of inclination and the tangent of an angle being positive or negative according to whether the angle is acute or obtuse, we can give the following procedure for finding relative maxima. If $y = f(x)$ is the equation of a curve to be tested for relative maximum, then

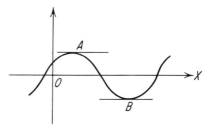

FIGURE 21-2

1. *Find* $f'(x)$. This gives a general expression for the slope.

2. *Set* $f'(x) = 0$ *and solve for* x. The values obtained are called *critical values of* x. They include all possible abscissas of relative maxima and perhaps other values of x.

3. *Find the value of* f' *for* x *less than and greater than the value for which* $f' = 0$ *but arbitrarily near that value.* This can be accomplished by evaluating $f'(x_1 - \varepsilon)$ and $f'(x_1 + \varepsilon)$ where $f'(x_1) = 0$ and $\varepsilon > 0$ is arbitrarily small.

4. *If* $f'(x_1 - \varepsilon) > 0, f'(x_1) = 0,$ *and* $f'(x_1 + \varepsilon) < 0,$ *then* $x = x_1$ *gives a relative maximum for* $y = f(x)$.

A point that is lower than any other point on a curve and in the vicinity of the point is called a *relative minimum*.

An argument similar to the one given above can be used to develop a procedure for finding any relative minimum that may exist. The first three steps are the same as those used in finding maxima and the fourth step is: *If* $f'(x_1 - \varepsilon) < 0,$ $f'(x_1) = 0,$ *and* $f'(x_1 + \varepsilon) > 0,$ *then* $x = x_1$ *gives a relative minimum value of* $y = f(x)$. The point B in Fig. 21.2 is a relative minimum since it is lower than any other point in its vicinity and on the curve.

If $f'(x_1) = 0$ and $f'(x_1 - \varepsilon)$ and $f'(x_1 + \varepsilon)$ are both positive or both negative, then $x = x_1$ does not give a maximum or minimum since in both cases there are points near $[x_1, f(x_1)]$ that are higher or lower than it.

EXAMPLE

1. Find the critical values of x if $y = f(x) = x^2 - 6x + 3$. Classify them as to maximum, minimum or neither.

SOLUTION

If
$$y = f(x) = x^2 - 6x + 3, \text{ then}$$
$$y' = f'(x) = 2x - 6$$
$$= 0 \text{ for } x = 3.$$

Consequently, $x = 3$ is the only critical value. We shall see if it gives a maximum, a minimum, or neither for $f(x)$. By substituting, we see that

$$f'(3 - \varepsilon) = 2(3 - \varepsilon) - 6 = -2\varepsilon < 0 \text{ for } \varepsilon > 0$$

and

$$f'(3 + \varepsilon) = 2(3 + \varepsilon) - 6 = 2\varepsilon > 0 \text{ for } \varepsilon > 0.$$

Therefore, there is a minimum value of $f(x)$ for $x = 3$ since the derivative changes from negative to zero to positive as x increases through the value for which the derivative is zero.

EXAMPLE

2. Find the values of x for which

$$y = f(x) = 4x^3 - 9x^2 - 12x + 5$$

is a maximum or minimum.

SOLUTION

If
$$y = f(x) = 4x^3 - 9x^2 - 12x + 5 \text{ as given, then}$$
$$y' = f'(x) = 12x^2 - 18x - 12$$
$$= 6(2x + 1)(x - 2)$$
$$= 0 \text{ for } x = -\frac{1}{2}, 2.$$

Therefore, we must test each of these values to see if it gives a maximum, a minimum, or neither for $f(x)$. Thus,

$$f'\left(-\frac{1}{2}-\varepsilon\right) = 6(-1-2\varepsilon+1)\left(-\frac{1}{2}-\varepsilon-2\right)$$

$$= 6(-2\varepsilon)\left(-\frac{1}{2}-\varepsilon-2\right) > 0$$

since 6 is positive and the other two factors are negative; furthermore,

$$f'\left(-\frac{1}{2}+\varepsilon\right) = 6(-1+2\varepsilon+1)\left(-\frac{1}{2}+\varepsilon-2\right)$$

$$= 6(2\varepsilon)(\varepsilon-2.5) < 0,$$

since the first two factors are positive and the last is negative. Therefore, $x = -\frac{1}{2}$ gives a maximum value for $f(x)$.

We shall now test $x = 2$ by the usual procedure. Thus,

$$f'(2-\varepsilon) = 6(4-2\varepsilon+1)(2-\varepsilon-2)$$
$$= 6(5-2\varepsilon)(-\varepsilon) < 0$$

and

$$f'(2+\varepsilon) = 6(4+2\varepsilon+1)(2+\varepsilon-2)$$
$$= 6(5+2\varepsilon)\,\varepsilon > 0.$$

Consequently, $f(x)$ has a minimum value for $x = 2$.

21.8. The Second Derivative

The derivative of a function was defined in Art. 21.2, and we shall now define the second derivative. The derivative of the derivative is called the *second derivative* and is denoted by $D_x^2 y$, $D_x^2 f$, $f''(x)$, y'', d^2y/dx^2, or d^2f/dx^2 if the function is $y = f(x)$.

EXAMPLE

1. If $\qquad f(x) = 2x^3 + x^2 - 4x + 6$, find $f'(x)$ and $f''(x)$.

SOLUTION

If we apply the formulas for the derivative of a polynomial, we find that

$$f'(x) = 6x^2 + 2x - 4$$

and

$$f''(x) = D_x f'(x) = 12x + 2$$

We can get a physical interpretation of the second derivative if we recall that the rate of change of velocity with respect to time is called acceleration just as the rate of change of distance with respect to time is called velocity.

Thus, if the distance from some selected point is given by $s(t)$, it not only follows that velocity is equal to ds/dt as pointed out in Art. 21.5 but also that $dv/dt = d^2s/dt^2$ is acceleration.

EXAMPLE

    2. If distance is given by $s(t) = t^3 - 4t^2 - 3$, find the acceleration for $t = 3$.

SOLUTION

    If we take the derivative of s with respect to t, we find that $v = ds/dt = 3t^2 - 8t$. Therefore,

$$\frac{dv}{dt} = \frac{d}{dt}(3t^2 - 8t)$$

$$= 6t - 8$$

is a formula for the acceleration. Consequently, the acceleration for $t = 3$ is $6(3) - 8 = 10$.

    In the last article, we found that $f(x)$ has a maximum for $x = x_1$ if $f'(x_1 - \varepsilon) > 0, f'(x_1) = 0$, and $f'(x_1 + \varepsilon) < 0$. Therefore, $f'(x)$ is a decreasing function; consequently, its derivative $f''(x)$ is negative in an interval about $x = x_1$. We can now state that *if $f'(x_1) = 0$ and $f''(x_1) < 0$, then $f(x)$ has a maximum for $x = x_1$.* By a similar argument, we can show that *if $f'(x_1) = 0$ and $f''(x_1) > 0$, then $f(x)$ has a minimum for $x = x_1$.*

EXAMPLE

    3. Find the values of x for which

$$y = f(x) = 4x^3 - 9x^2 - 12x + 5$$

is a maximum or minimum by use of the second derivative test.

SOLUTION

    All values of x for which $f(x)$ can be a maximum or a minimum are included among the roots of $f'(x)$; hence, we must find it. It is

$$f'(x) = 12x^2 - 18x - 12$$
$$= 6(2x + 1)(x - 2)$$

and is zero for $x = -\frac{1}{2}, 2$. We next find $f''(x)$ and evaluate it for $-\frac{1}{2}$ and 2. Thus,

$$f''(x) = 24x - 18,$$
$$f''(-\tfrac{1}{2}) = 24(-\tfrac{1}{2}) - 18 = -30 < 0,$$
$$f''(2) = 24(2) - 18 = 30 > 0.$$

Therefore, in keeping with the tests given above, we know that $f(x)$ has a maximum for $x = -\frac{1}{2}$ and a minimum for $x = 2$. This problem is the one solved as Example 2 in Art. 21.7 where we reach the same conclusions as here.

Exercise 21.3

Find the velocity and acceleration for the given value of t if s is as given in each of Problems 1 through 4.

1. $s(t) = t^3 - 12t + 5$, $t = 3$.

2. $s(t) = t^3 - 4t^2 - 3$, $t = 4$.

3. $s(t) = t^4 - 5t^3 + 13t$, $t = 2$.

4. $s(t) = t^5 - 3t^4 + 2t^3 - 4t$, $t = 2$.

Find the values of x for which $f(x)$ has a maximum or minimum in each of Problems 5 through 20. Use the first derivative test in Problems 5 through 12 and the second derivative test in 13 through 20.

5. $f(x) = x^2 - 4x + 2$.

6. $f(x) = x^2 - x + 2$.

7. $f(x) = -x^2 + 3x + 2$.

8. $f(x) = -x^2 + 5x + 3$.

9. $f(x) = 2x^3 - 9x^2 + 12x + 1$.

10. $f(x) = x^3 - 3x^2 - 9x + 7$.

11. $f(x) = -4x^3 + 9x^2 - 6x + 5$.

12. $f(x) = -x^3 + 2x^2 + 4x + 4$.

13. $f(x) = 2x^3 + 3x^2 - 12x + 4$.

14. $f(x) = 2x^3 - 9x^2 - 24x + 7$.

15. $f(x) = 4x^3 - x^2 - 4x + 3$.

16. $f(x) = x^3 - 4x^2 - 3x - 2$.

17. $f(x) = -x^3 - 3x^2 + 9x - 1$.

18. $f(x) = -2x^3 - 3x^2 + 36x - 41$.

19. $f(x) = -x^3 - x^2 + x + 17$.

20. $f(x) = -4x^3 + 3x^2 + 6x + 11$.

21.9. Applications of Maxima and Minima

In the last article, we saw how to find the maxima and minima values, if any, of a function. We shall now apply those principles to problems that arise in engineering, physics, and business. After the quantity that is to be a maximum or minimum is expressed as a function of some variable that enters into the problem, we proceed as in the last article; hence, our first task is to *express the quantity that is to be a maximum or minimum in terms of a variable that enters in the problem.* At times this is more readily done by expressing the desired quantity in terms of two variables, then finding a second relation between them. If this second relation is solved for one of the variables and the value thus found is substituted in the expression for the desired quantity, it is then in terms of one variable.

EXAMPLE

1. What positive number added to its reciprocal gives a minimum?

SOLUTION

If we represent the number by x, then its reciprocal is $1/x$. Consequently, the quantity $Q(x)$ that is to be a minimum is

$$Q(x) = x + \frac{1}{x}$$

$$= x + x^{-1}$$

Therefore, we must have

$$Q'(x) = 1 - \frac{1}{x^2}$$

$$= \frac{x^2 - 1}{x^2}$$

$$= 0$$

It is zero for $x = \pm 1$ but we can not use -1, since the problem calls for a positive number. We now know that if there is a number which satisfies the conditions of the problem it is 1 and must see if $Q(x)$ is a minimum for $x = 1$. We shall use the first derivative test. Thus,

$$Q'(1 - \varepsilon) = \frac{(1 - \varepsilon)^2 - 1}{(1 - \varepsilon)^2}$$

$$= \frac{1 - 2\varepsilon + \varepsilon^2 - 1}{(1 - \varepsilon)^2}$$

$$= \frac{\varepsilon(\varepsilon - 2)}{(1 - \varepsilon)^2} < 0,$$

since $\varepsilon > 0$, $\varepsilon - 2 < 0$, and $(1 - \varepsilon)^2 > 0$

$$Q'(1 + \varepsilon) = \frac{(1 + \varepsilon)^2 - 1}{(1 + \varepsilon)^2} > 0$$

since $(1 + \varepsilon)^2 > 0$. Therefore, the derivative changes from negative to zero to positive and the function is a minimum for $x = 1$.

EXAMPLE

2. What must be the ratio of the height h to radius r of a tin can (Fig. 21-3) in order to have the largest possible can from a given amount of material?

SOLUTION

The volume of the can is to be a maximum; hence, we must obtain an expression for it in terms of a variable that enters in the problem. We know that the volume is

(1) $V = \pi r^2 h.$

Since this is in terms of two variables, we must get another relation between them. This relation can be found by making use of the fact that the amount of material used in making the can is fixed; i.e., is a constant. The total surface of the can is made up of the two ends and the sides. Therefore,

$$(2) \qquad S = 2\pi r^2 + 2\pi rh$$

is a second relation between r and h. If solved for h, we have

$$h = \frac{S - 2\pi r^2}{2\pi r}$$

This for h in the expression for V gives

FIGURE 21-3

$$V(r) = \pi r^2 \left(\frac{S - 2\pi r^2}{2\pi r} \right)$$

$$= \frac{r}{2}(S - 2\pi r^2) \quad \text{removing common factors}$$

$$= \frac{1}{2} Sr - \pi r^3;$$

hence,

$$(3) \qquad V'(r) = \frac{1}{2} S - \pi 3 r^2$$

since S and π are constants.
Now $V'(r) = 0$ for

$$3\pi r^2 - \frac{1}{2} S = 3\pi r^2 - \frac{1}{2}(2\pi r^2 + 2\pi rh)$$

$$= 3\pi r^2 - \pi r^2 - \pi rh$$
$$= 2\pi r^2 - \pi rh$$
$$= \pi r(2r - h)$$
$$= 0 \text{ for } h = 2r$$

Therefore, if there is a maximum volume that can be obtained from a given amount of material it occurs for $h = 2r$. This can be seen to be a maximum from physical considerations but we shall be more formal by using the second derivative test. Thus, if we use $V'(r)$ as given in (3), we find that

$$V''(r) = 0 - 3\pi 2r, \quad S \text{ is a constant,}$$
$$= -6\pi r < 0$$

Therefore, the volume is a maximum if $h = 2r$.

Exercise 21.4

1. Divide 12 into two parts such that the sum of the squares of the parts is a minimum.

2. Divide 16 into two parts such that their product is a maximum.

3. Divide 24 into two parts such that the product of one and the cube of the other is as large as possible.

4. Separate 20 into two parts such that the cube of one part plus the square of the other is as small as possible.

5. What must be the relation between length and width of a rectangle so that the area is a maximum for a given perimeter?

6. What must be the relation between length and width of a rectangle so as to have a minimum perimeter for a specified area?

7. An isosceles triangle is inscribed in a semicircle of radius 5 inches. Find the base length for the maximum area if the vertex between the equal sides is at the center of the circle. The equation of the circle is $x^2 + y^2 = 25$.

8. The base of a right triangle is 10 inches and its altitude is 6 inches. Find the area of the largest rectangle that can be inscribed in the triangle if a side of the rectangle is along the base of the triangle.

9. A box is formed by cutting squares of equal size from the corners of a square piece of cardboard that is 12 inches on each side and then turning up the edges. What should be the edge of each square that is removed in order for the box to have a maximum volume?

10. Find the dimensions of a topless tank with a volume of 256 cubic feet and a square base if the amount of material used in constructing it is a minimum.

11. What are the coordinates of the point on $y = 4x - x^2$ that is closest to $(-1, 4)$?

12. Show that the largest rectangle that can be inscribed in the circle $x^2 + y^2 = a^2$ is a square.

13. A rectangular garden has one edge against a barn, but the other sides must be fenced. If 120 feet of fence are available, what should be the dimensions of the garden in order to have a maximum area?

14. Find the dimensions of the rectangular gutter of largest cross-sectional area that can be made from a piece of metal that is 12 inches wide provided it is made by turning up equal strips along each edge.

15. A gardener wants to enclose a rectangular plot and divide it into three equal pieces by cross fences that are parallel to the ends. What dimensions would enclose the largest area if the gardener has 720 feet of fence?

16. Show that a cylindrical tin can of specified volume with bottom but no top requires a minimum amount of material if its height is equal to the radius.

17. A group of students want to go on a special train to a ball game. The understanding is that the train will run if as many as 250 students sign up and that the cost will be $14 each if exactly 250 decide to go. It is further understood that the price will be reduced 4 cents for each student beyond 250 who goes. How many should sign up in order for total income to be largest?

18. If the speed of a point on a flywheel t seconds after starting is $s = -t^4 + 50t^2$, find the value of t for whicn s is a maximum.

19. The height of a ball t seconds after being thrown vertically upward was $h = 128t - 16t^2$. When was it the highest? How high did it rise?

20. If a manufacturer charges x per suit, he can sell $2005 - 30x$ of them per month. The cost of operating the plant for a month is $2400 plus $25x. Find the price that he should charge per suit to make a maximum profit. How many suits should he manufacture?

21. A Norman window consists of a rectangle surmounted by a semicircle. What should be the relation between base and altitude of the rectangle in order for the area of the window to be a maximum for a given perimeter?

22. A tank with top is made up of a hemisphere surmounted by a right circular cylinder. Determine the relation between radius and height of the cylinder that makes a maximum volume of the tank for a given amount of material.

23. A sheet of paper is to contain 21 square inches of printed material and is to have a 1 inch margin at the top and bottom and .75 inch at each side. Determine the dimensions of the page that will require the least amount of paper.

24. A figure is made up of a rectangle surmounted by an equilateral triangle, and one side of the triangle coincides with the upper base of the rectangle. How long should the side of the triangle be for a maximum area of the entire figure if its perimeter is $(24 - 4\sqrt{3})$ inches?

21.10. Derivative of a Product, a Quotient, and u^n

We shall obtain three new differentiation formulas in this article. They are for the derivative of a product, of a quotient, and of a function of a function. We shall first let u and v be differentiable functions of x and consider

(1) $y = uv$, the given function,

(2) $y + \Delta y = (u + \Delta u)(v + \Delta v)$, giving each function an increment,

(3) $\Delta y = (u + \Delta u)(v + \Delta v) - uv$, (2) — (1),

         $= u\Delta v + v\Delta u + \Delta u\Delta v$, expanding and collecting,

(4) and (5) $$\lim_{\Delta x \to 0} \frac{\Delta y}{\Delta x} = \lim_{\Delta x \to 0} \left(u \frac{\Delta v}{\Delta x} + v \frac{\Delta u}{\Delta x} + \frac{\Delta u}{\Delta x} \Delta v \right)$$

$$= uD_xv + vD_xu + (D_xu)(0)$$

$$= uD_xv + vD_xu$$

We now know that

$$D_x(uv) = uD_xv + vD_xu$$

and can put it in words by saying that the *derivative of the product of two functions is the first times the derivative of the second plus the second times the derivative of the first.*

EXAMPLE

1. Find the derivative of $(2x^3 - 1)(4x^2 - 5x)$ without expanding.

SOLUTION

The given function is a product and, differentiating it as such, we have

$$\frac{d(2x^3 - 1)(4x^2 - 5x)}{dx} = (2x^3 - 1) \frac{d(4x^2 - 5x)}{dx} + (4x^2 - 5x) \frac{d(2x^3 - 1)}{dx}$$

$$= (2x^3 - 1)(8x - 5) + (4x^2 - 5x)6x^2, \quad \text{differentiating,}$$
$$= 16x^4 - 10x^3 - 8x + 5 + 24x^4 - 30x^3, \quad \text{expanding,}$$
$$= 40x^4 - 40x^3 - 8x + 5, \quad \text{collecting,}$$

We shall now consider

(1) $$y = \frac{u}{v}, \quad \text{the given function,}$$

(2) $$y + \Delta y = \frac{u + \Delta u}{v + \Delta v}, \quad \text{giving each function an increment,}$$

(3) $$\Delta y = \frac{u + \Delta u}{v + \Delta v} - \frac{u}{v}, \quad (2) - (1),$$

$$= \frac{(u + \Delta u)v - (v + \Delta v)u}{v(v + \Delta v)}, \quad \text{getting a common denominator,}$$

$$= \frac{v\Delta u - u\Delta v}{v(v + \Delta v)}, \quad \text{expanding and collecting,}$$

(4) and (5) $\lim\limits_{\Delta x \to 0} \dfrac{\Delta v}{\Delta x} = \lim\limits_{\Delta x \to 0} \dfrac{\dfrac{v\Delta u}{\Delta x} - \dfrac{u\Delta v}{\Delta x}}{v(v + \Delta v)}$

$\qquad\qquad = \dfrac{vD_x u - uD_x v}{v^2}$, taking each limit.

We now know that

$$D_x\left(\frac{u}{v}\right) = \frac{vD_x u - uD_x v}{v^2}$$

and can put it in words as: *The derivative of a quotient is the denominator times the derivative of the numerator minus the numerator times the derivative of the denominator all divided by the square of the denominator.*

EXAMPLE

2. Find the derivative of

$$\frac{3x^2 - 2x}{x^3 + 3}.$$

SOLUTION

The given function is a quotient and, differentiating it as such, we have

$$D_x \frac{3x^2 - 2x}{x^3 + 3} = \frac{(x^3 + 3)D_x(3x^2 - 2x) - (3x^2 - 2x)D_x(x^3 + 3)}{(x^3 + 3)^2}$$

$$= \frac{(x^3 + 3)(6x - 2) - (3x^2 - 2x)3x^2}{(x^3 + 3)^2}, \quad \text{differentiating,}$$

$$= \frac{6x^4 - 2x^3 + 18x - 6 - 9x^4 + 6x^3}{(x^3 + 3)^2}, \quad \text{expanding,}$$

$$= \frac{-3x^4 + 4x^3 + 18x - 6}{(x^3 + 3)^2}, \quad \text{collecting}$$

We shall now consider the situation in which y depends on u and u depends on x. Consequently, if x is given an increment Δx, then u takes on an increment Δu, and this causes y to take on an increment Δy. Therefore, if u is a differentiable function of x and if $\Delta u \neq 0$, we have

$$\frac{\Delta y}{\Delta x} = \frac{\Delta y \Delta u}{\Delta x \Delta u}, \quad \text{multiplying} \quad \frac{\Delta y}{\Delta x} \quad \text{by} \quad \frac{\Delta u}{\Delta u},$$

$$= \frac{\Delta y \Delta u}{\Delta u \Delta x}, \quad \text{rearranging.}$$

Now making use of the fact that the limit of a product is the product of the limits, we have

$$\lim_{\Delta x \to 0} \frac{\Delta y}{\Delta x} = \lim_{\Delta x \to 0} \frac{\Delta y}{\Delta u} \frac{\Delta u}{\Delta x}$$

$$= \lim_{\Delta x \to 0} \frac{\Delta y}{\Delta u} \lim_{\Delta x \to 0} \frac{\Delta u}{\Delta x}$$

$$= \lim_{\Delta u \to 0} \frac{\Delta y}{\Delta u} \lim_{\Delta x \to 0} \frac{\Delta u}{\Delta x},$$ Δu must approach zero as Δx does since u is a differentiable function of x,

$$= D_u y D_x u$$

We now know that

$$\frac{dy}{dx} = \frac{dy}{du} \frac{du}{dx}$$

3. Find $\dfrac{dy}{dx}$ if $y = (2x^3 - 1)^4$.

SOLUTION

If we let $u = 2x^3 - 1$, we then have $y = u^4$. Consequently,

$$\frac{dy}{dx} = \frac{dy}{du} \frac{du}{dx}$$

become:

$$\frac{dy}{dx} = \frac{du^4}{du} \frac{d(2x^3 - 1)}{dx}$$
$$= 4u^3 6x^2$$
$$= 4(2x^3 - 1)^3 6x^2, \ u = 2x^3 - 1,$$
$$= 24x^2(2x^3 - 1)^3.$$

We could have been more direct about this differentiation by not making any formal substitutions. Thus,

$$\frac{d(2x^3 - 1)^4}{dx} = 4(2x^3 - 1)^3 \frac{d}{dx} (2x^3 - 1)$$

$$= 4(2x^3 - 1)^3 6x^2$$
$$= 24x^2(2x^3 - 1)^3$$

Exercise 21.5

Find $D_x y$ in each of Problems 1 through 20.

1. $y = (x^2 + 1)(3x - 2)$.

2. $y = (3x^4 + 2x)(x^2 - 1)$.

3. $y = (x^1 + 2x^2 - 3)(x^2 - 2x)$.

4. $y = (x^3 - 3x)(3x^2 + 4)$.

5. $y = \dfrac{x^2 - 3}{x + 2}$.

6. $y = \dfrac{2x + 5}{3x - 1}$.

7. $y = \dfrac{x^4 - 2x^2 + 5x}{2x - 7}$.

8. $y = \dfrac{3x^4 - 2x^2 - 5}{x^2 - 4}$.

9. $y = (2x - 1)^4$.

10. $y = (3x^2 - 4x + 7)^3$.

11. $y = \sqrt{x^2 - 2x + 3}$.

12. $y = (x^3 - 3x^2 + 6)^{2/3}$.

13. $y = (x^3 + 3x - 2)^{2/3}(3x - 1)^2$.

14. $y = \sqrt{x^4 - 2x^2 + 3}\,(2x^2 - 3)^3$.

15. $y = (2x - 1)^3(3x + 2)^2$.

16. $y = (x^2 - 3x)^2(3x^2 - 2)^3$.

17. $y = \dfrac{(3x^2 - 5)^3}{\sqrt{2x + 3}}$.

18. $y = \dfrac{(x^3 - x)^2}{\sqrt{8x - 1}}$.

19. $y = \dfrac{(2x^3 - 5x)^2}{\sqrt{4x - 3}}$.

20. $y = \dfrac{(5x + 1)^3}{\sqrt{6x + 5}}$.

Find the values of x for which each of the following is a maximum or a minimum.

21. $y = \dfrac{2x}{x^2 + 4}$.

22. $y = \dfrac{5x^2}{x^2 - x + 1}$.

23. $y = \dfrac{3x^2}{x^2 + x + 2}$.

24. $y = \dfrac{3x}{2x^2 - 3x + 6}$.

21.11. Finding an Integral of a Function

We have found a second operation that is associated with a given one on several occasions in our past work. For example, raising to a power and extracting roots are associated operations. Multiplication and division and also addition and subtraction are well known associated operations. We have seen how to find the derivatives of some functions and shall now be interested in finding the function when the derivative is given. This process is sometimes referred to as finding an antiderivative and at other times is called getting an integral or merely as integration. We shall say that $F(x) + c$ is an integral of $f(x)dx$ for C any constant if and only if $\dfrac{dF(x)}{dx} = f(x)$ and shall indicate this symbolically by writing

$$\int f(x)\,dx = F(x) + C$$

Since an interpretation of the derivative of a function is the slope of the curve, it follows that finding an integral is determining a function that has a given slope.

EXAMPLE

1. Find a curve which has a slope at all points on it equal to twice the x coordinate of the point.

SOLUTION

Our task is to find a curve $y = F(x) + C$ whose derivative is $2x$. Since the derivative of a power of x is obtained by using the exponent as a coefficient and decreasing the exponent by one and since the derivative of a constant is zero, it follows that

$$\int 2x\,dx = x^2 + C$$

This can be verified by use of the definition given above since

$$\frac{d}{dx}(x^2 + C) = 2x.$$

The function that we referred to in the definition as an integral of $f(x)dx$ is often called *the indefinite integral* because there is a value of it for each value assigned to the constant C. This constant is called the *constant of integration*. Figure 21-4

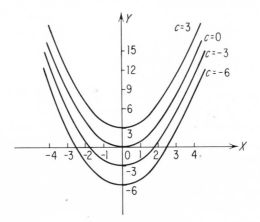

FIGURE 21-4

shows the graph of $y = \int 2x\,dx = x^2 + C$ for several values of C. The curves $y = x^2 + C$ are called a family of curves since they have a common property. The constant of integration can be determined so as to select the member of the family of curves that passes through any specified point. Thus if we want the member of the family $y = x^2 + C$ that passes through $(3, 6)$, we substitute $(3, 6)$ for (x, y) in the equation and find that $6 = 3^2 + C$; hence, $C = -3$, and $y = x^2 - 3$ is the member of the family that passes through $(3, 6)$. This is one of the curves shown in Fig. 21-4.

21.12. Some Integration Formulas

We are familiar with the fact that if $y = f(x)$, we then write $dy/dx = f'(x)$. At times, we shall want to use $dy = f'(x)dx$ and shall refer to it as the *differential of y*. Similarly, du is the differential of u. We shall now state the rules or theorems of integration that are necessary for us to be able to integrate a polynomial. Each one will be stated in symbols and in words, then illustrated and can be verified by differentiation. In each, u represents a function of x.

(1)
$$\int a\,du = a \int du$$

If translated into words, this states that *a constant factor may be introduced under or taken out from under the integral sign.* Thus,

$$\int 3x\,dx = 3 \int x\,dx$$

(2)
$$\int (u + v)\,dx = \int u\,dx + \int v\,dx$$

In words, this becomes: *the integral of a sum is the sum of the integrals.* Thus,

$$\int (x^2 + 7x)dx = \int x^2\,dx + \int 7x\,dx$$

(3)
$$\int u^n\,du = \frac{u^{n+1}}{n+1} + c, \qquad n \neq -1$$

If stated in words, this becomes: *the integral of a power of a function times its differential is the function with its exponent increased by 1 and divided by the new exponent with an arbitrary constant added.* Thus,

$$\int x^4\,dx = \frac{x^5}{5} + C, \qquad C \text{ an arbitrary constant.}$$

EXAMPLES

1.
$$\int 6x^2\,dx = 6 \int x^2\,dx = \frac{6x^3}{3} + C, \text{ by (1) and (3).}$$

2.
$$\int (x^3 + x^{-2})dx = \int x^3\,dx + \int x^{-2}dx, \text{ by (2)}$$

$$= \frac{x^4}{4} + \frac{x^{-1}}{-1} + C, \text{ by (3)}$$

$$= \frac{x^4}{4} - \frac{1}{x} + C.$$

3.
$$\int (5x^6 + 3x + 4)dx = \frac{5x^7}{7} + \frac{3x^2}{2} + 4x + C.$$

In performing the integration in this problem, we not only used our three integration formulas but also the fact that $4 = 4x^0$.

4.
$$\int (x^3 + 2x)^4(3x^2 + 2)dx = \frac{(x^3 + 2x)^5}{5} + C.$$

In this problem, we used (3) and also the fact that $(3x^2 + 2)dx$ is the differential of $x^3 + 2x$.

Exercise 21.6

Perform the indicated integrations in Problems 1 through 24 and verify your results by differentiation.

1. $\int x^2dx.$

2. $\int x^3dx.$

3. $\int x^{-3}dx.$

4. $\int x^7dx.$

5. $\int 5x^4dx.$

6. $\int 7x^{-8}dx.$

7. $\int {}'4x^2dx.$

8. $\int 3x^3dx.$

9. $\int (2x^4 - 5x^2)dx.$

10. $\int (5x^4 - 8x^3)dx.$

11. $\int (6x^2 - 7)dx.$

12. $\int (12x^3 - 9x^2)dx.$

13. $\int (x + \sqrt{x})dx.$

14. $\int (2\sqrt{x} - x^{-2})dx.$

15. $\int (x^{1/3} - x^{-1/3})dx.$

16. $\int (5x^{1/4} - 3x^{-4})dx.$

17. $\int (3x^2 - 4x)^3(6x - 4)dx.$

18. $\int (x^3 - 5x)^2(3x - 5)dx.$

19. $\int (x^4 - 2x^{-3})^{-2}(4x^3 + 6x^{-4})dx.$

20. $\int (x^2 - x^{-1})^{-3}(2x + x^{-2})dx.$

21. $\int (x^6 - x^4)^3(3x^5 - 2x^3)dx.$

22. $\int (x^4 + 2x^2)^2(x^3 + x)dx.$

23. $\int (7x^4 - 4x)^{1/2}(7x^3 - 1)dx.$

24. $\int (6x^5 - 4x^3)^{1/3}(5x^4 - 2x^2)dx.$

25. The slope of a curve is $dy/dx = 2x - 1$. Find its equation if it goes through (3, 4).

26. The slope of a curve is $dy/dx = 3x^2 - 4x$ and it passes through (2, 1). Find the equation.

27. The velocity in feet per second of a projectile is given by $v(t) = 128t - 18t^2$. How far is it from the starting point after two seconds?

28. A curve passes through the origin and has slope $dy/dx = 2x + 3$. Find the coordinates of its maximum.

21.13. Area by Integration

We shall now find a method for evaluating the area bounded by the X axis, $x = a$, $x = b$, and the curve $y = f(x)$ as shown in Fig. 21-5. We shall think of the area as being generated by an ordinate of variable length which reaches from the X axis to $y = f(x)$, begins at $x = a$ and moves to the right. Consequently, the area is a function of x since it depends on x. We shall call the area $A(x)$ and show that its derivative with respect to x is $f(x)$ and, as a result, its value can be found by integrating $f(x)$. We shall let x take on an increment Δx and let the corresponding change in A be ΔA. In Fig. 21-5, $y = f(x)$ is an increasing function of x and the increment ΔA is greater than $y\Delta x$ and less than $(y + \Delta y)x$. Symbolically,

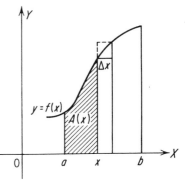

FIGURE 21-5

$$y\Delta x < \Delta A < (y + \Delta y)\Delta x$$

$$y < \frac{\Delta A}{\Delta x} < y + \Delta y, \text{ dividing by } \Delta x.$$

Now making use of the definition of a derivative and the fact that Δy approaches zero as Δx does if $y = f(x)$ is differentiable, we have

$$\lim_{\Delta x \to 0} \frac{\Delta A}{\Delta x} = \frac{dA}{dx} = y = f(x).$$

Consequently,

(1) $$A(x) = \int f(x)\, dx = F(x) + C.$$

Now, since the value of A is zero for the starting value a of x, we have $0 = F(a) + C$; hence, $C = -F(a)$. Therefore, (1) can be put in the form $A(x) = F(x) - F(a)$. Finally, the area under the curve from $x = a$ to $x = b$ is found by replacing x by b and is $A = F(b) - F(a)$. A similar argument can be given if $y = f(x)$ is a decreasing function. We can now say that *the area bounded by $x = a$, the X axis, $x = b$ and $y = f(x)$ is*

$$A = \int_a^b f(x)\, dx = F(x)\Big|_a^b = F(b) - F(a).$$

The symbol $\int_a^b f(x)dx$ is called *the definite integral of $f(x)$ from $x = a$ to $x = b$.* The numbers a and b are called the *lower* and *upper limits*, respectively.

EXAMPLE

1. Find the area bounded by the X axis, $x = -1$, $x = 2$ and the curve $y = x^2 + x + 1$.

SOLUTION

The area called for in this problem is shown in Fig. 21-6. If we apply the procedure developed above, we find that the area is

$$A = \int_{-1}^2 (x^2 + x + 1)dx$$

$$= \frac{x^3}{3} + \frac{x^2}{2} + x\Big|_{-1}^2$$

$$= \frac{8}{3} + \frac{4}{2} + 2 - \left(-\frac{1}{3} + \frac{1}{2} - 1\right)$$

$$= 7.5$$

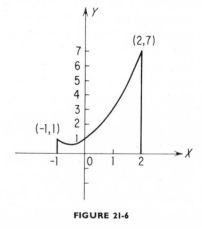

FIGURE 21-6

If part of the area is below and part is above the X axis, then the result of applying the above procedure will yield the difference between the two pieces of area. This can be seen by evaluating

$$\int_{-1}^{2} 2x\,dx = x^2 \Big|_{-1}^{2} = 3$$

and then finding the area of each triangle as indicated in Fig. 21-7. Their areas are $\frac{1}{2}(1)(2) = 1$ and $\frac{1}{2}(2)(4) = 4$ and the difference between them is 3 as given by integrating.

21.14. The Area Between Two Curves

The area bounded by two intersecting curves can be found in a manner similar to that used in the last article. The

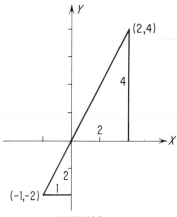

FIGURE 21-7

procedure consists of finding the points (a, c) and (b, d) of intersection of the two curves, then finding the area bounded by each curve, the X axis, $x = a$ and $x = b$, and finally getting the difference between these areas.

EXAMPLE

1. Find the area between $y = 8x - x^2$ and $y = 2x$.

SOLUTION

The two curves are shown in Fig. 21-8, and their points of intersection are labelled. These intersections were found by solving the two equations simultaneously. Thus, equating the expressions for y, we have

$$2x = 8x - x^2$$
$$x^2 - 6x = 0$$
$$x = 0, 6$$

The area under $y = 8x - x^2$ is

$$A_1 = \int_0^6 (8x - x^2)\,dx$$

$$= 4x^2 - \frac{x^3}{3} \Big|_0^6$$

$$= 144 - \frac{216}{3} = 72$$

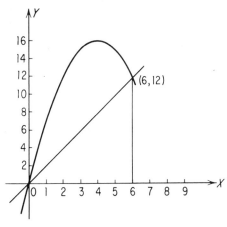

FIGURE 21-8

and that under $y = 2x$ is

$$A_2 = \int_0^6 2x \, dx$$

$$= x^2 \Big|_0^6$$

$$= 36.$$

Consequently, the area between the curves is

$$A = A_1 - A_2 = 72 - 36 = 36$$

Exercise 21.7

In each of Problems 1 through 16, find the area bounded by the X axis, the given ordinates, and the given curve. Sketch each curve as an aid in determining whether to perform one or more integrations.

1. $y = 2x + 3, x = 1, x = 3.$ **2.** $y = 3x - 4, x = 2, x = 5.$

3. $y = 2x - 1, x = 2, x = 6.$ **4.** $y = 4x - 3, x = 1, x = 4.$

5. $y = x^2 + 3x, x = 1, x = 3.$ **6.** $y = x^2 - 2x, x = 3, x = 5.$

7. $y = 3x - x^2, x = 0, x = 3.$ **8.** $y = 5x - x^2, x = 2, x = 4.$

9. $y = x^3 + 3x, x = 1, x = 4.$ **10.** $y = 2x^3 - x^2, x = 0, x = 3.$

11. $y = x^3 + 9x^2 + 36, x = -2, x = 2.$ **12.** $y = 12x - x^3, x = -1, x = 2.$

13. $y = 3x - 6, x = 1, x = 5.$ **14.** $y = 2x + 1, x = -3, x = 2.$

15. $y = x^2 - 4x + 3, x = 2, x = 4.$ **16.** $y = x^2 - 6x + 8, x = 1, x = 3.$

In each of Problems 17 through 24, sketch the curves and find the area bounded by them.

17. $y = x^2 - 2x, y = x + 4.$ **18.** $y = x^2 + 5x, y = 2x - 2.$

19. $y = x^2 - 3x, y = x + 12.$ **20.** $y = x^2 - 5x, y = 3x - 7.$

21. $2y = x^2, y = 4x - 6.$ **22.** $y = x^2 - 10x + 9, y = 2 - 2x.$

23. $y = 8x - x^2, 3y = x^2.$ **24.** $y = -x^2 + 6x + 8$
$\qquad\qquad\qquad\qquad\qquad 3y = x^2 + 2x + 4.$

APPENDIX

Table I. Natural Trigonometric Functions

To Four Places

Angle	Sin	Cos	Tan	Cot	
0° 00′	.0000	1.0000	.0000		**90° 00′**
10	029	000	029	343.8	50
20	058	000	058	171.9	40
30	.0087	1.0000	.0087	114.6	30
40	116	.9999	116	85.94	20
50	145	999	145	68.75	10
1° 00′	.0175	.9998	.0175	57.29	**89° 00′**
10	204	998	204	49.10	50
20	233	997	233	42.96	40
30	.0262	.9997	.0262	38.19	30
40	291	996	291	34.37	20
50	320	995	320	31.24	10
2° 00′	.0349	.9994	.0349	28.64	**88° 00′**
10	378	993	378	26.43	50
20	407	992	407	24.54	40
30	.0436	.9990	.0437	22.90	30
40	465	989	466	21.47	20
50	494	988	495	20.21	10
3° 00′	.0523	.9986	.0524	19.08	**87° 00′**
10	552	985	553	18.07	50
20	581	983	582	17.17	40
30	.0610	.9981	.0612	16.35	30
40	640	980	641	15.60	20
50	669	978	670	14.92	10
4° 00′	.0698	.9976	.0699	14.30	**86° 00′**
10	727	974	729	13.73	50
20	756	971	758	13.20	40
30	.0785	.9969	.0787	12.71	30
40	814	967	816	12.25	20
50	843	964	846	11.83	10
5° 00′	.0872	.9962	.0875	11.43	**85° 00′**
10	901	959	904	11.06	50
20	929	957	934	10.71	40
30	.0958	.9954	.0963	10.39	30
40	987	951	992	10.08	20
50	.1016	948	.1022	9.788	10
6° 00′	.1045	.9945	.1051	9.514	**84° 00′**
10	074	942	080	9.255	50
20	103	939	110	9.010	40
30	.1132	.9936	.1139	8.777	30
40	161	932	169	8.556	20
50	190	929	198	8.345	10
7° 00′	.1219	.9925	.1228	8.144	**83° 00′**
10	248	922	257	7.953	50
20	276	918	287	7.770	40
30	.1305	.9914	.1317	7.596	30
40	334	911	346	7.429	20
50	363	907	376	7.269	10
8° 00′	.1392	.9903	.1405	7.115	**82° 00′**
10	421	899	435	6.968	50
20	449	894	465	6.827	40
30	.1478	.9890	.1495	6.691	30
40	507	886	524	6.561	20
50	536	881	554	6.435	10
9° 00′	.1564	.9877	.1584	6.314	**81° 00′**
	Cos	Sin	Cot	Tan	Angle

Table I.   Natural Trigonometric Functions (cont.)
To Four Places

Angle	Sin	Cos	Tan	Cot	
9° 00′	.1564	.9877	.1584	6.314	81° 00′
10	593	872	614	197	50
20	622	868	644	084	40
30	.1650	.9863	.1673	5.976	30
40	679	858	703	871	20
50	708	853	.733	769	10
10° 00′	.1736	.9848	.1763	5.671	80° 00′
10	765	843	793	576	50
20	794	838	823	485	40
30	.1822	.9833	.1853	5.396	30
40	851	827	883	309	20
50	880	822	914	226	10
11° 00′	.1908	.9816	.1944	5.145	79° 00′
10	937	811	974	066	50
20	965	805	.2004	4.989	40
30	.1994	.9799	.2035	4.915	30
40	.2022	793	065	843	20
50	051	787	095	773	10
12° 00′	.2079	.9781	.2126	4.705	78° 00′
10	108	775	156	638	50
20	136	769	186	574	40
30	.2164	.9763	.2217	4.511	30
40	193	757	247	449	20
50	221	750	278	390	10
13° 00′	.2250	.9744	.2309	4.331	77° 00′
10	278	737	339	275	50
20	306	730	370	219	40
30	.2334	.9724	.2401	4.165	30
40	363	717	432	113	20
50	391	710	462	061	10
14° 00′	.2419	.9703	.2493	4.011	76° 00′
10	447	696	524	3.962	50
20	476	689	555	914	40
30	.2504	.9681	.2586	3.867	30
40	532	674	617	821	20
50	560	667	648	776	10
15° 00′	.2588	.9659	.2679	3.732	75° 00′
10	616	652	711	689	50
20	644	644	742	647	40
30	.2672	.9636	.2773	3.606	30
40	700	628	805	566	20
50	728	621	836	526	10
16° 00′	.2756	.9613	.2867	3.487	74° 00′
10	784	605	899	450	50
20	812	596	931	412	40
30	.2840	.9588	.2962	3.376	30
40	868	580	994	340	20
50	896	572	.3026	305	10
17° 00′	.2924	.9563	.3057	3.271	73° 00′
10	952	555	089	237	50
20	979	546	121	204	40
30	.3007	.9537	.3153	3.172	30
40	035	528	185	140	20
50	062	520	217	108	10
18° 00′	.3090	.9511	.3249	3.078	72° 00′
	Cos	Sin	Cot	Tan	Angle

Table I. Natural Trigonometric Functions (cont.)
To Four Places

Angle	Sin	Cos	Tan	Cot	
18° 00′	.3090	.9511	.3249	3.078	**72° 00′**
10	118	502	281	047	50
20	145	492	314	018	40
30	.3173	.9483	.3346	2.989	30
40	201	474	378	960	20
50	228	465	411	932	10
19° 00′	.3256	.9455	.3443	2.904	**71° 00′**
10	283	446	476	877	50
20	311	436	508	850	40
30	.3338	.9426	.3541	2.824	30
40	365	417	574	798	20
50	393	407	607	773	10
20° 00′	.3420	.9397	.3640	2.747	**70° 00′**
10	448	387	673	723	50
20	475	377	706	699	40
30	.3502	.9367	.3739	2.675	30
40	529	356	772	651	20
50	557	346	805	628	10
21° 00′	.3584	.9336	.3839	2.605	**69° 00′**
10	611	325	872	583	50
20	638	315	906	560	40
30	.3665	.9304	.3939	2.539	30
40	692	293	973	517	20
50	719	283	.4006	496	10
22° 00′	.3746	.9272	.4040	2.475	**68° 00′**
10	773	261	074	455	50
20	800	250	108	434	40
30	.3827	.9239	.4142	2.414	30
40	854	228	176	394	20
50	881	216	210	375	10
23° 00′	.3907	.9205	.4245	2.356	**67° 00′**
10	934	194	279	337	50
20	961	182	314	318	40
30	.3987	.9171	.4348	2.300	30
40	.4014	159	383	282	20
50	041	147	417	264	10
24° 00′	.4067	.9135	.4452	2.246	**66° 00′**
10	094	124	487	229	50
20	120	112	522	211	40
30	.4147	.9100	.4557	2.194	30
40	173	088	592	177	20
50	200	075	628	161	10
25° 00′	.4226	.9063	.4663	2.145	**65° 00′**
10	253	051	699	·128	50
20	279	038	734	112	40
30	.4305	.9026	.4770	2.097	30
40	331	013	806	081	20
50	358	001	841	066	10
26° 00′	.4384	.8988	.4877	2.050	**64° 00′**
10	410	975	913	035	50
20	436	962	950	020	40
30	.4462	.8949	.4986	2.006	30
40	488	936	.5022	1.991	20
50	514	923	059	977	10
27° 00′	.4540	.8910	.5095	1.963	**63° 00′**
	Cos	Sin	Cot	Tan	Angle

Table I. Natural Trigonometric Functions (cont.)

To Four Places

Angle	Sin	Cos	Tan	Cot	
27° 00′	.4540	.8910	.5095	1.963	63° 00′
10	566	897	132	949	50
20	592	884	169	935	40
30	.4617	.8870	.5206	1.921	30
40	643	857	243	907	20
50	669	843	280	894	10
28° 00′	.4695	.8829	.5317	1.881	62° 00′
10	720	816	354	868	50
20	746	802	392	855	40
30	.4772	.8788	.5430	1.842	30
40	797	774	467	829	20
50	823	760	505	816	10
29° 00′	.4848	.8746	.5543	1.804	61° 00′
10	874	732	581	792	50
20	899	718	619	780	40
30	.4924	.8704	.5658	1.767	30
40	950	689	696	756	20
50	975	675	735	744	10
30° 00′	.5000	.8660	.5774	1.732	60° 00′
10	025	646	812	720	50
20	050	631	851	709	40
30	.5075	.8616	.5890	1.698	30
40	100	601	930	686	20
50	125	587	969	675	10
31° 00′	.5150	.8572	.6009	1.664	59° 00′
10	175	557	048	653	50
20	200	542	088	643	40
30	.5225	.8526	.6128	1.632	30
40	250	511	168	621	20
50	275	496	208	611	10
32° 00′	.5299	.8480	.6249	1.600	58° 00′
10	324	465	289	590	50
20	348	450	330	580	40
30	.5373	.8434	.6371	1.570	30
40	398	418	412	560	20
50	422	403	453	550	10
33° 00′	.5446	.8387	.6494	1.540	57° 00′
10	471	371	536	530	50
20	495	355	577	520	40
30	.5519	.8339	.6619	1.511	30
40	544	323	661	501	20
50	568	307	703	1.492	10
34° 00′	.5592	.8290	.6745	1.483	56° 00′
10	616	274	787	473	50
20	640	258	830	464	40
30	.5664	.8241	.6873	1.455	30
40	688	225	916	446	20
50	712	208	959	437	10
35° 00′	.5736	.8192	.7002	1.428	55° 00′
10	760	175	046	419	50
20	783	158	089	411	40
30	.5807	.8141	.7133	1.402	30
40	831	124	177	393	20
50	854	107	221	385	10
36° 00′	.5878	.8090	.7265	1.376	54° 00′
	Cos	Sin	Cot	Tan	Angle

Table I. Natural Trigonometric Functions (cont.)

To Four Places

Angle	Sin	Cos	Tan	Cot	
36° 00'	.5878	.8090	.7265	1.376	**54° 00'**
10	901	073	310	368	50
20	925	056	355	360	40
30	.5948	.8039	.7400	1.351	30
40	972	021	445	343	20
50	995	004	490	335	10
37° 00'	.6018	.7986	.7536	1.327	**53° 00'**
10	041	969	581	319	50
20	065	951	627	311	40
30	.6088	.7934	.7673	1.303	30
40	111	916	720	295	20
50	134	898	766	288	10
38° 00'	.6157	.7880	.7813	1.280	**52° 00'**
10	180	862	860	272	50
20	202	844	907	265	40
30	.6225	.7826	.7954	1.257	30
40	248	808	.8002	250	20
50	271	790	050	242	10
39° 00'	.6293	.7771	.8098	1.235	**51° 00'**
10	316	753	146	228	50
20	338	735	195	220	40
30	.6361	.7716	.8243	1.213	30
40	383	698	292	206	20
50	406	679	342	199	10
40° 00'	.6428	.7660	.8391	1.192	**50° 00'**
10	450	642	441	185	50
20	472	623	491	178	40
30	.6494	.7604	.8541	1.171	30
40	517	585	591	164	20
50	539	566	642	157	10
41° 00'	.6561	.7547	.8693	1.150	**49° 00'**
10	583	528	744	144	50
20	604	509	796	137	40
30	.6626	.7490	.8847	1.130	30
40	648	470	899	124	20
50	670	451	952	117	10
42° 00'	.6691	.7431	.9004	1.111	**48° 00'**
10	713	412	057	104	50
20	734	392	110	098	40
30	.6756	.7373	.9163	1.091	30
40	777	353	217	085	20
50	799	333	271	079	10
43° 00'	.6820	.7314	.9325	1.072	**47° 00'**
10	841	294	380	066	50
20	862	274	435	060	40
30	.6884	.7254	.9490	1.054	30
40	905	234	545	048	20
50	926	214	601	042	10
44° 00'	.6947	.7193	.9657	1.036	**46° 00'**
10	967	173	713	030	50
20	988	153	770	024	40
30	.7009	.7133	.9827	1.018	30
40	030	112	884	012	20
50	050	092	942	006	10
45° 00'	.7071	.7071	1.000	1.000	**45° 00'**
	Cos	Sin	Cot	Tan	Angle

Table II. Powers and Roots

No.	Sq.	Sq. Root	Cube	Cube Root	No.	Sq.	Sq. Root	Cube	Cube Root
1	1	1.000	1	1.000	51	2,601	7.141	132,651	3.708
2	4	1.414	8	1.260	52	2,704	7.211	140,608	3.732
3	9	1.732	27	1.442	53	2,809	7.280	148,877	3.756
4	16	2.000	64	1.587	54	2,916	7.348	157,464	3.780
5	25	2.236	125	1.710	55	3,025	7.416	166,375	3.803
6	36	2.449	216	1.817	56	3,136	7.483	175,616	3.826
7	49	2.646	343	1.913	57	3,249	7.550	185,193	3.848
8	64	2.828	512	2.000	58	3,364	7.616	195,112	3.871
9	81	3.000	729	2.080	59	3,481	7.681	205,379	3.893
10	100	3.162	1,000	2.154	60	3,600	7.746	216,000	3.915
11	121	3.317	1,331	2.224	61	3,721	7.810	226,981	3.936
12	144	3.464	1,728	2.289	62	3,844	7.874	238,328	3.958
13	169	3.606	2,197	2.351	63	3,969	7.937	250,047	3.979
14	196	3.742	2,744	2.410	64	4,096	8.000	262,144	4.000
15	225	3.873	3,375	2.466	65	4,225	8.062	274,625	4.021
16	256	4.000	4,096	2.520	66	4,356	8.124	287,496	4.041
17	289	4.123	4,913	2.571	67	4,489	8.185	300,763	4.062
18	324	4.243	5,832	2.621	68	4,624	8.246	314,432	4.082
19	361	4.359	6,859	2.668	69	4,761	8.307	328,509	4.102
20	400	4.472	8,000	2.714	70	4,900	8.367	343,000	4.121
21	441	4.583	9,261	2.759	71	5,041	8.426	357,911	4.141
22	484	4.690	10,648	2.802	72	5,184	8.485	373,248	4.160
23	529	4.796	12,167	2.844	73	5,329	8.544	389,017	4.179
24	576	4.899	13,824	2.884	74	5,476	8.602	405,224	4.198
25	625	5.000	15,625	2.924	75	5,625	8.660	421,875	4.217
26	676	5.099	17,576	2.962	76	5,776	8.718	438,976	4.236
27	729	5.196	19,683	3.000	77	5,929	8.775	456,533	4.254
28	784	5.291	21,952	3.037	78	6,084	8.832	474,552	4.273
29	841	5.385	24,389	3.072	79	6,241	8.888	493,039	4.291
30	900	5.477	27,000	3.107	80	6,400	8.944	512,000	4.309
31	961	5.568	29,791	3.141	81	6,561	9.000	531,441	4.327
32	1,024	5.657	32,768	3.175	82	6,724	9.055	551,368	4.344
33	1,089	5.745	35,937	3.208	83	6,889	9.110	571,787	4.362
34	1,156	5.831	39,304	3.240	84	7,056	9.165	592,704	4.380
35	1,225	5.916	42,875	3.271	85	7,225	9.220	614,125	4.397
36	1,296	6.000	46,656	3.302	86	7,396	9.274	636,056	4.414
37	1,369	6.083	50,653	3.332	87	7,569	9.327	658,503	4.431
38	1,444	6.164	54,872	3.362	88	7,744	9.381	681,472	4.448
39	1,521	6.245	59,319	3.391	89	7,921	9.434	704,969	4.465
40	1,600	6.325	64,000	3.420	90	8,100	9.487	729,000	4.481
41	1,681	6.403	68,921	3.448	91	8,281	9.539	753,571	4.498
42	1,764	6.481	74,088	3.476	92	8,464	9.592	778,688	4.514
43	1,849	6.557	79,507	3.503	93	8,649	9.644	804,357	4.531
44	1,936	6.633	85,184	3.530	94	8,836	9.695	830,584	4.547
45	2,025	6.708	91,125	3.557	95	9,025	9.747	857,375	4.563
46	2,116	6.782	97,336	3.583	96	9,216	9.798	884,736	4.579
47	2,209	6.856	103,823	3.609	97	9,409	9.849	912,673	4.595
48	2,304	6.928	110,592	3.634	98	9,604	9.899	941,192	4.610
49	2,401	7.000	117,649	3.659	99	9,801	9.950	970,299	4.626
50	2,500	7.071	125,000	3.684	100	10,000	10.000	1,000,000	4.642

Table III.   Common Logarithms of Numbers

N	0	1	2	3	4	5	6	7	8	9
10	0000	0043	0086	0128	0170	0212	0253	0294	0334	0374
11	0414	0453	0492	0531	0569	0607	0645	0682	0719	0755
12	0792	0828	0864	0899	0934	0969	1004	1038	1072	1106
13	1139	1173	1206	1239	1271	1303	1335	1367	1399	1430
14	1461	1492	1523	1553	1584	1614	1644	1673	1703	1732
15	1761	1790	1818	1847	1875	1903	1931	1959	1987	2014
16	2041	2068	2095	2122	2148	2175	2201	2227	2253	2279
17	2304	2330	2355	2380	2405	2430	2455	2480	2504	2529
18	2553	2577	2601	2625	2648	2672	2695	2718	2742	2765
19	2788	2810	2833	2856	2878	2900	2923	2945	2967	2989
20	3010	3032	3054	3075	3096	3118	3139	3160	3181	3201
21	3222	3243	3263	3284	3304	3324	3345	3365	3385	3404
22	3424	3444	3464	3483	3502	3522	3541	3560	3579	3598
23	3617	3636	3655	3674	3692	3711	3729	3747	3766	3784
24	3802	3820	3838	3856	3874	3892	3909	3927	3945	3962
25	3979	3997	4014	4031	4048	4065	4082	4099	4116	4133
26	4150	4166	4183	4200	4216	4232	4249	4265	4281	4298
27	4314	4330	4346	4362	4378	4393	4409	4425	4440	4456
28	4472	4487	4502	4518	4533	4548	4564	4579	4594	4609
29	4624	4639	4654	4669	4683	4698	4713	4728	4742	4757
30	4771	4786	4800	4814	4829	4843	4857	4871	4886	4900
31	4914	4928	4942	4955	4969	4983	4997	5011	5024	5038
32	5051	5065	5079	5092	5105	5119	5132	5145	5159	5172
33	5185	5198	5211	5224	5237	5250	5263	5276	5289	5302
34	5315	5328	5340	5353	5366	5378	5391	5403	5416	5428
35	5441	5453	5465	5478	5490	5502	5514	5527	5539	5551
36	5563	5575	5587	5599	5611	5623	5635	5647	5658	5670
37	5682	5694	5705	5717	5729	5740	5752	5763	5775	5786
38	5798	5809	5821	5832	5843	5855	5866	5877	5888	5899
39	5911	5922	5933	5944	5955	5966	5977	5988	5999	6010
40	6021	6031	6042	6053	6064	6075	6085	6096	6107	6117
41	6128	6138	6149	6160	6170	6180	6191	6201	6212	6222
42	6232	6243	6253	6263	6274	6284	6294	6304	6314	6325
43	6335	6345	6355	6365	6375	6385	6395	6405	6415	6425
44	6435	6444	6454	6464	6474	6484	6493	6503	6513	6522
45	6532	6542	6551	6561	6571	6580	6590	6599	6609	6618
46	6628	6637	6646	6656	6665	6675	6684	6693	6702	6712
47	6721	6730	6739	6749	6758	6767	6776	6785	6794	6803
48	6812	6821	6830	6839	6848	6857	6866	6875	6884	6893
49	6902	6911	6920	6928	6937	6946	6955	6964	6972	6981
50	6990	6998	7007	7016	7024	7033	7042	7050	7059	7067
N	0	1	2	3	4	5	6	7	8	9

Table III.   Common Logarithms of Numbers (cont.)

N	0	1	2	3	4	5	6	7	8	9
50	6990	6998	7007	7016	7024	7033	7042	7050	7059	7067
51	7076	7084	7093	7101	7110	7118	7126	7135	7143	7152
52	7160	7168	7177	7185	7193	7202	7210	7218	7226	7235
53	7243	7251	7259	7267	7275	7284	7292	7300	7308	7316
54	7324	7332	7340	7348	7356	7364	7372	7380	7388	7396
55	7404	7412	7419	7427	7435	7443	7451	7459	7466	7474
56	7482	7490	7497	7505	7513	7520	7528	7536	7543	7551
57	7559	7566	7574	7582	7589	7597	7604	7612	7619	7627
58	7634	7642	7649	7657	7664	7672	7679	7686	7694	7701
59	7709	7716	7723	7731	7738	7745	7752	7760	7767	7774
60	7782	7789	7796	7803	7810	7818	7825	7832	7839	7846
61	7853	7860	7868	7875	7882	7889	7896	7903	7910	7917
62	7924	7931	7938	7945	7952	7959	7966	7973	7980	7987
63	7993	8000	8007	8014	8021	8028	8035	8041	8048	8055
64	8062	8069	8075	8082	8089	8096	8102	8109	8116	8122
65	8129	8136	8142	8149	8156	8162	8169	8176	8182	8189
66	8195	8202	8209	8215	8222	8228	8235	8241	8248	8254
67	8261	8267	8274	8280	8287	8293	8299	8306	8312	8319
68	8325	8331	8338	8344	8351	8357	8363	8370	8376	8382
69	8388	8395	8401	8407	8414	8420	8426	8432	8439	8445
70	8451	8457	8463	8470	8476	8482	8488	8494	8500	8506
71	8513	8519	8525	8531	8537	8543	8549	8555	8561	8567
72	8573	8579	8585	8591	8597	8603	8609	8615	8621	8627
73	8633	8639	8645	8651	8657	8663	8669	8675	8681	8686
74	8692	8698	8704	8710	8716	8722	8727	8733	8739	8745
75	8751	8756	8762	8768	8774	8779	8785	8791	8797	8802
76	8808	8814	8820	8825	8831	8837	8842	8848	8854	8859
77	8865	8871	8876	8882	8887	8893	8899	8904	8910	8915
78	8921	8927	8932	8938	8943	8949	8954	8960	8965	8971
79	8976	8982	8987	8993	8998	9004	9009	9015	9020	9025
80	9031	9036	9042	9047	9053	9058	9063	9069	9074	9079
81	9085	9090	9096	9101	9106	9112	9117	9122	9128	9133
82	9138	9143	9149	9154	9159	9165	9170	9175	9180	9186
83	9191	9196	9201	9206	9212	9217	9222	9227	9232	9238
84	9243	9248	9253	9258	9263	9269	9274	9279	9284	9289
85	9294	9299	9304	9309	9315	9320	9325	9330	9335	9340
86	9345	9350	9355	9360	9365	9370	9375	9380	9385	9390
87	9395	9400	9405	9410	9415	9420	9425	9430	9435	9440
88	9445	9450	9455	9460	9465	9469	9474	9479	9484	9489
89	9494	9499	9504	9509	9513	9518	9523	9528	9533	9538
90	9542	9547	9552	9557	9562	9566	9571	9576	9581	9586
91	9590	9595	9600	9605	9609	9614	9619	9624	9628	9633
92	9638	9643	9647	9652	9657	9661	9666	9671	9675	9680
93	9685	9689	9694	9699	9703	9708	9713	9717	9722	9727
94	9731	9736	9741	9745	9750	9754	9759	9763	9768	9773
95	9777	9782	9786	9791	9795	9800	9805	9809	9814	9818
96	9823	9827	9832	9836	9841	9845	9850	9854	9859	9863
97	9868	9872	9877	9881	9886	9890	9894	9899	9903	9908
98	9912	9917	9921	9926	9930	9934	9939	9943	9948	9952
99	9956	9961	9965	9969	9974	9978	9983	9987	9991	9996
N	0	1	2	3	4	5	6	7	8	9

Table IV. Logarithms of Functions

Angle	L. Sin.	L. Tan.	L. Cot.	L. Cos.	
0° 00′	———	———	———	.0000	90° 00′
10	7.4637	7.4637	2.5363	.0000	50
20	7.7648	7.7648	2.2352	.0000	40
30	7.9408	7.9408	2.0591	.0000	30
40	8.0658	8.0658	1.9342	.0000	20
50	8.1627	8.1627	1.8373	.0000	10
1° 00′	8.2419	8.2419	1.7581	9.9999	89° 00′
10	8.3088	8.3089	1.6911	9.9999	50
20	8.3668	8.3669	1.6331	9.9999	40
30	8.4179	8.4181	1.5819	9.9999	30
40	8.4637	8.4638	1.5362	9.9998	20
50	8.5050	8.5053	1.4947	9.9998	10
2° 00′	8.5428	8.5431	1.4569	9.9997	88° 00′
10	8.5776	8.5779	1.4221	9.9997	50
20	8.6097	8.6101	1.3899	9.9996	40
30	8.6397	8.6401	1.3599	9.9996	30
40	8.6677	8.6682	1.3318	9.9995	20
50	8.6940	8.6945	1.3055	9.9995	10
3° 00′	8.7188	8.7194	1.2806	9.9994	87° 00′
10	8.7423	8.7429	1.2571	9.9993	50
20	8.7645	8.7652	1.2348	9.9993	40
30	8.7857	8.7865	1.2135	9.9992	30
40	8.8059	8.8067	1.1933	9.9991	20
50	8.8251	8.8261	1.1739	9.9990	10
4° 00′	8.8436	8.8446	1.1554	9.9989	86° 00′
10	8.8613	8.8624	1.1376	9.9989	50
20	8.8783	8.8795	1.1205	9.9988	40
30	8.8946	8.8960	1.1040	9.9987	30
40	8.9104	8.9118	1.0882	9.9986	20
50	8.9256	8.9272	1.0728	9.9985	10
5° 00′	8.9403	8.9420	1.0580	9.9983	85° 00′
10	8.9545	8.9563	1.0437	9.9982	50
20	8.9682	8.9701	1.0299	9.9981	40
30	8.9816	8.9836	1.0164	9.9980	30
40	8.9945	8.9966	1.0034	9.9979	20
50	9.0070	9.0093	.9907	9.9977	10
6° 00′	9.0192	9.0216	.9784	9.9976	84° 00′
10	9.0311	9.0336	.9664	9.9975	50
20	9.0426	9.0453	.9547	9.9973	40
30	9.0539	9.0567	.9433	9.9972	30
40	9.0648	9.0678	.9322	9.9971	20
50	9.0755	9.0786	.9214	9.9969	10
7° 00′	9.0859	9.0891	.9109	9.9968	83° 00′
10	9.0961	9.0995	.9005	9.9966	50
20	9.1060	9.1096	.8904	9.9964	40
30	9.1157	9.1194	.8806	9.9963	30
40	9.1252	9.1291	.8709	9.9961	20
50	9.1345	9.1385	.8615	9.9959	10
8° 00′	9.1436	9.1478	.8522	9.9958	82° 00′
10	9.1525	9.1569	.8431	9.9956	50
20	9.1612	9.1658	.8342	9.9954	40
30	9.1697	9.1745	.8255	9.9952	30
40	9.1781	9.1831	.8169	9.9950	20
50	9.1863	9.1915	.8085	9.9948	10
9° 00′	9.1943	9.1997	.8003	9.9946	81° 00′
	L. Cos.	L. Cot.	L. Tan.	L. Sin.	Angle

Table IV. Logarithms of Functions (cont.)

Angle	L. Sin.	L. Tan.	L. Cot.	L. Cos.	
9° 00'	9.1943	9.1997	.8003	9.9946	81° 00'
10	9.2022	9.2078	.7922	9.9944	50
20	9.2100	9.2158	.7842	9.9942	40
30	9.2176	9.2236	.7764	9.9940	30
40	9.2251	9.2313	.7687	9.9938	20
50	9.2324	9.2389	.7611	9.9936	10
10° 00'	9.2397	9.2463	.7537	9.9934	80° 00'
10	9.2468	9.2536	.7464	9.9931	50
20	9.2538	9.2609	.7391	9.9929	40
30	9.2606	9.2680	.7320	9.9927	30
40	9.2674	9.2750	.7250	9.9924	20
50	9.2740	9.2819	.7181	9.9922	10
11° 00'	9.2806	9.2887	.7113	9.9919	79° 00'
10	9.2870	9.2953	.7047	9.9917	50
20	9.2934	9.3020	.6980	9.9914	40
30	9.2997	9.3085	.6915	9.9912	30
40	9.3058	9.3149	.6851	9.9909	20
50	9.3119	9.3212	.6788	9.9907	10
12° 00'	9.3179	9.3275	.6725	9.9904	78° 00'
10	9.3238	9.3336	.6664	9.9901	50
20	9.3296	9.3397	.6603	9.9899	40
30	9.3353	9.3458	.6542	9.9896	30
40	9.3410	9.3517	.6483	9.9893	20
50	9.3466	9.3576	.6424	9.9890	10
13° 00'	9.3521	9.3634	.6366	9.9887	77° 00'
10	9.3575	9.3691	.6309	9.9884	50
20	9.3629	9.3748	.6252	9.9881	40
30	9.3682	9.3804	.6196	9.9878	30
40	9.3734	9.3859	.6141	9.9875	20
50	9.3786	9.3914	.6086	9.9872	10
14° 00'	9.3837	9.3968	.6032	9.9869	76° 00'
10	9.3887	9.4021	.5979	9.9866	50
20	9.3937	9.4074	.5926	9.9863	40
30	9.3986	9.4127	.5873	9.9859	30
40	9.4035	9.4178	.5822	9.9856	20
50	9.4083	9.4230	.5770	9.9853	10
15° 00'	9.4130	9.4281	.5719	9.9849	75° 00'
10	9.4177	9.4331	.5669	9.9846	50
20	9.4223	9.4381	.5619	9.9843	40
30	9.4269	9.4430	.5570	9.9839	30
40	9.4314	9.4479	.5521	9.9836	20
50	9.4359	9.4527	.5473	9.9832	10
16° 00'	9.4403	9.4575	.5425	9.9828	74° 00'
10	9.4447	9.4622	.5378	9.9825	50
20	9.4491	9.4669	.5331	9.9821	40
30	9.4533	9.4716	.5284	9.9817	30
40	9.4576	9.4762	.5238	9.9814	20
50	9.4618	9.4808	.5192	9.9810	10
17° 00'	9.4659	9.4853	.5147	9.9806	73° 00'
10	9.4700	9.4898	.5102	9.9802	50
20	9.4741	9.4943	.5057	9.9798	40
30	9.4781	9.4987	.5013	9.9794	30
40	9.4821	9.5031	.4969	9.9790	20
50	9.4861	9.5075	.4925	9.9786	10
18° 00'	9.4900	9.5118	.4882	9.9782	72° 00'
	L. Cos.	L. Cot.	L. Tan.	L. Sin.	Angle

Table IV.   *Logarithms of Functions* (cont.)

Angle	L. Sin.	L. Tan.	L. Cot.	L. Cos.	
18° 00′	9.4900	9.5118	.4882	9.9782	72° 00′
10	9.4939	9.5161	.4839	9.9778	50
20	9.4977	9.5203	.4797	9.9774	40
30	9.5015	9.5245	.4755	9.9770	30
40	9.5052	9.5287	.4713	9.9765	20
50	9.5090	9.5329	.4671	9.9761	10
19° 00′	9.5126	9.5370	.4630	9.9757	71° 00′
10	9.5163	9.5411	.4589	9.9752	50
20	9.5199	9.5451	.4549	9.9748	40
30	9.5235	9.5491	.4509	9.9743	30
40	9.5270	9.5531	.4469	9.9739	20
50	9.5306	9.5571	.4429	9.9734	10
20° 00′	9.5341	9.5611	.4389	9.9730	70° 00′
10	9.5375	9.5650	.4350	9.9725	50
20	9.5409	9.5689	.4311	9.9721	40
30	9.5443	9.5727	.4273	9.9716	30
40	9.5477	9.5766	.4234	9.9711	20
50	9.5510	9.5804	.4196	9.9706	10
21° 00′	9.5543	9.5842	.4158	9.9702	69° 00′
10	9.5576	9.5879	.4121	9.9697	50
20	9.5609	9.5917	.4083	9.9692	40
30	9.5641	9.5954	.4046	9.9687	30
40	9.5673	9.5991	.4009	9.9682	20
50	9.5704	9.6028	.3972	9.9677	10
22° 00′	9.5736	9.6064	.3936	9.9672	68° 00′
10	9.5767	9.6100	.3900	9.9667	50
20	9.5798	9.6136	.3864	9.9661	40
30	9.5828	9.6172	.3828	9.9656	30
40	9.5859	9.6208	.3792	9.9651	20
50	9.5889	9.6243	.3757	9.9646	10
23° 00′	9.5919	9.6279	.3721	9.9640	67° 00′
10	9.5948	9.6314	.3686	9.9635	50
20	9.5978	9.6348	.3652	9.9629	40
30	9.6007	9.6383	.3617	9.9624	30
40	9.6036	9.6417	.3583	9.9618	20
50	9.6065	9.6452	.3548	9.9613	10
24° 00′	9.6093	9.6486	.3514	9.9607	66° 00′
10	9.6121	9.6520	.3480	9.9602	50
20	9.6149	9.6553	.3447	9.9596	40
30	9.6177	9.6587	.3413	9.9590	30
40	9.6205	9.6620	.3380	9.9584	20
50	9.6232	9.6654	.3346	9.9579	10
25° 00′	9.6259	9.6687	.3313	9.9573	65° 00′
10	9.6286	9.6720	.3280	9.9567	50
20	9.6313	9.6752	.3248	9.9561	40
30	9.6340	9.6785	.3215	9.9555	30
40	9.6366	9.6817	.3183	9.9549	20
50	9.6392	9.6850	.3150	9.9543	10
26° 00′	9.6418	9.6882	.3118	9.9537	64° 00′
10	9.6444	9.6914	.3086	9.9530	50
20	9.6470	9.6946	.3054	9.9524	40
30	9.6495	9.6977	.3023	9.9518	30
40	9.6521	9.7009	.2991	9.9512	20
50	9.6546	9.7040	.2960	9.9505	10
27° 00′	9.6570	9.7072	.2928	9.9499	63° 00′
	L. Cos.	L. Cot.	L. Tan.	L. Sin.	Angle

Table IV.   Logarithms of Functions (cont.)

Angle	L. Sin.	L. Tan.	L. Cot.	L. Cos.	
27° 00'	9.6570	9.7072	.2928	9.9499	**63 ° 00**
10	9.6595	9.7103	.2897	9.9492	50
20	9.6620	9.7134	.2866	9.9486	40
30	9.6644	9.7165	.2835	9.9479	30
40	9.6668	9.7196	.2804	9.9473	20
50	9.6692	9.7226	.2774	9.9466	10
28° 00'	9.6716	9.7257	.2743	9.9459	**62° 00'**
10	9.6740	9.7287	.2713	9.9453	50
20	9.6763	9.7317	.2683	9.9446	40
30	9.6787	9.7348	.2652	9.9439	30
40	9.6810	9.7378	.2622	9.9432	20
50	9.6833	9.7408	.2592	9.9425	10
29° 00'	9.6856	9.7438	.2562	9.9418	**61° 00'**
10	9.6878	9.7467	.2533	9.9411	50
20	9.6901	9.7497	.2503	9.9404	40
30	9.6923	9.7526	.2474	9.9397	30
40	9.6946	9.7556	.2444	9.9390	20
50	9.6968	9.7585	.2415	9.9383	10
30° 00'	9.6990	9.7614	.2386	9.9375	**60° 00'**
10	9.7012	9.7644	.2356	9.9368	50
20	9.7033	9.7673	.2327	9.9361	40
30	9.7055	9.7701	.2299	9.9353	30
40	9.7076	9.7730	.2270	9.9346	20
50	9.7097	9.7759	.2241	9.9338	10
31° 00'	9.7118	9.7788	.2212	9.9331	**59° 00'**
10	9.7139	9.7816	.2184	9.9323	50
20	9.7160	9.7845	.2155	9.9315	40
30	9.7181	9.7873	.2127	9.9308	30
40	9.7201	9.7902	.2098	9.9300	20
50	9.7222	9.7930	.2070	9.9292	10
32° 00'	9.7242	9.7958	.2042	9.9284	**58° 00'**
10	9.7262	9.7986	.2014	9.9276	50
20	9.7282	9.8014	.1986	9.9268	40
30	9.7302	9.8042	.1958	9.9260	30
40	9.7322	9.8070	.1930	9.9252	20
50	9.7342	9.8097	.1903	9.9244	10
33° 00'	9.7361	9.8125	.1875	9.9236	**57° 00'**
10	9.7380	9.8153	.1847	9.9228	50
20	9.7400	9.8180	.1820	9.9219	40
30	9.7419	9.8208	.1792	9.9211	30
40	9.7438	9.8235	.1765	9.9203	20
50	9.7457	9.8263	.1737	9.9194	10
34° 00'	9.7476	9.8290	.1710	9.9186	**56° 00'**
10	9.7494	9.8317	.1683	9.9177	50
20	9.7513	9.8344	.1656	9.9169	40
30	9.7531	9.8371	.1629	9.9160	30
40	9.7550	9.8398	.1602	9.9151	20
50	9.7568	9.8425	.1575	9.9142	10
35° 00'	9.7586	9.8452	.1548	9.9134	**55° 00'**
10	9.7604	9.8479	.1521	9.9125	50
20	9.7622	9.8506	.1494	9.9116	40
30	9.7640	9.8533	.1467	9.9107	30
40	9.7657	9.8559	.1441	9.9098	20
50	9.7675	9.8586	.1414	9.9089	10
36° 00'	9.7692	9.8613	.1387	9.9080	**54° 00'**
	L. Cos.	L. Cot.	L. Tan.	L. Sin.	Angle

Table IV.   Logarithms of Functions (cont.)

Angle	L. Sin.	L. Tan.	L. Cot.	L. Cos.	
36° 00′	9.7692	9.8613	.1387	9.9080	54° 00′
10	9.7710	9.8639	.1361	9.9070	50
20	9.7727	9.8666	.1334	9.9061	40
30	9.7744	9.8692	.1308	9.9052	30
40	9.7761	9.8718	.1282	9.9042	20
50	9.7778	9.8745	.1255	9.9033	10
37° 00′	9.7795	9.8771	.1229	9.9023	53° 00′
10	9.7811	9.8797	.1203	9.9014	50
20	9.7828	9.8824	.1176	9.9004	40
30	9.7844	9.8850	.1150	9.8995	30
40	9.7861	9.8876	.1124	9.8985	20
50	9.7877	9.8902	.1098	9.8975	10
38° 00′	9.7893	9.8928	.1072	9.8965	52° 00′
10	9.7910	9.8954	.1046	9.8955	50
20	9.7926	9.8980	.1020	9.8945	40
30	9.7941	9.9006	.0994	9.8935	30
40	9.7957	9.9032	.0968	9.8925	20
50	9.7973	9.9058	.0942	9.8915	10
39° 00	9.7989	9.9084	.0916	9.8905	51° 00′
10	9.8004	9.9110	.0890	9.8895	50
20	9.8020	9.9135	.0865	9.8884	40
30	9.8035	9.9161	.0839	9.8874	30
40	9.8050	9.9187	.0813	9.8864	20
50	9.8066	9.9212	.0788	9.8853	10
40° 00′	9.8081	9.9238	.0762	9.8843	50° 00′
10	9.8096	9.9264	.0736	9.8832	50
20	9.8111	9.9289	.0711	9.8821	40
30	9.8125	9.9315	.0685	9.8810	30
40	9.8140	9.9341	.0659	9.8800	20
50	9.8155	9.9366	.0634	9.8789	10
41° 00′	9.8169	9.9392	.0608	9.8778	49° 00′
10	9.8184	9.9417	.0583	9.8767	50
20	9.8198	9.9443	.0557	9.8756	40
30	9.8213	9.9468	.0532	9.8745	30
40	9.8227	9.9494	.0506	9.8733	20
50	9.8241	9.9519	.0481	9.8722	10
42° 00′	9.8255	9.9544	.0456	9.8711	48° 00′
10	9.8269	9.9570	.0430	9.8699	50
20	9.8283	9.9595	.0405	9.8688	40
30	9.8297	9.9621	.0379	9.8676	30
40	9.8311	9.9646	.0354	9.8665	20
50	9.8324	9.9671	.0329	9.8653	10
43° 00′	9.8338	9.9697	.0303	9.8641	47° 00′
10	9.8351	9.9722	.0278	9.8629	50
20	9.8365	9.9747	.0253	9.8618	40
30	9.8378	9.9772	.0228	9.8606	30
40	9.8391	9.9798	.0202	9.8594	20
50	9.8405	9.9823	.0177	9.8582	10
44° 00′	9.8418	9.9848	.0152	9.8569	46° 00′
10	9.8431	9.9874	.0126	9.8557	50
20	9.8444	9.9899	.0101	9.8545	40
30	9.8457	9.9924	.0076	9.8532	30
40	9.8469	9.9949	.0051	9.8520	20
50	9.8482	9.9975	.0025	9.8507	10
45° 00′	9.8495	.0000	.0000	9.8495	45° 00′
	L. Cos.	L. Cot.	L. Tan.	L. Sin.	Angle

Table V. Amount of 1 at Compound Interest

$$(1 + i)^n$$

n	1½%	2%	2½%	3%	4%	5%	6%	n
1	1.0150	1.0200	1.0250	1.0300	1.0400	1.0500	1.0600	1
2	1.0302	1.0404	1.0506	1.0609	1.0816	1.1025	1.1236	2
3	1.0457	1.0612	1.0769	1.0927	1.1249	1.1576	1.1910	3
4	1.0614	1.0824	1.1038	1.1255	1.1699	1.2155	1.2625	4
5	1.0773	1.1041	1.1314	1.1593	1.2167	1.2763	1.3382	5
6	1.0934	1.1262	1.1597	1.1941	1.2653	1.3401	1.4185	6
7	1.1098	1.1487	1.1887	1.2299	1.3159	1.4071	1.5036	7
8	1.1265	1.1717	1.2184	1.2668	1.3686	1.4775	1.5938	8
9	1.1434	1.1951	1.2489	1.3048	1.4233	1.5513	1.6895	9
10	1.1605	1.2190	1.2801	1.3440	1.4802	1.6289	1.7908	10
11	1.1779	1.2434	1.3121	1.3842	1.5395	1.7103	1.8983	11
12	1.1956	1.2682	1.3449	1.4258	1.6010	1.7959	2.0122	12
13	1.2136	1.2936	1.3785	1.4685	1.6651	1.8856	2.1329	13
14	1.2318	1.3195	1.4130	1.5126	1.7317	1.9799	2.2609	14
15	1.2502	1.3459	1.4483	1.5580	1.8009	2.0789	2.3966	15
16	1.2690	1.3728	1.4845	1.6047	1.8730	2.1829	2.5404	16
17	1.2880	1.4002	1.5216	1.6528	1.9479	2.2920	2.6928	17
18	1.3073	1.4282	1.5597	1.7024	2.0258	2.4066	2.8543	18
19	1.3270	1.4568	1.5987	1.7535	2.1068	2.5270	3.0256	19
20	1.3469	1.4859	1.6386	1.8061	2.1911	2.6533	3.2071	20
21	1.3671	1.5157	1.6796	1.8603	2.2788	2.7860	3.3996	21
22	1.3876	1.5460	1.7216	1.9161	2.3699	2.9253	3.6035	22
23	1.4084	1.5769	1.7646	1.9736	2.4647	3.0715	3.8197	23
24	1.4295	1.6084	1.8087	2.0328	2.5633	3.2251	4.0489	24
25	1.4509	1.6406	1.8539	2.0938	2.6658	3.3864	4.2919	25
26	1.4727	1.6734	1.9003	2.1566	2.7725	3.5557	4.5494	26
27	1.4948	1.7069	1.9478	2.2213	2.8834	3.7335	4.8223	27
28	1.5172	1.7410	1.9965	2.2879	2.9987	3.9201	5.1117	28
29	1.5400	1.7758	2.0464	2.3566	3.1187	4.1161	5.4184	29
30	1.5631	1.8114	2.0976	2.4273	3.2434	4.3219	5.7435	30
31	1.5865	1.8476	2.1500	2.5001	3.3731	4.5380	6.0881	31
32	1.6103	1.8845	2.2038	2.5751	3.5081	4.7649	6.4534	32
33	1.6345	1.9222	2.2589	2.6523	3.6484	5.0032	6.8406	33
34	1.6590	1.9607	2.3153	2.7319	3.7943	5.2533	7.2510	34
35	1.6839	1.9999	2.3732	2.8139	3.9461	5.5160	7.6861	35
36	1.7091	2.0399	2.4325	2.8983	4.1039	5.7918	8.1473	36
37	1.7348	2.0807	2.4933	2.9852	4.2681	6.0814	8.6361	37
38	1.7608	2.1223	2.5557	3.0748	4.4388	6.3855	9.1543	38
39	1.7872	2.1647	2.6196	3.1670	4.6164	6.7048	9.7035	39
40	1.8140	2.2080	2.6851	3.2620	4.8010	7.0400	10.2857	40
41	1.8412	2.2522	2.7522	3.3599	4.9931	7.3920	10.9029	41
42	1.8688	2.2972	2.8210	3.4607	5.1928	7.7616	11.5570	42
43	1.8969	2.3432	2.8915	3.5645	5.4005	8.1497	12.2505	43
44	1.9253	2.3901	2.9638	3.6715	5.6165	8.5572	12.9855	44
45	1.9542	2.4379	3.0379	3.7816	5.8411	8.9850	13.7646	45
46	1.9835	2.4866	3.1139	3.8950	6.0748	9.4343	14.5905	46
47	2.0133	2.5363	3.1917	4.0119	6.3178	9.9060	15.4659	47
48	2.0435	2.5871	3.2715	4.1323	6.5705	10.4013	16.3939	48
49	2.0741	2.6388	3.3533	4.2562	6.8333	10.9213	17.3775	49
50	2.1052	2.6916	3.4371	4.3839	7.1067	11.4674	18.4202	50

Table VI. Present Value of I at Compound Interest

$$(1 + i)^{-n}$$

n	1½%	2%	2½%	3%	4%	5%	6%	n
1	0.9852	0.9804	0.9756	0.9709	0.9615	0.9524	0.9434	1
2	0.9707	0.9612	0.9518	0.9426	0.9246	0.9070	0.8900	2
3	0.9563	0.9423	0.9286	0.9151	0.8890	0.8638	0.8396	3
4	0.9422	0.9238	0.9056	0.8885	0.8548	0.8227	0.7921	4
5	0.9283	0.9057	0.8839	0.8626	0.8219	0.7835	0.7473	5
6	0.9145	0.8880	0.8623	0.8375	0.7903	0.7462	0.7050	6
7	0.9010	0.8706	0.8413	0.8131	0.7599	0.7107	0.6651	7
8	0.8877	0.8535	0.8207	0.7894	0.7307	0.6768	0.6274	8
9	0.8746	0.8368	0.8007	0.7664	0.7026	0.6446	0.5919	9
10	0.8617	0.8203	0.7812	0.7441	0.6756	0.6139	0.5584	10
11	0.8489	0.8043	0.7621	0.7224	0.6496	0.5847	0.5268	11
12	0.8364	0.7885	0.7436	0.7014	0.6246	0.5568	0.4970	12
13	0.8240	0.7730	0.7254	0.6810	0.6006	0.5303	0.4688	13
14	0.8118	0.7579	0.7077	0.6611	0.5775	0.5051	0.4423	14
15	0.7999	0.7430	0.6905	0.6419	0.5553	0.4810	0.4173	15
16	0.7880	0.7284	0.6736	0.6232	0.5339	0.4581	0.3936	16
17	0.7764	0.7142	0.6572	0.6050	0.5134	0.4363	0.3714	17
18	0.7649	0.7002	0.6412	0.5874	0.4936	0.4155	0.3503	18
19	0.7536	0.6864	0.6255	0.5703	0.4746	0.3957	0.3305	19
20	0.7425	0.6730	0.6103	0.5537	0.4564	0.3769	0.3118	20
21	0.7315	0.6598	0.5954	0.5375	0.4388	0.3589	0.2942	21
22	0.7207	0.6468	0.5809	0.5219	0.4220	0.3418	0.2775	22
23	0.7100	0.6342	0.5667	0.5067	0.4057	0.3256	0.2618	23
24	0.6995	0.6217	0.5529	0.4919	0.3901	0.3101	0.2470	24
25	0.6892	0.6095	0.5394	0.4776	0.3751	0.2953	0.2330	25
26	0.6790	0.5976	0.5262	0.4637	0.3607	0.2812	0.2198	26
27	0.6690	0.5859	0.5134	0.4502	0.3468	0.2678	0.2074	27
28	0.6591	0.5744	0.5009	0.4371	0.3335	0.2551	0.1956	28
29	0.6494	0.5631	0.4887	0.4243	0.3207	0.2429	0.1846	29
30	0.6398	0.5521	0.4767	0.4120	0.3083	0.2314	0.1741	30
31	0.6303	0.5412	0.4651	0.4000	0.2965	0.2204	0.1643	31
32	0.6210	0.5306	0.4538	0.3883	0.2851	0.2099	0.1550	32
33	0.6118	0.5202	0.4427	0.3770	0.2741	0.1999	0.1462	33
34	0.6028	0.5100	0.4319	0.3660	0.2636	0.1904	0.1379	34
35	0.5939	0.5000	0.4214	0.3554	0.2534	0.1813	0.1301	35
36	0.5851	0.4902	0.4111	0.3450	0.2437	0.1727	0.1227	36
37	0.5764	0.4806	0.4011	0.3350	0.2343	0.1644	0.1158	37
38	0.5679	0.4712	0.3913	0.3252	0.2253	0.1566	0.1092	38
39	0.5595	0.4619	0.3817	0.3158	0.2166	0.1491	0.1031	39
40	0.5513	0.4529	0.3724	0.3066	0.2083	0.1420	0.0972	40
41	0.5431	0.4440	0.3633	0.2976	0.2003	0.1353	0.0917	41
42	0.5351	0.4353	0.3545	0.2890	0.1926	0.1288	0.0865	42
43	0.5272	0.4268	0.3458	0.2805	0.1852	0.1227	0.0816	43
44	0.5194	0.4184	0.3374	0.2724	0.1780	0.1169	0.0770	44
45	0.5117	0.4102	0.3292	0.2644	0.1712	0.1113	0.0727	45
46	0.5042	0.4022	0.3211	0.2567	0.1646	0.1060	0.0685	46
47	0.4967	0.3943	0.3133	0.2493	0.1583	0.1009	0.0647	47
48	0.4894	0.3865	0.3057	0.2420	0.1522	0.0961	0.0610	48
49	0.4821	0.3790	0.2982	0.2350	0.1463	0.0916	0.0575	49
50	0.4750	0.3715	0.2909	0.2281	0.1407	0.0872	0.0543	50

Table VII. Amount of I Per Period

$$s_{\overline{n}|i} = \frac{(1 + i)^n - 1}{i}$$

n	$\frac{5}{12}\%$	$\frac{1}{2}\%$	1%	$1\frac{1}{4}\%$	$1\frac{1}{2}\%$	2%	3%	n
1	1.0000	1.0000	1.0000	1.0000	1.0000	1.0000	1.0000	1
2	2.0042	2.0050	2.0100	2.0125	2.0150	2.0200	2.0300	2
3	3.0125	3.0150	3.0301	3.0377	3.0452	3.0604	3.0909	3
4	4.0251	4.0301	4.0604	4.0756	4.0909	4.1216	4.1836	4
5	5.0418	5.0503	5.1010	5.1266	5.1523	5.2040	5.3091	5
6	6.0628	6.0755	6.1520	6.1907	6.2296	6.3081	6.4684	6
7	7.0881	7.1059	7.2135	7.2680	7.3230	7.4343	7.6625	7
8	8.1176	8.1414	8.2857	8.3589	8.4328	8.5830	8.8923	8
9	9.1515	9.1821	9.3685	9.4634	9.5593	9.7546	10.1591	9
10	10.1896	10.2280	10.4622	10.5817	10.7027	10.9497	11.4639	10
11	11.2321	11.2792	11.5668	11.7139	11.8633	12.1687	12.8078	11
12	12.2789	12.3356	12.6825	12.8604	13.0412	13.4121	14.1920	12
13	13.3300	13.3972	13.8093	14.0211	14.2368	14.6803	15.6178	13
14	14.3856	14.4642	14.9474	15.1964	15.4504	15.9739	17.0863	14
15	15.4455	15.5365	16.0969	16.3863	16.6821	17.2934	18.5989	15
16	16.5099	16.6142	17.2579	17.5912	17.9324	18.6393	20.1569	16
17	17.5786	17.6973	18.4304	18.8111	19.2014	20.0121	21.7616	17
18	18.6519	18.7858	19.6147	20.0462	20.4894	21.4123	23.4144	18
19	19.7296	19.8797	20.8109	21.2968	21.7967	22.8406	25.1169	19
20	20.8118	20.9791	22.0190	22.5630	23.1237	24.2974	26.8704	20
21	21.8985	22.0840	23.2392	23.8450	24.4705	25.7833	28.6765	21
22	22.9898	23.1944	24.4716	25.1431	25.8376	27.2990	30.5368	22
23	24.0856	24.3104	25.7163	26.4574	27.2251	28.8450	32.4529	23
24	25.1859	25.4320	26.9735	27.7881	28.6335	30.4219	34.4265	24
25	26.2909	26.5591	28.2432	29.1354	30.0630	32.0303	36.4593	25
26	27.4004	27.6919	29.5256	30.4996	31.5140	33.6709	38.5530	26
27	28.5146	28.8304	30.8209	31.8809	32.9867	35.3443	40.7096	27
28	29.6334	29.9745	32.1291	33.2794	34.4815	37.0512	42.9309	28
29	30.7569	31.1244	33.4504	34.6954	35.9987	38.7922	45.2189	29
30	31.8850	32.2800	34.7849	36.1291	37.5387	40.5681	47.5754	30
31	33.0179	33.4414	36.1327	37.5807	39.1018	42.3794	50.0027	31
32	34.1554	34.6086	37.4941	39.0504	40.6883	44.2270	52.5028	32
33	35.2978	35.7817	38.8690	40.5386	42.2986	46.1116	55.0778	33
34	36.4448	36.9606	40.2577	42.0453	43.9331	48.0338	57.7302	34
35	37.5967	38.1454	41.6603	43.5709	45.5921	49.9945	60.4621	35
36	38.7533	39.3361	43.0769	45.1155	47.2760	51.9944	63.2759	36
37	39.9148	40.5328	44.5076	46.6794	48.9851	54.0343	66.1742	37
38	41.0811	41.7354	45.9527	48.2926	50.7199	56.1149	69.1594	38
39	42.2523	42.9441	47.4123	49.8862	52.4807	58.2372	72.2342	39
40	43.4283	44.1588	48.8864	51.4895	54.2679	60.4020	75.4013	40
41	44.6093	45.3796	50.3752	53.1332	56.0819	62.6100	78.6633	41
42	45.7952	46.6065	51.8790	54.7973	57.9231	64.8622	82.0232	42
43	46.9860	47.8396	53.3978	56.4823	59.7920	67.1595	85.4839	43
44	48.1818	49.0788	54.9318	58.1883	61.6889	69.5027	89.0484	44
45	49.3825	50.3242	56.4811	59.9157	63.6142	71.8927	92.7199	45
46	50.5883	51.5758	58.0459	61.6646	65.5684	74.3306	96.5015	46
47	51.7991	52.8337	59.6263	63.4354	67.5519	76.8172	100.3965	47
48	53.0149	54.0978	61.2226	65.2284	69.5652	79.3535	104.4084	48
49	54.2358	55.3683	62.8348	67.0437	71.6087	81.9406	108.5406	49
50	55.4618	56.6452	64.4632	68.8818	73.6828	84.5794	112.7969	50

Table VIII.   Present Value of 1 Per Period

$$a_{\overline{n}|i} = \frac{1 - (1 + i)^{-n}}{i}$$

n	5/12%	1/2%	1%	1¼%	1½%	2%	3%	n
1	0.9959	0.9950	0.9901	0.9877	0.9852	0.9804	0.9709	1
2	1.9876	1.9851	1.9704	1.9631	1.9559	1.9416	1.9135	2
3	2.9752	2.9702	2.9410	2.9265	2.9122	2.8839	2.8286	3
4	3.9587	3.9505	3.9020	3.8781	3.8544	3.8077	3.7171	4
5	4.9381	4.9259	4.8534	4.8178	4.7826	4.7135	4.5797	5
6	5.9135	5.8964	5.7955	5.7460	5.6972	5.6014	5.4172	6
7	6.8848	6.8621	6.7282	6.6627	6.5982	6.4720	6.2303	7
8	7.8521	7.8230	7.6517	7.5681	7.4859	7.3255	7.0197	8
9	8.8153	8.7791	8.5660	8.4623	8.3605	8.1622	7.7861	9
10	9.7746	9.7304	9.4713	9.3455	9.2222	8.9826	8.5302	10
11	10.7299	10.6770	10.3676	10.2178	10.0711	9.7868	9.2526	11
12	11.6812	11.6189	11.2551	11.0793	10.9075	10.5753	9.9540	12
13	12.6286	12.5562	12.1337	11.9302	11.7315	11.3484	10.6350	13
14	13.5721	13.4887	13.0037	12.7706	12.5434	12.1062	11.2961	14
15	14.5116	14.4166	13.8651	13.6005	13.3432	12.8493	11.9379	15
16	15.4472	15.3399	14.7179	14.4203	14.1313	13.5777	12.5611	16
17	16.3790	16.2586	15.5623	15.2299	14.9076	14.2919	13.1661	17
18	17.3069	17.1728	16.3983	16.0295	15.6726	14.9920	13.7535	18
19	18.2309	18.0824	17.2260	16.8193	16.4262	15.6785	14.3238	19
20	19.1511	18.9874	18.0456	17.5993	17.1686	16.3514	14.8775	20
21	20.0675	19.8880	18.8570	18.3697	17.9001	17.0112	15.4150	21
22	20.9801	20.7841	19.6604	19.1306	18.6208	17.6580	15.9369	22
23	21.8889	21.6757	20.4558	19.8820	19.3309	18.2922	16.4436	23
24	22.7939	22.5629	21.2434	20.6242	20.0304	18.9139	16.9355	24
25	23.6952	23.4456	22.0232	21.3572	20.7196	19.5235	17.4131	25
26	24.5927	24.3240	22.7952	22.0813	21.3986	20.1210	17.8768	26
27	25.4865	25.1980	23.5596	22.7963	22.0676	20.7069	18.3270	27
28	26.3766	26.0677	24.3164	23.5025	22.7267	21.2813	18.7641	28
29	27.2630	26.9330	25.0658	24.2000	23.3761	21.8444	19.1885	29
30	28.1457	27.7941	25.8077	24.8889	24.0158	22.3965	19.6004	30
31	29.0248	28.6508	26.5423	25.5693	24.6461	22.9377	20.0004	31
32	29.9002	29.5033	27.2696	26.2413	25.2671	23.4683	20.3888	32
33	30.7720	30.3515	27.9897	26.9050	25.8790	23.9886	20.7658	33
34	31.6402	31.1955	28.7027	27.5605	26.4817	24.4986	21.1318	34
35	32.5047	32.0354	29.4086	28.2079	27.0756	24.9986	21.4872	35
36	33.3657	32.8710	30.1075	28.8473	27.6607	25.4888	21.8323	36
37	34.2231	33.7025	30.7995	29.4788	28.2371	25.9695	22.1672	37
38	35.0770	34.5299	31.4847	30.1025	28.8051	26.4406	22.4925	38
39	35.9273	35.3531	32.1630	30.7185	29.3646	26.9026	22.8082	39
40	36.7740	36.1722	32.8347	31.3269	29.9158	27.3555	23.1148	40
41	37.6173	36.9873	33.4997	31.9278	30.4590	27.7995	23.4124	41
42	38.4571	37.7983	34.1581	32.5213	30.9941	28.2348	23.7014	42
43	39.2933	38.6053	34.8100	33.1075	31.5212	28.6616	23.9819	43
44	40.1261	39.4082	35.4555	33.6864	32.0406	29.0800	24.2543	44
45	40.9555	40.2072	36.0945	34.2582	32.5523	29.4902	24.5187	45
46	41.7814	41.0022	36.7272	34.8229	33.0565	29.8923	24.7754	46
47	42.6039	41.7932	37.3537	35.3806	33.5532	30.2866	25.0247	47
48	43.4230	42.5803	37.9740	35.9315	34.0426	30.6731	25.2667	48
49	44.2387	43.3635	38.5881	36.4755	34.5247	31.0521	25.5017	49
50	45.0509	44.1428	39.1961	37.0129	34.9997	31.4236	25.7298	50

ANSWERS

5. 7, 9.   **6.** −7, −3.   **7.** 1, 1.   **9.** 10, 18.

10. 12, 2.   **11.** −6, −6.   **13.** 2, −2.   **14.** −2, 2.

15. 0, 0.   **17.** −24.   **18.** −60.   **19.** 24.

21. 20.   **22.** 25.   **23.** 15.   **25.** −28.

26. −9.   **27.** 21.   **29.** 6.   **30.** 1.

31. 4.

1. $\frac{4}{6}$.   **2.** $\frac{15}{25}$.   **3.** $\frac{20}{15}$.   **5.** $\frac{2}{5}$.

6. $\frac{2}{3}$.   **7.** $\frac{7}{4}$.   **9.** $\frac{5}{7}$.   **10.** $\frac{3}{1}$.

11. $\frac{9}{6}$.   **13.** $\frac{5}{14}$.   **14.** $\frac{12}{4}$.   **15.** $\frac{27}{6}$.

17. $\frac{3}{4}$.   **18.** $\frac{3}{5}$.   **19.** $\frac{5}{9}$.   **21.** $\frac{5}{7}$.

22. $\frac{3}{4}$.   **23.** $\frac{4}{5}$.   **25.** $\frac{7}{11}$.   **26.** $\frac{4}{7}$.

27. $\frac{2}{3}$.   **29.** $\frac{21}{20}$.   **30.** $\frac{22}{45}$.   **31.** $\frac{39}{140}$.

33. $\frac{4}{3}$.   **34.** $\frac{3}{10}$.   **35.** $\frac{5}{14}$.   **37.** $\frac{7}{15}$.

38. $\frac{5}{22}$.   **39.** $\frac{3}{10}$.   **41.** $\frac{14}{15}$.   **42.** $\frac{68}{91}$.

43. $\frac{52}{45}$.   **45.** $\frac{1}{4}$.   **46.** $\frac{4}{3}$.   **47.** $\frac{3}{2}$.

49. $\frac{1}{7}$.   **50.** $\frac{2}{9}$.   **51.** 4.

347

EXERCISE 1.3, PAGE 13

1. $\frac{29}{21}$. **2.** $\frac{17}{20}$. **3.** $\frac{23}{40}$. **5.** $\frac{8}{15}$.

6. $1\frac{7}{30}$. **7.** $\frac{1}{63}$. **9.** $\frac{3}{2}$. **10.** $\frac{1}{4}$.

11. $\frac{1}{10}$. **13.** $-\frac{1}{12}$. **14.** $\frac{17}{18}$. **15.** $\frac{21}{20}$.

17. $\frac{2}{3}$. **18.** $\frac{2}{5}$. **19.** $\frac{1}{4}$. **21.** $\frac{13}{30}$.

22. $\frac{37}{105}$. **23.** $\frac{47}{60}$. **25.** $1\frac{3}{5}$. **26.** $1\frac{11}{12}$.

27. $2\frac{1}{8}$. **29.** $1\frac{2}{3}$. **30.** $\frac{7}{10}$. **31.** $7\frac{2}{3}$.

33. $1\frac{17}{60}$. **34.** $-\frac{5}{6}$. **35.** $\frac{5}{7}$. **36.** $-\frac{2}{5}$.

EXERCISE 1.4, PAGE 18

1. 27.167. **2.** 34.14. **3.** 27.371. **5.** 11.346.

6. 64.974. **7.** 65.358. **9.** 906.1128. **10.** 4361. 604.

11. 157.7730. **13.** 1.82. **14.** 24.5. **15.** 7.21.

17. 9.8196. **18.** 186.4252. **19.** $-.148823$. **21.** 35.9.

22. 4.80. **23.** .208. **25.** 4.52. **26.** 1.14.

27. 10.7. **29.** $16.93. **30.** 2016 miles

31. $10.50. **33.** $2.20. **34.** $.90. **35.** $13.61.

EXERCISE 1.5, PAGE 21

1. 27. **2.** 119. **3.** 6.2. **5.** 13.

6. $5.0(10^2)$. **7.** $(6.18)(10^3)$. **9.** 3.3.

10. 2.6. **11.** 4.0. **13.** 92.

14. 1.5. **15.** 2.33. **17.** 6.55.

18. 80.3. **19.** 84.4. **21.** 2.06.

22. 6.3. **23.** 10.7. **25.** 20.73.

26. 41.26. **27.** -2.0. **29.** 3.2.

30. 4.0. **31.** 255.8. **33.** $8.35(10^7)$.

34. $1.03(10^7)$. **35.** $1.5(10^6)$. **37.** $5.27(10)$.

38. $2.746(10^2)$. **39.** $7.43(10^2)$.

EXERCISE 1.6, PAGE 25

1. .03, .22. **2.** .008, .91. **3.** .367, .829.

5. 7%, 63%. **6.** 15%, 71%. **7.** 8.6%, 62.9%.

9. $63. **10.** $30. **11.** $11.25.

13. $8.40. **14.** $23.79. **15.** $25.48.

17. 5%. **18.** 6%. **19.** 6%.

21. 6.09%. **22.** 5.56%. **23.** 4.05%.

25. $105, $244.50, $69. **26.** $46.72, $27.74, $107.31.

27. $23.93, $49.16, $20.88. **29.** $572.08.

30. $1264.44. **31.** $1847.01.

33. $723.68, $82.43. **34.** $1578.95, $180.28.

35. $2563.67, $288.47. **37.** $363.

38. $47.60. **39.** $72.80.

EXERCISE 1.7, PAGE 32

1. 5, VII.
3. 60, LIII.
6. 91, DCCXLII.
17. 213.
19. 199.
22. 20,2211₃.
25. 1202₅.
27. 269₁₁.
30. 545.
33. 20343.
35. 34,344.

2. 100, CI.
5. 4, XCII.
7. 612, CMIXX.
18. 223.
21. 2066₇.
23. 1457₁₂.
26. 2417₉.
29. 555.
31. 1240.
34. 14,145.

EXERCISE 2.1, PAGE 35

1. 9.

2. 612.

3. 11.2 seconds, 24.64 seconds.
5. 10,920.
6. $1136.07.
7. 45.
9. 400 feet, 784 feet, 1600 feet.
10. 2023 cubic inches, 1,785,214 cubic feet.
11. 33.5104 cubic inches, 904.7808 cubic inches, 268,083, 200,000 cubic miles.
13. $967.74.
14. 31, 120.
15. 80, 63.
17. 78.20 square centimeters.
18. 9660 cubic centimeters.
19. 29.1 square feet.

EXERCISE 2.2, PAGE 42

1. 1, 7, −17.
3. 8, −7, −26.5.
6. 2, 5, 7.5.
9. 0, $-\frac{1}{2}$, no number.
11. 0, $-\frac{9}{8}$, no number.
14. 18, −2, 42.
17. $\frac{1}{9}$, $-\frac{1}{4}$, $-\frac{31}{25}$.
19. 0, 0, 5.
22. 0, no number, $\frac{55}{12}$.
25. $\frac{64}{63}$.
27. $\frac{675}{833}$.
30. -12, $-\frac{221}{6}$.

2. −14, 11, 26.
5. 2, $\frac{11}{5}$, 4.5.
7. 5, 3, 0.
10. 0, $\frac{15}{7}$, no number.
13. 11, 1, 11.
15. −2, 4, −2.
18. $-\frac{77}{25}$, $-\frac{22}{9}$, $-\frac{9}{8}$.
21. 0, $-\frac{15}{7}$, $-\frac{16}{7}$.
23. 0, no number, $-\frac{19}{18}$.
26. $\frac{144}{25}$.
29. 26, $-\frac{2}{3}$.
31. 23, $\frac{31}{100}$.

3. First, fourth, fourth, third, fourth.

5. −.5

7. 2.

10. −6.

13. −2.3, 1.3.

15. −.4, 5.4.

18. −1.0, .3.

21. −2.9, .4.

23. 1.1, −.5.

6. 1.7.

9. 7.5.

11. −3.3.

14. .4, 2.6.

17. −.9, 2.4.

19. −1.8, .8.

22. 1.2, 2.5.

1. Any four, any three, any two, or any one of the elements a, b, c, d, e, is a proper subset of $\{a, b, c, d, e\}$.

2. A proper subset is obtained by omitting any one, any two, any three, any four, or any five of the given elements.

3. A proper subset is obtained by omitting one or more of rational numbers.

5. A is a subset of B since each element of A is also an element of B.

6. A is not a subset of B since John Jones is not and was not a Vice-President of the U.S.

7. A is not a subset of B since $A \supset B$.

In 9, 10, 11, 13, 14, 15, 17, and 19, an answer is given but there are others.

9. $\{x \mid x$ is a resident of the U.S. $\}$.

10. $\{0, 1, 3, 5, 7, 9, \}$.

11. $\{a, c, e, g, i, b\}$.

13. A President in the given set whose name did not contain a or e; Woodrow Wilson.

14. The members of the set that are divisible by 3; 6, 9, 12.

15. $\{x \mid x$ is between 0 and 1 $\}$.

17. 1   3   5   7   9
   2   5   8 11 23.

18. A one to one correspondence cannot be set up since the two sets do not have the same number of elements.

19. 2   4   6...$2n$...
   3   6   9...$3n$....

21. Yes.

22. They are identical since each element of each is an element of the other.

23. The first set contains the names of all men who were President of the U.S. during the period specified in the second set.

1. $\{2, 3, 4, 6, 8, 9, 10, 12\}$.

3. $\{2, 3, 4, 5, 6, 8, 10, 12\}$.

2. $\{2, 3, 4, 5, 6, 9, 12\}$.

5. $\{6, 12\}$.

6. {3}.   **7.** {2, 4}.
9. {2, 4, 6, 8, 10, 12}.   **10.** {2, 3, 4, 5, 6, 8, 10, 12}.
11. {2, 4, 6, 8, 10, 12}.
13. {(−2, 4), (−2, 5), (−2, 6), (0, 4), (0, 5), (0, 6), (5, 4), (5, 5), (5, 6)}.
14. {(4, −2), (4, 4), (4, 6), (5, −2), (5, 4), (5, 6), (6, −2), (6, 4), (6, 6)}.
15. {(−2, −2), (−2, 4), (−2, 6), (0, −2), (0, 4), (0, 6), (5, −2), (5, 4), (5, 6)}.
17. {(−2, −2), (−2, 4), (−2, 5), (−2, 6), (0, −2), (0, 4), (0, 5), (0, 6), (5, −2),
         (5, 4), (5, 5), (5, 6)}.
18. {(−2, −2), (−2, 4), (−2, 6), (0, −2), (0, 4), (0, 6), (4, −2), (4, 4), (4, 6),
         (5, −2), (5, 4), (5, 6), (6, −2), (6, 4,) (6, 6,)}
19. {(−2, 4), (−2, 6), (0, 4,), (0, 6), (5, 4,) (5, 6)}.
21. {−2, 0, 4, 5, 6}.   **22.** {−2, 4, 6}.
23. {−2, 5}.   **25.** {−2, 0, 4, 5, 6}.
26. {−2, 0, 4, 5, 6}.   **27.** ∅.
29. {−2, 4, 5, 6}.   **30.** {4, 6}.
31. {(5, −2), (5, 0), (5, 4), (5, 5), (5, 6)}.
33. {(5, 4), (5, 6)}.
34. {(−2, 4), (0, 4), (5, 4), (−2, 6), (0, 6), (5, 6)}.
35. {−2, 4, 5, 6}.

EXERCISE 4.1, PAGE 58

1. 4.   **2.** −6.   **3.** −7.
5. 2.   **6.** 3.   **7.** −2.
9. 5.   **10.** 4.   **11.** −7.
13. $\frac{2}{3}$.   **14.** $-\frac{1}{2}$.   **15.** $-\frac{3}{5}$.
17. $\frac{1}{9}$.   **18.** $\frac{1}{7}$.   **19.** $\frac{1}{11}$.
21. 3.   **22.** 3.   **23.** 5.
25. $\frac{13}{20}$.   **26.** $-\frac{1}{42}$.   **27.** $\frac{3}{22}$.
29. $\frac{26}{3}$.   **30.** $\frac{1}{2}$.   **31.** $\frac{10}{7}$.
33. 15.   **34.** 14.   **35.** 21.
37. 30.   **38.** 42.   **39.** 110.
41. $\frac{3}{4}$.   **42.** $\frac{5}{2}$.   **43.** $\frac{4}{3}$.

EXERCISE 4.2, PAGE 61

1. (1.9, .4).   **2.** (3.3, .5).   **3.** Inconsistent.
5. Dependent.   **6.** Inconsistent.   **7.** (−1.6, .8).
9. (.8, −1.2).   **10.** (2, −4).   **11.** Dependent.
13. Inconsistent.   **14.** Dependent.   **15.** (0, 1).
17. Dependent.   **18.** Inconsistent.   **19.** (3.9, −3.1).
21. (.8, .1).   **22.** (1.9, −.8).   **23.** Dependent.
25. Inconsistent.   **26.** Dependent.   **27.** (2.8, −1.5).
29. (2.8, 2.2).   **30.** (.1, −.5).   **31.** Inconsistent.

EXERCISE 4.3, PAGE 65

1. $4x + 6y$.	**2.** $15x - 20y$.	**3.** $14x - 35y$.
5. $-6x + y$.	**6.** $-22x + 68y$.	**7.** $-3x - 30y$.
9. $(2, 1)$.	**10.** $(3, -1)$.	**11.** $(2, -3)$.
13. $(-2, 3)$.	**14.** $(5, -4)$.	**15.** $(-2, -2)$.
17. $(\frac{1}{2}, \frac{1}{3})$.	**18.** $(\frac{2}{3}, -\frac{1}{2})$.	**19.** $(\frac{1}{2}, \frac{1}{4})$.
21. $(3, 2)$.	**22.** $(1, -3)$.	**23.** $(2, -2)$.
25. $(3, -4)$.	**26.** $(2, 0)$.	**27.** $(0, -1)$.
29. $(\frac{1}{6}, \frac{1}{3})$.	**30.** $(\frac{1}{2}, \frac{3}{4})$.	**31.** $(\frac{1}{2}, -\frac{1}{2})$.
33. $(2, -3)$.	**34.** $(-1, -2)$.	**35.** $(3, -4)$.
37. $(2, -1)$.	**38.** $(-2, 3)$.	**39.** $(3, \frac{1}{2})$.
41. $(\frac{1}{2}, -\frac{1}{3})$.	**42.** $(-\frac{1}{3}, \frac{1}{4})$.	**43.** $(\frac{2}{3}, \frac{3}{5})$.

EXERCISE 4.4, PAGE 67

1. $(2, 1, 3)$.	**2.** $(1, 2, -1)$.	**3.** $(-2, 1, -1)$.
5. $(3, 1, 0)$.	**6.** $(2, -2, 1)$.	**7.** $(-5, 4, 1)$.
9. $(1, -1, 2)$.	**10.** $(1, 0, -2)$.	**11.** $(7, 8, -6)$.
13. $(2, -3, -1)$.	**14.** $(1, 2, -3)$.	**15.** $(5, 4, -6)$.
17. $(1, 0, -2)$.	**18.** $(2, -3, 7)$.	**19.** $(5, 5, -5)$.
21. $(\frac{1}{2}, \frac{1}{4}, -\frac{1}{4})$.	**22.** $(\frac{1}{6}, \frac{1}{6}, \frac{1}{3})$.	**23.** $(\frac{1}{5}, \frac{1}{10}, -\frac{1}{5})$.

EXERCISE 4.5, PAGE 71

1. 28, 15.	**2.** 31, 14.	**3.** \$32, \$74.
5. \$7300, \$9400.	**6.** 80, 70.	
7. 108, 14.	**9.** 8 years, 4 years.	
10. 310, 450.	**11.** 18 years, 6 years.	

13. John earned \$500, Fred earned \$600.

14. \$30, \$35.	**15.** 32 calves, 47 pigs.
17. 22 pounds.	**18.** 6.
19. 9 days, 800 miles.	**21.** \$1200 at 5%, \$1700 at 6%.

22. \$2200 at 4%, \$1800 at 5%.

23. \$1400 at 5%, \$2800 at 6%.

25. \$500 at 4%, \$700 at 5%, \$1200 at 6%.

26. \$600 at 3%, \$1800 at 4%, \$1300 at 5%.

27. \$500 at 4%, \$1300 at 5%, \$2300 at 6%.

29. 50 half dollars, 64 quarters, 30 dimes.

30. 24 quarters, 34 dimes, 72 nickels.

31. 7 were 16 years old, 13 were 17 years old, 10 were 18 years old.

EXERCISE 4.6, PAGE 78

1. 5.

2. -5.

3. 6.

5. 2.

6. -10.

7. 4.

9. 18.

10. -41.

11. -16.

13. -22.

14. 34.

15. 0.

17. (2, 1).

18. (3, -1).

19. (3, 2).

21. ($\frac{1}{2}$, -1).

22. (1, $-\frac{1}{3}$).

23. ($\frac{1}{2}$, $\frac{1}{4}$).

25. (1, 2, 1).

26. (2, -1, -3).

27. (4, 2, -3).

29. ($\frac{1}{2}$, $-\frac{1}{4}$, 1).

30. ($\frac{1}{3}$, $\frac{1}{4}$, $\frac{1}{12}$).

31. ($\frac{1}{2}$, $\frac{1}{4}$, $\frac{1}{6}$).

EXERCISE 5.1, PAGE 84

1. x^7.

2. x^8.

3. a^4.

5. $6y^3$.

6. $35y^3$.

7. $6y^5$.

9. x^3.

10. x^5.

11. x^5.

13. $3x^6$.

14. $3x^3$.

15. $2x^4$.

17. $5a^5 - 15a^4 + 20a^3$.

18. $-6a^4 + 21a^3 - 12a^2$.

19. $6a^7 - 8a^6 + 14a^5$.

21. $6x^3 - 5x^2y + 4xy^2 - y^3$.

22. $6x^3 - 17x^2y + 8xy^2 + 6y^3$.

23. $5a^3 - 12a^2b - 19ab^2 - 6b^3$.

25. $2x^4 - 11x^3 + 27x^2 - 31x + 10$.

26. $6x^4 - 25x^3 + 20x^2 + 13x - 12$.

27. $10x^4 + 29x^3 - 16x^2 - 23x + 12$.

29. $3x^3 - 4x^2 + 2x$.

30. $3x^2 + 2x - 5$.

31. $6x^3 - 4x^2 + 2x$.

33. $3x - 1$.

34. $2x - 1$.

35. $3x + 1$.

37. $2x^2 - 3x - 1$.

38. $3x^2 + 2x - 1$.

39. $2x^2 + 3x - 1$.

EXERCISE 5.2, PAGE 86

1. $x^2 - 9$.

2. $x^2 - 25$.

3. $w^2 - 4$.

5. $4x^2 - y^2$.

6. $9a^2 - b^2$.

7. $16x^2 - y^2$.

9. $9x^2 - 4y^2$.

10. $4x^2 - 25y^2$.

11. $25x^2 - 9y^2$.

13. $x^2 - 6x + 9$.

14. $x^2 - 8x + 16$.

15. $x^2 + 10x + 25$.

17. $9a^2 - 6a + 1$.

18. $4a^2 + 4a + 1$.

19. $25w^2 + 10w + 1$.

21. $9x^2 - 12xy + 4y^2$.

22. $4a^2 - 20ab + 25b^2$.

23. $25s^2 - 30st + 9t^2$.

25. $x^2 + x - 6$.

26. $a^2 - 7a + 12$.

27. $b^2 + 8b + 15$.

29. $6x^2 + x - 1$.

30. $12a^2 - 7a + 1$.

31. $10b^2 + 7b + 1$.

33. $6a^2 + 5a - 6$.

34. $10x^2 - 31x + 15$.

35. $8b^2 + 26b + 15$.

37. $6a^2 + 5ab - 6b^2$.

38. $20x^2 + xb - 12b^2$.

39. $12x^2 - 31xy + 20y^2$.

EXERCISE 5.3, PAGE 89

1. $(x + 2)(x - 2)$.
2. $(x + 3)(x - 3)$.
3. $(x + 4)(x - 4)$.
5. $(2a + b)(2a - b)$.
6. $(3a + b)(3a - b)$.
7. $(6b + c)(6b - c)$.
9. $(2x + 3y)(2x - 3y)$.
10. $(5x + 4y)(5x - 4y)$.
11. $(3x + 7y)(3x - 7y)$.
13. $(x - 2)^2$.
14. $(x - 3)^2$.
15. $(x + 5)^2$.
17. $(3t + 1)^2$.
18. $(5y + 1)^2$.
19. $(4a - 1)^2$.
21. $(3x - 2y)^2$.
22. $(2x + 5y)^2$.
23. $(5s + 3t)^2$.
25. $(x + 3)(x - 2)$.
26. $(a - 3)(a - 4)$.
27. $(x + 5)(x + 3)$.
29. $(3x + 1)(2x + 1)$.
30. $(6a - 1)(a - 1)$.
31. $(5b + 1)(2b + 1)$.
33. $(2a + 3)(3a - 2)$.
34. $(5a - 3)(2a - 5)$.
35. $(4a + 3)(2a + 5)$.
37. Not factorable.
38. Not factorable.
39. $(x + 6)(x + 2)$.
41. $(2x - 3)^2$.
42. $(3x + 5)^2$.
43. Not factorable.
45. Not factorable.
46. Not factorable.
47. $(3x - 4y)(2x - y)$.

EXERCISE 5.4, PAGE 93

1. x^6.
2. x^{12}.
3. x^{20}.
5. $9a^2$.
6. $8b^3$.
7. $36a^2$.
9. $512x^6$.
10. $81x^8$.
11. a^6b^9.
13. $(a^3)^2, (a^2)^3$.
14. $(a^5)^3, (a^3)^5$.
15. $(a^3)^3$.
17. $(a^6)^2, (a^2)^6, (a^3)^4, (a^4)^3$.
18. $(a^{15})^2, (a^2)^{15}, (a^{10})^3, (a^3)^{10}, (a^6)^5, (a^5)^6$.
19. $(a^9)^3, (a^3)^9$.
21. $(a - b)(a^2 + ab + b^2)$.
22. $(x - y)(x^2 + xy + y^2)$.
23. $(c + d)(c^2 - cd + d^2)$.
25. $(2c + 3d)(4c^2 - 6cd + 9d^2)$.
26. $(3a + 4d)(9a^2 - 12ad + 16d^2)$.
27. $(5a - 2b)(25a^2 + 10ab + 4b^2)$.
29. $(a - b - c)(a^2 - 2ab + b^2 + ac - bc + c^2)$.
30. $(2a - b + 2c)(4a^2 - 4ab + b^2 - 4ac + 2bc + 4c^2)$.
31. $(3a + 2b + c)(9a^2 - 6ab - 3ac + 4b^2 + 4bc + c^2)$.
33. $a(x - 2)(x - 3)$.
34. $a(y - 1)(y - 2)$.
35. $a(x + a)^2$.
37. $(x + 3 + 2y)(x + 3 - 2y)$.
38. $(a - 2 + 3b)(a - 2 - 3b)$.
39. $(2x + 2y - 3)(2x - 2y + 3)$.
41. $(3a - 2b)(2 - b)$.
42. $(2a + 3)(a + 2b)$.
43. $(3a - 2)(2a - 3b)$.
45. $(a - b)(a - b + c)$.
46. $(x - 2y)(x - 2y + z)$.
47. $(a + b)(a^2 - a - ab - b + b^2)$.
49. $(x + 2)(2x - 1 + y)$.
50. $(3x + 2)(x - 1 - y)$.
51. $(2x - 3)(2x - 5 + y)$.

Exercise 6.1, page 97

1. a^8/b^6.

2. a^{11}/b^2.

3. a^3/b.

5. $3c^5/d^9$.

6. $c^7/6d^5$.

7. $2c^4d^2$.

9. $2x/3$.

10. $2b^4y^2/15$.

11. $2c^2/3w^3$.

13. $\frac{5}{6}$.

14. $\frac{2}{3}$.

15. 4.

17. $\frac{3}{2}$.

18. $\frac{3}{4}$,

19. 5.

21. $1/(3x + y)(2x + 5y)$.

22. $(3a + 4b)(4a - 3b)$.

23. $(3x - 2a)/(5x + a)$.

25. $xy/(x + y)$.

26. $6/(x + 3y)x$.

27. $\frac{2}{3}$.

29. $(x + 3)/(x + 4)$.

30. $(x + 4)/(x + 5)$.

31. $(x + 2)/(2x + 1)$.

33. $(x - 1)(x^2 - x - 1)/(x + 3)(x - 3)(x + 1)$.

34. $(x^2 + 2x + 3)(x - 2)/(x^2 + x + 5)(x - 1)$.

35. $(x - 2)/(2x - 1)$.

37. $a/(a + 2b)$.

38. $(a + 3b)/b$.

39. $(a^2 + 2ab + 4b^2)/(4a^2 - 2ab + b^2)$.

41. $(x^2 - 3x + 9)/(x^2 - x + 1)$.

42. $(x^2 - 5x + 25)/(x^2 + 2x + 4)$.

43. $(2x - 1)(2x + 1)/(x - 4)(x^2 - 4x + 16)$.

Exercise 6.2, page 102

1. $(16x - 14)/3x(x - 2)$.

2. $(x + 1)/2x(3x - 1)$.

3. $(5x - 1)/2x(x + 3)$.

5. $2(x^2 + 2x - 5)/(x + 2)(2x + 1)$.

6. $2(3x^2 - 13)/(x - 1)(2x + 3)$.

7. $-2(6x - 1)(x + 1)/(4x + 1)(3x + 2)$.

9. $(7x + 13)/(x + 1)^2(2x + 5)$.

10. $-2(13x - 7)/(3x - 1)^2(x - 3)$.

11. $(-2x^2 + x + 7)/(3x + 2)(2x - 1)^2$.

13. $(3x^2 - x + 2)/(x + 1)(x - 1)(2x - 1)$.

14. $(10x^2 + 27x + 7)/(3x - 1)(x + 2)(2x + 3)$.

15. $-6(6x + 1)/(2x - 1)(2x + 3)(2x - 3)$.

17. $(3x^3 - 6x^2 + 4x - 4)/(x - 1)(x + 2)(x - 2)$.

18. $(5x^3 - 19x^2 + 39x - 63)/(x + 3)(2x - 3)(x - 3)$.

19. $(7x^3 + 29x^2 + 7x - 18)/(4x + 1)(x + 4)(x + 3)$.

21. 1.

22. 2.

23. 3.

25. 2.

26. 1.

27. -2.

29. 0.

30. 1.

31. -1.

33. $-.5$.

34. 1.75.

35. 0.

37. -1.

38. 0.

39. 2.

9. 13.   **10.** 17.   **11.** 5.
13. ± 4.   **14.** $\pm 8.$ **15.** ± 24.
17. -5.   **18.** -8.   **19.** -24.
21. $y = 24, r = 26$.   **22.** $y = -8, r = 10$.
23. $y = -7.5, r = 8.5$.   **25.** $x = -24, r = 26$.
26. $x = -3.5, r = 12.5$.   **27.** $x = 3, r = 5$.
29. $y = 3, r = 5$.   **30.** $y = -24, r = 25$.
31. $x = 6, r = 10$.

In each of the first 12 problems, the values of the functions are given in the order sin A, cos A, tan A, cot A, sec A, csc A.

1. $\frac{12}{13}, \frac{5}{13}, \frac{12}{5}, \frac{5}{12}, \frac{13}{5}, \frac{13}{12}$.

2. $\frac{3}{5}, \frac{4}{5}, \frac{3}{4}, \frac{4}{3}, \frac{5}{4}, \frac{5}{3}$.

3. $\frac{15}{17}, \frac{8}{17}, \frac{15}{8}, \frac{8}{15}, \frac{17}{8}, \frac{17}{15}$.

5. $-\frac{5}{13}, \frac{12}{13}, -\frac{5}{12}, -\frac{12}{5}, \frac{13}{12}, -\frac{13}{5}$.

6. $\frac{24}{25}, -\frac{7}{25}, -\frac{24}{7}, -\frac{7}{24}, -\frac{25}{7}, \frac{25}{24}$.

7. $-\frac{15}{17}, -\frac{8}{17}, \frac{15}{8}, \frac{8}{15}, -\frac{17}{8}, -\frac{17}{15}$.

9. $\frac{2}{3}, -\sqrt{5}/3, -2/\sqrt{5}, -\sqrt{5}/2, -3/\sqrt{5}, \frac{3}{2}$.

10. $-\sqrt{11}/6, \frac{5}{6}, -\sqrt{11}/5, -5/\sqrt{11}, \frac{6}{5}, -6/\sqrt{11}$.

11. $-\frac{2}{5}, \sqrt{21}/5, -2/\sqrt{21}, -\sqrt{21}/2, 5/\sqrt{21}, -\frac{5}{2}$.

13. Fourth.   **14.** Third.   **15.** Fourth.
17. Fourth.   **18.** Third.   **19.** Second.

The order is the same in the answers below as in 1 through 12, and the given value is also included.

21. $\frac{3}{5}, \pm\frac{4}{5}, \pm\frac{3}{4}, \pm\frac{4}{3}, \pm\frac{5}{4}, \frac{5}{3}$.

22. $\pm\frac{12}{13}, \frac{5}{13}, \pm\frac{12}{5}, \pm\frac{5}{12}, \frac{13}{5}, \pm\frac{13}{12}$.

23. $\pm\frac{24}{25}, \frac{7}{25}, \pm\frac{24}{7}, \pm\frac{7}{24}, \frac{25}{7}, \pm\frac{25}{24}$,

25. $\pm\frac{15}{17}, -\frac{8}{17}, \pm\frac{15}{8}, \pm\frac{8}{15}, -\frac{17}{8}, \pm\frac{17}{15}$.

26. $\pm\frac{5}{13}, -\frac{12}{13}, \pm\frac{5}{12}, \pm\frac{12}{5}, -\frac{13}{12}, \pm\frac{13}{5}$.

27. $-\frac{4}{5}, \pm\frac{3}{5}, \pm\frac{4}{3}, \pm\frac{3}{4}, \pm\frac{5}{3}, -\frac{5}{4}$.

29. $\pm\frac{7}{25}, \pm\frac{24}{25}, \frac{7}{24}, \frac{24}{7}, \pm\frac{25}{24}, \pm\frac{25}{7}$.

30. $\pm\frac{5}{13}, \pm\frac{12}{13}, \frac{12}{5}, \frac{5}{12}, \pm\frac{13}{12}, \pm\frac{13}{5}$.

31. $\pm\frac{3}{5}, \mp\frac{4}{5}, -\frac{3}{4}, -\frac{4}{3}, \mp\frac{5}{4}, \pm\frac{5}{3}$.

33. $-\frac{5}{13}, -\frac{12}{13}, \frac{5}{12}, \frac{12}{5}, -\frac{13}{12}, -\frac{13}{5}$.

34. $\frac{8}{17}, -\frac{15}{17}, -\frac{8}{15}, -\frac{15}{8}, -\frac{17}{15}, \frac{17}{8}$.

35. $-\frac{24}{25}, -\frac{7}{25}, \frac{24}{7}, \frac{7}{24}, -\frac{25}{7}, -\frac{25}{24}$.

EXERCISE 7.3, PAGE 115

In Problems 1 through 8, the answers are given in the usual order.

1. $1/\sqrt{2}, -1/\sqrt{2}, -1, -1, -\sqrt{2}, \sqrt{2}; -\sqrt{3}/2, -\frac{1}{2}, -\sqrt{3}, -1/\sqrt{3}, -2, -2/\sqrt{3}.$

2. $-\frac{1}{2}, -\sqrt{3}/2, 1/\sqrt{3}, \sqrt{3}, -2/\sqrt{3}, -2; 1, 0,$ no number, 0, no number, 1.

3. $-1/\sqrt{2}, -1/\sqrt{2}, 1, 1, -\sqrt{2}, -\sqrt{2}; -1, 0,$ no number, 0, no number, $-1.$

5. $\frac{1}{2}, -\sqrt{3}/2, -1/\sqrt{3}, -\sqrt{3}, -2/\sqrt{3}, 2; 0, 1, 0,$ no number, 1, no number.

6. $-1/\sqrt{2}, 1/\sqrt{2}, -1, -1, \sqrt{2}, -\sqrt{2}; -\sqrt{3}/2, \frac{1}{2}, -\sqrt{3}, -1/\sqrt{3}, 2, -2/\sqrt{3}$

7. $-\frac{1}{2}, \sqrt{3}/2, -1/\sqrt{3}, -\sqrt{3}, 2/\sqrt{3}, -2; \sqrt{3}/2, -\frac{1}{2}, -\sqrt{3}, -1/\sqrt{3}, -2, 2/\sqrt{3}.$

9. $1.$ **10.** $(1-\sqrt{3})/2\sqrt{2}.$ **11.** $(\sqrt{3}-1)2\sqrt{2}.$

13. $(\sqrt{3}+1)/2\sqrt{2}.$ **14.** $(1+\sqrt{3})/2\sqrt{2}.$ **15.** $-1.$

EXERCISE 8.1, PAGE 119

1. .5348.	**2.** .8774.	**3.** .2711.
5. .9925.	**6.** .7400.	**7.** .9110.
9. 2.177.	**10.** .1793.	**11.** .7826.
13. 1°30′.	**14.** 10°30′.	**15.** 21°50′.
17. 38°40′.	**18.** 39°30′.	**19.** 52°.
21. 65°30′.	**22.** 68°50′.	**23.** 78°20′.
25. 6.926.	**26.** .2333.	**27.** .8966.
29. .7775.	**30.** .7104.	**31.** .7375.
33. .4784.	**34.** .9335.	**35.** .2016.
37. 1°17′.	**38.** 3°49′.	**39.** 12°32′.
41. 30°57′.	**42.** 35°9′.	**43.** 43°21′.
45. 58°27′.	**46.** 71°34′.	**47.** 66°55′.

EXERCISE 8.2, PAGE 124

1. $B = 35°40′, a = 230, b = 165.$
2. $B = 57°30′, a = 307, b = 482.$
3. $A = 17°10′, a = 12.2, b = 39.6.$
5. $A = 36°20′, B = 53°40′, b = 418.$
6. $A = 41°20′, B = 48°40′, b = .733.$
7. $A = 68°20′, B = 21°40′, a = 2.58.$
9. $A = 41°, B = 49°, c = 669.$
10. $A = 50°20′, B = 39°40′, c = 670.$
11. $A = 18°30′, B = 71°30′, c = .0353.$
13. $B = 48°50′, b = 27.3, c = 36.3.$
14. $B = 28°, b = 286, c = 608.$
15. $A = 16°30′, a = 127, c = 446.$
17. $A = 39°10′, b = 21.2, c = 27.4.$
18. $A = 32°50′, b = .598, c = .712.$
19. $B = 53°, a = 2.09, c = 3.47.$

21. $A = 27°47'$, $a = 3762$, $b = 7141$.
22. $A = 50°6'$, $a = 4557$, $b = 3812$.
23. $B = 62°22'$, $a = 440.5$, $b = 841.5$.
25. $A = 43°14'$, $B = 46°46'$, $a = 66.98$.
26. $A = 47°20'$, $B = 42°40'$, $a = .6301$.
27. $A = 46°1'$, $B = 43°59'$, $b = 3971$.
29. $A = 38°50'$, $B = 51°10'$, $c = 4420$.
30. $A = 50°24'$, $B = 39°36'$, $c = .6486$.
31. $A = 58°41'$, $B = 31°19'$, $c = 8.639$.
33. $A = 37°24'$, $a = 4588$, $c = 7559$.
34. $A = 52°30'$, $a = 6148$, $c = 7750$.
35. $B = 66°4'$, $b = .5308$, $c = .5807$.
37. $B = 52°23'$, $a = 46.34$, $c = 75.91$.
38. $B = 26°16'$, $a = 6641$, $c = 7406$.
39. $A = 61°42'$, $b = .4370$, $c = .9219$.

EXERCISE 8.3, PAGE 127

1. 182 inches, 53°40′.
2. 385 feet, 714 feet.
3. 13.08 centimeters, 32.74 centimeters.
5. 48.4 feet.
6. 291 feet.
7. 523 feet.
9. 505 feet.
10. 389 feet.
11. 20 feet.
13. 926 miles, N52°30′W.
14. 447 miles, S79°10′W.
15. 625 miles, N62°50′W.
17. 41°, 299 feet.
18. 420 miles, S80°E.
19. 389.2 feet, 1058 feet.

EXERCISE 9.1, PAGE 133

1. (2, 405°), (−2, 225°), (−2, −135°); (5, 500°), (−5, 320°), (−5, −40°); (5, 575°), (−5, 395°), (−5, 35°); (4, 290°), (−4, 110°), (−4, −250°).
2. (3, 420°), (−3, 240°), (−3, −120°); (−2, 490°), (2, 310°), (2, −50°); (4, 590°), (−4, 410°), (−4, 50°); (5, 320°), (−5, 140°), (−5, −220°).
3. (4, 475°), (−4, 295°), (−4, −65°); (3, 325°), (−3, 145°), (−3, −215°); (6, 360°), (−6, 180°), (−6, −180°); (2, 230°), (−2, 50°), (−2, −310°).
5. (5, 400°), (−5, 220°), (−5, −140°); (3, 315°), (−3, 135°), (−3, −225°); (−2, 300°), (2, 120°), (2, −240°); (5, 675°), (−5, 495°), (−5, 135°).
6. (−2, 390°), (2, 210°), (2, −150°); (2, 330°), (−2, 150°), (−2, −210°); (4, 450°), (−4, 270°), (−4, −90°); (3, 600°), (−3, 420°), (−3, 60°).
7. (−3, 630°), (3, 450°), (3, 90°); (7, 165°), (−7, −15°), (−7, −375°); (−4, 250°), (4, 70°), (4, −290°); (2, 570°), (−2, 390°), (−2, 30°).
9. 69°10′, 35°, 70°, 35°.
10. 40°, 40°, 50°, 49°40′.
11. 20°, 80°, 59°30′, 20°.
13. sin 45°.
14. −cos 40°30′.
15. −tan 54°40′.
17. sec 80°.
18. csc 39°20′.
19. −cos 50°20′.
21. cot 35°50′.

22. sec 40°.

23. csc 80°.

25. −tan 14°50′.

26. cot 45°.

27. sec 70°.

29. sin 49°30′.

30. cos 35°30′.

31. −tan 30°.

33. −.6472.

34. −.6539.

35. −2.675.

37. .6338.

38. −.4950.

39. .8195.

EXERCISE 9.2, PAGE 139

The intercepts are given first; following the semicolon are points or lines of symmetry as found by applying the tests given in Section 9.5. There may be symmetry that does not show up with these tests.

1. (4, 0°), (4, 90°), (4, 180°), (4, 270°); pole, both axes.

2. (3, 0°), (3, 90°), (3, 180°), (3, 270°); pole, both axes.

3. (−2, 0°), (−2, 90°), (−2, 180°), (−2, 270°); pole, both axes.

5. The pole; pole.   **6.** The pole; pole.

7. The pole; pole.

9. (4, 0°), (0, 90°), (−4, 180°), (0, 270°); polar axis.

10. (0, 0°), (2, 90°), (0, 180°), (−2, 270°); normal axis.

11. (0, 0°), (6, 90°), (0, 180°), (−2, 270°); normal axis.

13. (0, 0°), (0, 90°), (0, 180°), (0, 270°); pole.

14. (0, 0°), (0, 90°), (0, 180°), (0, 270°); pole.

15. (1, 0°), (1, 90°), (1, 180°), (1, 270°); pole, both axes.

17. (1, 0°), (0, 90°), (−1, 180°), (0, 270°); polar axis.

18. (1, 0°), (0, 90°), (−1, 180°), (0, 270°), polar axis.

19. (0, 0°), (1, 90°), (0, 180°), (−1, 270°); normal axis.

21. (−2, 0°), (0, 90°), (−2, 180°), (−4, 270°); normal axis.

22. (4, 0°), (2, 90°), (0, 180°), (2, 270°); polar axis.

23. (−4, 0°), (−1, 90°), (−4, 180°), (−7, 270°); normal axis.

25. (1, 0°), (−1, 180°); polar axis.

26. (0, 0°), (0, 180°); pole.

27. (0, 90°), (0, 270°); pole.

29. (0, 0°), (±1, 90°), (0, 180°); normal axis.

30. (±1, 0°), (0, 90°), (0, 270°); polar axis.

31. (±1, 0°), (±1, 90°), (±1, 180°), (±1, 270°); pole, both axes.

33. (2, 0°), (1, 90°), (2, 180°); normal axis.

34. (2, 0°), (2, 180°), (1, 270°); normal axis.

35. (1.5, 0°), (3, 90°), (3, 270°); polar axis.

37. (−4, 0°), (4, 90°), ($\frac{4}{3}$, 180°), (4, 270°); polar axis.

38. (6, 0°), (1.5, 90°), (6, 180°), (−3, 270°); normal axis.

39. ($\frac{1}{3}$, 0°), (.5, 90°), ($\frac{1}{3}$, 180°), ($\frac{1}{4}$, 270°); normal axis.

EXERCISE 9.3, PAGE 141

Each odd numbered equation is followed by the same equation in terms of another coordinate system.

EXERCISE 10.1, PAGE 146

1. $\frac{1}{8}$. **2.** $\frac{1}{25}$. **3.** $\frac{1}{6}$. **5.** $\frac{1}{125}$.

6. $\frac{1}{243}$. **7.** $\frac{1}{36}$. **9.** $\frac{1}{9}$. **10.** 16.

11. 5. **13.** 8. **14.** $\frac{1}{9}$. **15.** $\frac{1}{16}$.

17. $1/a^3$. **18.** $1/x^2$. **19.** $1/y$. **21.** $1/a^6$.

22. $1/b$. **23.** $1/b^3$. **25.** x. **26.** y^3.

27. $1/a^5$. **29.** b^5/a^3. **30.** $1/ab^2$. **31.** $1/a^7b$.

33. a^4/b^8. **34.** $1/b^6$. **35.** c^4/d^2. **37.** $1/a^6b^9$.

38. $1/a^2b^4$. **39.** a^6b^3. **41.** $(b-a)/ab$. **42.** $(b^2+a^2)/ab$.

43. $2ab$. **45.** $(b^2-a)/ab$. **46.** $(b^2-a^3)/a^2b^2$.

47. $a(b+a^2)/b$. **49.** $ab/(b+a)$. **50.** a.

51. $(a-b)(a+b)/ab$.

EXERCISE 10.2, PAGE 149

1. 3. **2.** 11. **3.** 17. **5.** 3.

6. 7. **7.** 3. **9.** -4. **10.** -3.

11. -2. **13.** 3.8. **14.** .53. **15.** 73.

17. 8.73. **18.** 19.5. **19.** 16.8. **21.** 2.76.

22. 6.18. **23.** 5.30. **25.** 1.237. **26.** 9.016.

27. 24.67. **29.** 61.1. **30.** 41. **31.** 17.1 feet.

33. 18.42. **34.** 18.4 **35.** 38.7 feet.

EXERCISE 10.3, PAGE 152

1. 3. **2.** 2. **3.** 3. **5.** 64.

6. 9. **7.** 32. **9.** $\frac{64}{125}$. **10.** $\frac{9}{4}$.

11. $\frac{243}{32}$. **13.** $\frac{1}{2}$. **14.** $\frac{1}{5}$. **15.** $\frac{1}{3}$.

17. $\frac{1}{25}$. **18.** $\frac{1}{216}$. **19.** $\frac{1}{8}$. **21.** $a^{7/12}$.

22. $b^{7/10}$. **23.** $c^{8/15}$. **25.** $a^{1/6}$. **26.** $c^{1/12}$.

27. $d^{3/20}$. **29.** $a^{1/4}/a$. **30.** $a^{1/3}/a$. **31.** $a^{1/2}/a$.

33. $a^{1/2}b^{1/3}/ab$. **34.** $a^{1/3}b^{1/4}/ab$. **35.** $a^{1/4}b^{1/2}/ab$.

37. $3a^{1/3}b^{1/5}$. **38.** $3a^{1/2}b^{1/4}$. **39.** $5x^{1/3}y^{1/5}$.

41. $25ab^{1/2}/b$. **42.** $8x^{3/5}/y$. **43.** $a^{2/3}b^{4/5}/8ab$.

45. x/y. **46.** $a^{1/3}/b$. **47.** $a^{1/2}b/4a^2$.

49. $ab^{2/3}/b^2$. **50.** $a^2c^{2/3}/4b$. **51.** $b^{1/3}c^{1/4}/3a$.

53. $9c^{1/2}/ac^2$. **54.** $16a^3c^{2/3}/c^4$. **55.** $2a^2b^{1/3}c^{1/2}/bc^2$.

EXERCISE 10.4, PAGE 154

1. 5. **2.** 3. **3.** 4. **5.** $2\sqrt[3]{3}$.

6. $3\sqrt[4]{2}$. **7.** $2\sqrt[5]{3}$. **9.** $b\sqrt[5]{a}$. **10.** $ab\sqrt{a}$.

11. $ab\sqrt[3]{b}$. **13.** $2ab\sqrt{b}$. **14.** $3a^2b\sqrt{b}$. **15.** $4ab^2\sqrt{a}$.

17. $2xy\sqrt[3]{3y}$. **18.** $3x^2y^3\sqrt[3]{3y}$. **19.** $2xy\sqrt[4]{2y^2}$. **21.** $\frac{y}{a}\sqrt[4]{a^2y}$.

22. $\frac{x}{y}\sqrt[5]{x^3y^4}$. **23.** $\frac{y}{x}\sqrt{xy}$. **25.** $\frac{2}{x}\sqrt[5]{x}$. **26.** $\frac{2}{xy}\sqrt{2y}$.

27. $\dfrac{3}{x}\sqrt[3]{2x}$.   **29.** $b^2\sqrt[6]{a^5b^5}$.   **30.** $a^2b\sqrt[6]{a^5b^4}$.   **31.** $a^2b\sqrt[12]{a^5b^7}$.

33. $b\sqrt[4]{a}$.   **34.** $\dfrac{a^2}{b^2}\sqrt[6]{ab^5}$.   **35.** $\dfrac{1}{ab}\sqrt[15]{a^{11}b^4}$.

37. $\dfrac{3y^2}{x}$.   **38.** $\dfrac{7b^3}{a}$.   **39.** $\dfrac{2a}{3b^2}\sqrt[3]{a^2b^2}$.

41. $\sqrt[6]{xy}$.   **42.** $\dfrac{b}{a^2}\sqrt[6]{ab}$.   **43.** $\dfrac{b}{a^3}\sqrt[15]{a^{13}b^8}$.

Exercise 10.5, page 156

1. $9\sqrt{3}$.   **2.** $\sqrt{5}$.   **3.** $7\sqrt{2}$.

5. $\sqrt{3}-\sqrt{2}$.   **6.** $4\sqrt{2}-\sqrt{5}$.   **7.** $3\sqrt{2}-4\sqrt{3}$.

9. $4\sqrt{2}-7\sqrt[3]{3}$.   **10.** $-2\sqrt{3}-\sqrt[3]{2}$.   **11.** $6\sqrt[3]{4}-\sqrt{6}$.

13. $(y^2+xy-x^2)\sqrt{xy}$.   **14.** $(y+2x+xy)\sqrt{xy}$.

15. $(x+y)(x+2y)\sqrt{y}$.   **17.** $(y^2+y)\sqrt{x}-(x^2+x)\sqrt{y}$.

18. $y(x^2+3y)\sqrt{xy}+xy(2xy-3)\sqrt{x}$.

19. $2x\sqrt[3]{xy^2}+3xy\sqrt[3]{x^2y}-2x\sqrt{y}-3y\sqrt{x}$.

21. $\dfrac{4x^2-xy+4y}{4x}\sqrt{2xy}$.   **22.** $\dfrac{9xy+6x-y^3}{3xy^2}\sqrt{3xy}$.

23. $\dfrac{4x-8y-1}{4xy}\sqrt[3]{xy^2}$.   **25.** $\dfrac{x-2y}{xy}\sqrt{xy}+\dfrac{x-y^2}{xy}\sqrt[3]{xy}$.

26. $\dfrac{x+2y^2}{xy}\sqrt{2x}+\dfrac{y-2x}{y^2}\sqrt[3]{2x}$.   **27.** $\dfrac{3x+y^2}{3xy}\sqrt[3]{3x}+\dfrac{1}{3x}\sqrt{x}$.

Exercise 11.1, page 159

1. $5^3=125$.   **2.** $3^4=81$.   **3.** $8^2=64$.

5. $4^{2.5}=32$.   **6.** $9^{1.5}=27$.   **7.** $27^{4/3}=81$.

9. $3^{-2}=\frac{1}{9}$.   **10.** $4^{-3}=\frac{1}{64}$.   **11.** $2^{-5}=\frac{1}{32}$.

13. $\log_2 64=6$.   **14.** $\log_3 81=4$.   **15.** $\log_7 343=3$.

17. $\log_9 27=1.5$.   **18.** $\log_{16} 32=1.25$.   **19.** $\log_{32} 16=.8$.

21. $\log_2 \frac{1}{8}=-3$.   **22.** $\log_5 \frac{1}{625}=-4$.

23. $\log_6 \frac{1}{216}=-3$.   **25.** 9.

26. 125.   **27.** 1.   **29.** 4.

30. 2.   **31.** 27.   **33.** $\frac{1}{4}$.

34. $\frac{1}{27}$.   **35.** $\frac{1}{625}$.   **37.** 3.

38. 3.   **39.** 4.   **41.** 1.5.

42. $\frac{4}{3}$.   **43.** 1.5.   **45.** 5.

46. 2.   **47.** 3.   **49.** 9.

50. 8.   **51.** 16.

EXERCISE 11.2, PAGE 163

1. 2.5132.

2. 1.8543.

3. 0.9355.

5. 0.2405.

6. 2.7657.

7. 1.3927.

9. 1.3766.

10. 0.9571.

11. 2.9425.

13. $9.7818 - 10.$

14. $8.2480 - 10.$

15. $7.8494 - 10.$

17. $7.3243 - 10.$

18. $9.9217 - 10.$

19. $8.8597 - 10.$

21. $8.7419 - 10.$

22. $7.0755 - 10.$

23. $9.8312 - 10.$

25. 13.2.

26. 277.

27. 7.43.

29. 7.22.

30. 40.5.

31. 486.

33. .914.

34. .0156.

35. .0192.

37. 52.3.

38. 768.

39. 6.93.

41. 175.

42. 8.44.

43. 47.6.

45. 1.98.

46. 34.2.

47. 778.

EXERCISE 11.3, PAGE 168

1. $\log a + \log b.$

2. $\log c + \log d.$

3. $\log p + \log a + \log r.$

5. $\log a - \log b.$

6. $\log k - \log t.$

7. $\log k - \log(a + 2).$

9. $b \log a.$

10. $k \log c.$

11. $p \log(x - 1).$

13. $\log b + \log d - \log s.$

14. $\log a - \log s - \log t.$

15. $\log a + \log m - \log b - \log t.$

17. $t \log a + \log y.$

18. $3 \log b + a \log k.$

19. $\log s + \frac{1}{3} \log t.$

21. $\log a k.$

22. $\log 5b.$

23. $\log abc$

25. $\log \dfrac{A}{S}.$

26. $\log \dfrac{B}{3}.$

27. $\log \dfrac{AK}{T}.$

29. $\log \dfrac{M^a}{N}.$

30. $\log \dfrac{P}{M^a}.$

31. $\log \dfrac{a^2}{(b - 1)^3}.$

33. 181.

34. 243.

35. 8.51.

37. 7.27.

38. 0.727.

39. 30.0.

41. 3.32.

42. 3.63.

43. 1.24.

45. 299.

46. 52.3.

47. .932.

49. .964.

50. 21.2.

51. .353.

EXERCISE 11.4, PAGE 171

1. 18.15.

2. 2.227.

3. 636.7.

5. .5282.

6. .02652.

7. .001507.

9. 1.5845.

10. .8279.

11. 3.6987.

13. $9.8967 - 10.$

14. $8.1626 - 10.$

15. $7.4125 - 10.$

17. 772.2.

18. 45.77.

19. 382.0.

21. 1.988.

22. 2.097.

23. .003162.

25. 2.919.	**26.** 4.239.	**27.** .1334.
29. 22.51.	**30.** 23.11.	**31.** 9.054.
33. .8830.	**34.** .7812.	**35.** 181.6.
37. 4.636.	**38.** 1.705.	**39.** 1.421.

EXERCISE 12.1, PAGE 174

1. 9.4269 − 10.	**2.** 9.9765 − 10.
3. 9.6850 − 10.	**5.** 9.8884 − 10.
6. 9.9392 − 10.	**7.** 9.0093 − 10.
9. 0.0557.	**10.** 9.9006 − 10.
11. 9.9817 − 10.	**13.** 3°40′.
14. 7°30′.	**15.** 14°30′.
17. 30°30′.	**18.** 34°30′.
19. 47°50′.	**21.** 56°50′.
22. 68°.	**23.** 74°20′.
25. 9.1684 − 10.	**26.** 0.4995.
27. 9.5667 − 10.	**29.** 0.1292.
30. 9.8314 − 10.	**31.** 9.7908 − 10.
33. 9.9701 − 10.	**34.** 9.4520 − 10.
35. 0.7567.	**37.** 22°54′.
38. 32°53′.	**39.** 78°25′.
41. 17°29′.	**42.** 50°34′.
43. 79°10′.	**45.** 39°56′.
46. 31°35′.	**47.** 1°56′.

EXERCISE 12.2, PAGE 179

1. $b = 485$, $c = 538$, $C = 87°20′$.
2. $a = 238$, $c = 199$, $A = 85°50′$.
3. $a = 52.2$, $b = 69.7$, $B = 76°10′$.
5. $a = 66.6$, $c = 53.8$, $A = 105°10′$.
6. $a = 3.95$, $b = 5.27$, $B = 97°20′$.
7. $b = .570$, $c = .637$, $C = 97°40′$.
9. $a = 8.09$, $c = 5.79$, $B = 83°40′$.
10. $a = 38.1$, $b = 69.7$, $C = 106°10′$.
11. $b = .639$, $c = .835$, $A = 81°50′$.
13. $b = 5.873$, $c = 8.522$, $C = 67°55′$.
14. $a = 5195$, $c = 4166$, $A = 75°17′$.
15. $a = 4433$, $b = 4637$, $B = 67°59′$.
17. $b = 4.787$, $c = 8.058$, $A = 94°34′$.
18. $a = 5159$, $c = 6694$, $B = 97°39′$.
19. $a = .5370$, $b = .7000$, $C = 94°18′$.

21. 42.0 centimeters.   **22.** 363 varas, 441 varas.
23. 138 feet.   **25.** 107 feet.
26. 91.8 feet, 120 feet.
27. Wind, 22.3 miles per hour; ground speed 239 miles per hour.

Exercise 12.3, page 183

1. No solution.   **2.** No solution.
3. No solution.
5. One solution, $A = 90°$, $B = 26°50'$, $b = 440$.
6. One solution, $A = 90°$, $C = 53°$, $c = 247$.
7. One solution, $A = 38°50'$, $C = 99°20'$, $c = 1.14(10^3)$.
9. One solution, $B = 39°50'$, $C = 22°50'$, $c = 65.5$.
10. One solution , $C = 34°30'$, $A = 11°20'$, $a = 73.7$.
11. No solution.
13. Two solutions, $B = 44°40'$, $C = 95°50'$, $c = 46.5$, $B' = 135°20'$, $C' = 5°10'$, $c' = 4.21$.
14. Two solutions, $C = 50°30'$, $A = 82°40'$, $a = 5.37$, $C' = 129°30'$, $A' = 3°40'$, $a' = .346$.
15. Two solutions, $A = 67°$, $B = 55°20'$, $b = .490$, $A' = 113°$, $B' = 9°20'$, $b' = .0966$.
17. 752 miles, S12°10'W.
18. 20 minutes.
19. 510 miles, N83°50'W.

Exercise 12.4, page 186

1. 48°.   **2.** 79°.   **3.** 50°.   **5.** 98°.
6. 93°.   **7.** 108°   **9.** 129°30'.   **10.** 119°20'.
11. 98°50'.   **13.** 41.   **14.** 66.   **15.** 40.
17. 45.   **18.** 88.   **19.** 83.   **21.** 245.
22. 521.   **23.** 611.   **25.** 105°20', 74°40'.
26. 42 inches, 22 inches.   **27.** 223 mph, S21°10'W.

Exercise 12.5, page 189

1. 5.5(10²).   **2.** 1.4(10²).   **3.** 1.2(10³).
5. 1.2(10³).   **6.** 1.6(10³).   **7.** 3.0(10³).
9. 766.   **10.** 2.69(10⁴).   **11.** 1.47(10⁵).
13. 3.3(10²).   **14.** 98.   **15.** 9.2(10)².
17. 4.73(10⁴).   **18.** 6.45(10⁴).   **19.** 5.09(10⁴).
21. 563.0.   **22.** 32.52.   **23.** 1.174(10⁵).

Exercise 13.1, page 191

1. ± 2.   **2.** ± 1.   **3.** ± 5.   **5.** $\pm\frac{4}{3}$.
6. $\pm\frac{3}{2}$.   **7.** $\pm\frac{5}{4}$. **9.** 1, 2.   **10.** 2, 3.
11. 2, -4.   **13.** -2, 5.   **14.** -3, 7.   **15.** -4, -5.
17. 2, $\frac{5}{3}$. **18.** -1, $\frac{3}{2}$. **19.** -3, $\frac{2}{3}$. **21.** -2, $\frac{1}{2}$.
22. 3, $-\frac{2}{3}$, **23.** 5, $-\frac{3}{2}$. **25.** $\frac{2}{3}$, $\frac{3}{2}$. **26.** $\frac{3}{4}$, $-\frac{4}{3}$.
27. $-\frac{2}{5}$, $\frac{5}{2}$.   **29.** $\frac{2}{3}$, $\frac{3}{4}$.   **30.** $\frac{3}{5}$, $\frac{5}{4}$.
31. $-\frac{3}{2}$, $\frac{4}{5}$.   **33.** a, b.   **34.** a, $-b/2$.
35. b, d/c.

EXERCISE 13.2, PAGE 194

1. 2, 3.   **2.** 3, −2.   **3.** 1, 4.   **5.** −4, 2.
6. −5, 3.   **7.** 4, 7.   **9.** 2, $\frac{1}{2}$.   **10.** 3, $\frac{2}{3}$.
11. 5, $\frac{1}{4}$.   **13.** $\frac{3}{4}$, $\frac{4}{3}$.   **14.** $\frac{2}{5}$, $\frac{5}{3}$.

15. −$\frac{2}{3}$, −$\frac{3}{5}$.   **17.** 2 ± $\sqrt{3}$.   **18.** 3 ± $\sqrt{2}$.

19. 1 ± $\sqrt{5}$.   **21.** 2 ± $\sqrt{7}$.   **22.** 3 ± $\sqrt{3}$.

23. 2 ± $\sqrt{2}$.   **25.** 1 ± i.   **26.** 2 ± i.

27. 1 ± 2i.   **29.** $\frac{1}{2}(1 \pm i)$.   **30.** $\frac{1}{3}(2 \pm i)$.
31. $\frac{2}{3}(1 \pm 2i)$.   **33.** a, −b.   **34.** $a + b$, a.
35. a, a^2.

EXERCISE 13.3, PAGE 196

1. 2, 3,   **2.** 3, 4,   **3.** 2, 5.   **5.** 5, −2.
6. −7, 3.   **7.** −2, −6.   **9.** 2, $\frac{1}{3}$.   **10.** 3, $\frac{1}{2}$.
11. 1, $\frac{1}{4}$.   **13.** −4, $\frac{2}{3}$.   **14.** −2, $\frac{3}{4}$.   **15.** 3, −$\frac{2}{5}$.
17. $\frac{2}{3}$, $\frac{3}{2}$.   **18.** $\frac{3}{4}$, −$\frac{4}{3}$.   **19.** $\frac{3}{7}$, $\frac{7}{2}$.
21. −$\frac{2}{7}$, −$\frac{7}{3}$.   **22.** $\frac{2}{5}$, −$\frac{3}{4}$.   **23.** −$\frac{3}{4}$, −$\frac{2}{3}$.

25. 1 ± $\sqrt{2}$.   **26.** 2 ± $\sqrt{3}$.   **27.** 2 ± $\sqrt{5}$.

29. $\frac{1}{2}(2 \pm \sqrt{6})$.   **30.** $\frac{1}{3}(3 \pm \sqrt{5})$.   **31.** $\frac{2}{3}(3 \pm \sqrt{2})$.

33. ±2i.   **34.** ±5i.   **35.** ±4i.

37. 3 ± $i\sqrt{2}$.   **38.** 2 ± $i\sqrt{3}$.   **39.** −1 ± $i\sqrt{5}$.

41. −a, b.   **42.** $a + b$, −a.   **43.** $a − b$, b.

EXERCISE 13.4, PAGE 199

1. ±2, ±1.   **2.** ±3, ±1.   **3.** ±3, ±i.
5. 1, 2.   **6.** 1, −3.   **7.** 2, −3.
9. $\frac{1}{2}$, −$\frac{1}{5}$.   **10.** −$\frac{1}{3}$, $\frac{1}{4}$.   **11.** −1, $\frac{1}{7}$.
13. ±3.   **14.** ±1.   **15.** 1, −2.
17. 1, ±2, −3.   **18.** ±1, −2, −4.   **19.** 1, 2, 3, 4.
21. ±$\frac{1}{2}$, 2, 3.   **22.** −$\frac{1}{3}$, ±1, $\frac{5}{3}$,   **23.** −$\frac{5}{2}$, ±1, −$\frac{1}{2}$.
25. 1, $\frac{7}{5}$.   **26.** −$\frac{9}{7}$, −$\frac{8}{5}$.   **27.** −$\frac{10}{7}$, −$\frac{4}{5}$.
29. 0, $\frac{7}{3}$.   **30.** −$\frac{1}{4}$, −1.   **31.** −5, −1.

EXERCISE 13.5, PAGE 202

1. 2, -1. **2.** 1, 5. **3.** -1, 0. **5.** 3.
6. 5. **7.** -2. **9.** 6. **10.** -4, 0.
11. No solution. **13.** $\frac{3}{2}$.
14. $\frac{2}{3}, \frac{1}{3}$. **15.** $\frac{1}{2}$. **17.** 1. **18.** 1, $-\frac{3}{2}$.
19. 3, $\frac{14}{3}$. **21.** 1, $\frac{11}{3}$. **22.** -2, -1. **23.** $\frac{1}{2}$.
25. 2. **26.** -1, $\frac{1}{2}$. **27.** 0, $-\frac{16}{11}$. **29.** $\frac{1}{2}$.
30. $\frac{2}{3}$. **31.** 5, $\frac{85}{9}$. **33.** $-\frac{1}{3}$. **34.** -3.
35. 1. **37.** 0. **38.** No solution. **39.** $\frac{3}{2}$.

EXERCISE 13.6, PAGE 205

1. 14, 15. **2.** 15, 17. **3.** 14, 16. **5.** 8, 13.
6. 14, 21. **7.** 18, 22. **9.** 12. **10.** 7.
11. 7. **13.** 11 feet, 13 feet.
14. 15 feet, 21 feet. **15.** 25 feet, 8 feet.
17. 16 miles per hour. **18.** 225 miles per hour.
19. Jim drove 45 miles per hour, Joe drove 40 miles per hour.
21. 1.5 miles, .9 mile. **22.** 4 feet by 12 feet.
23. 4.

EXERCISE 14.2, PAGE 213

1. $x > 2$. **2.** $x > 3$. **3.** $x > -1$.
5. $x < -5$. **6.** $x < -3$. **7.** $x < 2$.
9. $x < -1$. **10.** $x < -4$. **11.** $x < 3$.
13. $x > 2$. **14.** $x > 2$. **15.** $x > 3$.
17. $-4 < x < 1$. **18.** $-\frac{7}{3} < x < 3$. **19.** $-\frac{18}{5} < x < 2$.
21. $x < -\frac{2}{3}$ and $x > 2$. **22.** $x < -4$ and $x > -3$.
23. $x < -3$ and $x > \frac{4}{3}$.

EXERCISE 14.3, PAGE 215

1. $x < -1$ and $x > \frac{5}{2}$. **2.** $x < -\frac{4}{3}$ and $x > 2$.
3. $x < -4$ and $x > -\frac{3}{2}$. **5.** $-\frac{1}{2} < x < \frac{7}{3}$.
6. $\frac{5}{4} < x < \frac{3}{2}$. **7.** $-\frac{5}{2} < x < -\frac{1}{3}$.
9. All real values of x. **10.** All real values of x.
11. No real values of x. **13.** $x < -3$ and $x > 2.4$.
14. $x < -1.6$ and $x > .4$. **15.** $x < -1.4$ and $x > 1.3$.
17. $-2.7 < x < 1.5$. **18.** $-1.6 < x < 1.8$.
19. $-.5 < x < 3.6$.

EXERCISE 15.1, PAGE 218

1. $368.97.
3. $1695.60.
6. $1258.87.
9. $210.72.
11. $309.89.
14. $2001.96.
17. $878.78.
19. $2127.13.
22. $8.6198(10⁹).
25. $18,365.60.
27. $2645.20.

2. $670.05.
5. $1047.55.
7. $3915.43.
10. $540.54.
13. $957.07.
15. $2699.32.
18. $6990. 34.
21. $3.6364(10⁸).
23. $5.5577(10⁵).
26. $1470.01

EXERCISE 15.2, PAGE 222

Each fractional part of a day is considered a full day.

1. 34.834 years; 34 years, 10 months, 3 days.
2. 43.109 years; 43 years, 1 month, 10 days.
3. 10.825 years; 10 years, 9 months, 27 days.
5. 21.442 years; 21 years, 5 months, 10 days.
6. 44.713 years; 44 years, 8 months, 17 days.
7. 35.956 years; 35 years; 11 months, 15 days.
9. 24.198 years; 24 years, 2 months, 12 days.
10. 22.749 years; 22 years. 9 months.
11. 23.733 years; 23 years, 8 months, 24 days.
13. 2.16%.
15. 1.96%.
18. 1.96%.
21. 5.98%.
23. 2.50%.
26. 9.727 years.
29. 3.24%.
31. 2.92%.

14. 1.55%.
17. 4.16%.
19. 3.94%.
22. 5.38%.
25. 17.669 years.
27. 9.093 years.
30. 5.46%.

EXERCISE 15.3, PAGE 225

1. $378.15.
3. $531.43.
6. $1854.72.
9. $347.32.
11. $835.81.
14. $375.08.
17. $353.05.
19. $234.33.
22. $64.20.
25. $13,985.90.
27. $2258.92.

2. $504.73.
5. $495.64.
7. $1574.70.
10. $518.21.
13. $305.37.
15. $489.60.
18. $379.69.
21. $689.22.
23. $3153.78.
26. $13,166.50.

EXERCISE 15.4, PAGE 227

1. $752.59.

2. $934.84.

3. $1895.59.

5. $1705.02.

6. $1851.22.

7. $1751.54.

9. $1258.73.

10. $674.31.

11. $1302.61.

13. $1313.13.

14. $1150.12.

15. $6651.73.

17. $9505.

18. $4340.60.

19. $5942.47.

EXERCISE 16.1, PAGE 231

1. 2, 6, 18, 54, 162.

2. 3, 6, 12, 24.

3. 1, 2, 4, 8, 16, 32.

5. 64, 32, 16, 8, 4.

6. 243, 81, 27, 9, 3, 1.

7. $\frac{1}{4}$, $-\frac{1}{2}$, 1, -2.

9. 24.

10. 486.

11. 9.

13. 96.

14. 81.

15. 486.

17. 121.

18. 126.

19. 252.

21. 126.

22. 242.

23. 63.5.

25. 93.

26. 252.

27. 781.2.

29. 10.83.

30. 20.41.

31. 20.61.

EXERCISE 16.2, PAGE 235

1. $3763.02, $4635.12.

2. $7161.90, $12,558.45.

3. $9493.50, $15,305.65.

5. $2605.38, $3180.89.

6. $2307.78, $3782.08.

7. $3615.81, $4326.97.

9. $2211.77, $5391.95.

10. $2058.38, $9023.75.

11. $4374.07, $19,175.47.

13. $1469.92, $2882.03.

14. $1708.83, $2755.02.

15. $1730.84, $2706.93.

17. $3032.00, $12,529.01.

18. $2453.85, $6348.28.

19. $2042.56, $4173.91.

21. $5427.88.

22. $59,510.

23. $2680.96.

25. $2385.67.

26. $8062.76.

27. $36,733.56.

29. The first one.

30. The single payment.

31. $631.21.

EXERCISE 16.3, PAGE 237

1. $176.25.

2. $83.92.

3. $126.92.

5. $399.25.

6. $700.11.

7. $645.80.

9. $36.82.

10. $16.61.

11. $80.84.

13. $240.86.

14. $224.10.

15. $124.57.

17. $1829.38.

18. $168.72.

19. $112.56.

21. $176.35.

22. $104.12.

23. $145.21.

25. $611.57.

26. $2604.96.

27. $2291.07.

EXERCISE 16.4, PAGE 240

1. 5.89%.   **2.** 33.40%.   **3.** 17.80%.
5. 8.56%.   **6.** 13.42%.   **7.** 29.89%.
9. 6.37%.   **10.** 12.99%.   **11.** 14.69%.
13. $28.14.   **14.** $38.73.   **15.** $36.43.
17. $121.70.   **18.** $86.43.   **19.** $80.32.
21. 9.14%.   **22.** 2.65%.   **23.** 12.61%.
25. $109.68.   **26.** $4585.45.   **27.** $89.95.

EXERCISE 17.1, PAGE 244

1. $\frac{1}{4}$.   **2.** $\frac{15}{2}$,   **3.** $\frac{16}{3}$.
5. $\frac{4}{1}$.   **6.** $\frac{7}{2}$.   **7.** $\frac{3}{1}$.
9. 44 miles/1 hour.   **10.** $.40/1 pound.
11. $635/1 month.   **13.** 28 students/1 class.
14. 51 women/50 men.   **15.** 315 persons/1 square mile.
17. 1.   **18.** 4.   **19.** 3.
21. 3.   **22.** $\frac{15}{7}$. **23.** 4.
25. ±4.   **26.** ±6.   **27.** ±5.
29. 18.   **30.** 2.   **31.** 12.5.
33. ±9.   **34.** ±8.   **35.** ±9.
37. $\frac{28}{15}$.   **38.** $\frac{5}{12}$.   **39.** $\frac{3}{40}$.
41. 20, 8.   **42.** 17.5 feet.   **43.** $1.85.

EXERCISE 17.2, PAGE 248

1. $s = kt$, $p = k/q$, $b = kcd$, $h = kj/m$.
2. 6.   **3.** 4.   **5.** 48.
6. 2.   **7.** 36.   **9.** 31.4.
10. 784 feet.   **11.** 17.496 horsepower.
13. 3 pounds.   **14.** $\frac{1}{25}$ seconds.   **15.** 18 feet.
17. 365 days.   **18.** 16 seconds.   **19.** 343 cubic inches.
21. 12 candlepower.   **22.** 163 pounds.   **23.** 375 ohms.
25. 1004.8 cubic inches.   **26.** 37.125 tons.
27. $48.   **29.** 135 pounds.   **30.** 60 cubic feet.
31. 800 pounds.
33. The kinetic energy traveling at 40 miles per hour is 16 times that at 10 miles per hour.
34. The resistance is 4 times as much at 60 miles per hour as at 30 miles per hour.
35. The weight of the first is 4.913 times that of the second.
37. The heat of the first is twice that of the second.
38. The resistance of the first is 7.2 times that of the second.
39. The attraction between the first two is $\frac{2}{3}$ that between the second two spheres.

Exercise 18.1, page 254

13. $-\frac{5}{3}$.   **14.** $\frac{15}{2}$.   **15.** -2.

17. $x^2 - 6x + y^2 - 8y + 16 = 0$.

18. $x^2 - 4x + y^2 + 6y = 12$.

19. $x + 5y = 10$.

Exercise 18.2, page 257

1. $(7, 0)$.   **2.** $(3, 7)$.   **3.** $(-1, 9)$.

5. $(-2, -10)$.   **6.** $(-9, -11)$.   **7.** $(7, -5)$.

9. $(5, a - 3)$.   **10.** $(-a, 1 + 2a)$.   **11.** $(a + 2, 2a - 3)$.

13. $(5, 1)$.   **14.** $(1, 1)$.   **15.** $(4, -4)$.

17. $(2, -2)$.   **18.** $(7, -5)$.   **21.** $(-4, 10), (2, -2), (8, -4)$.

22. $(-2, 4), (4, 6), (2, -4)$.   **23.** $(1, -7)$.

Exercise 18.3, page 263

1. 1.   **2.** $-\frac{2}{3}$.   **3.** $\frac{1}{5}$.

5. $-\frac{5}{3}$.   **6.** $-\frac{5}{11}$.   **7.** $-\frac{6}{5}$.

9. Parallel.   **10.** Perpendicular.   **11.** Neither.

13. Neither.   **14.** Parallel.   **15.** Parallel.

17. Parallel.   **18.** Parallel.   **19.** Perpendicular.

21. 6.   **22.** 4.   **23.** 2.

25. 1.   **26.** 20.5.   **27.** 6.

29. 18.5.   **30.** 13.5.   **31.** 32.

35. -2.

Exercise 19.1, page 269

1. Mean is 32, median is 34, mode is 23.

2. The mean, median, and mode are each 71.

3. Mean is 1356, median is 930, there is no mode.

5. Mean is $5937.50, median is $4700, mode is $3700. The mean is not a representative average since one salary is so much greater than most others. The mode is not satisfactory since it is very near the lowest salary.

6. The mean is 13 ounces, the median is 8 ounces, the mode is 6 ounces. The mean is not a representative average because one fish was so much larger than any others. The mode is not representative because it is the smallest weight.

7. Mean is $33, median is $33, mode is $35.

9. Mean is 31.5, median is 29, mode is 29.

10. Mean is $40\frac{5}{7}$, median is 40, mode is 25.

11. Mean is $693.50, median is $709.50, mode is $629.50.

EXERCISE 19.2, PAGE 273

1. 8.0°.
3. 12.5.
6. 14.5.
9. 6000, 8550, 8910.
11. 1600, 2280, 24.

2. $354.
5. 10.4°.
7. $320.
10. 2000, 2850, 2970.

EXERCISE 19.3, PAGE 278

1. .63.
5. .84.
9. .67.
13. .72.

2. −.11.
6. .39.
10. −.026.
14. .41.

3. .71.
7. .71.
11. .57.
15. .75.

EXERCISE 20.1, PAGE 283

1. 720.
3. 39, 826, 800.
6. 24.
9. 1620.
11. 1296.
14. 336.
21. 3024.
23. 15,120.
26. 358,800.
29. 43,200.
31. 622,080.
34. 40.

2. 40,320.
5. 6.
7. 20!/9!.
10. 20,580.
13. 360.
15. 6720.
22. 120.
25. 720.
27. 4096.
30. 75.
33. 120.
35. 720.

EXERCISE 20.2, PAGE 286

5. 15,504.
7. 28.
10. 52!/39!13!.
13. 24.
15. 45.
18. 387,600.
21. 9344.
23. 14,570.
26. 45,728,592,000.
29. 1050.
31. 150.

6. 4945.
9. 3,921,225.
11. 2808(143³).
14. 4320.
17. 20.
19. 203.
22. 27,804.
25. 1260.
27. 196, 614, 432.
30. 10,080.

EXERCISE 20.3, PAGE 289

1. $\frac{1}{2}$.

2. $\frac{1}{4}$, $\frac{1}{2}$.

3. $\frac{1}{8}$, $\frac{1}{8}$, $\frac{3}{8}$.

5. $\frac{1}{6}$, $\frac{1}{2}$.

6. $\frac{1}{18}$, $\frac{1}{6}$, $\frac{1}{18}$.

7. $\frac{1}{36}$, $\frac{1}{18}$, $\frac{1}{9}$.

9. $\frac{1}{13}$, $\frac{1}{4}$, $\frac{1}{2}$.

10. $\frac{1}{5525}$.

11. $\frac{1}{64}$, $\frac{1}{2197}$, $\frac{1}{16}$,

13. $\frac{1}{4}$, $\frac{5}{12}$, $\frac{1}{3}$, $\frac{7}{12}$.

14. $\frac{5}{33}$, $\frac{7}{22}$,

15. $\frac{1}{12}$, $\frac{1}{12}$.

17. $\frac{5}{11}$.

18. $\frac{1}{14}$, $\frac{1}{56}$.

19. $\frac{1}{105}$.

21. $\frac{11}{4165}$, $\frac{1}{270}$, 725, $\frac{46}{833}$.

22. 13! 39!/52!.

23. $\frac{1}{2}$.

25. $\frac{1}{7}$.

26. $\frac{5}{51}$, $\frac{35}{306}$.

27. $\frac{5}{204}$, $\frac{35}{272}$.

EXERCISE 20.4, PAGE 294

1. $\frac{7}{10}$.

2. $\frac{11}{36}$.

3. $\frac{43}{77}$.

5. $\frac{1}{2}$.

6. $\frac{3}{4}$.

7. $\frac{1}{5}$.

9. $\frac{2}{11}$.

10. $\frac{1}{3}$.

11. $\frac{3}{5}$. $\frac{2}{15}$, $\frac{1}{5}$.

13. $\frac{1}{216}$.

14. $\frac{9}{25}$.

15. $\frac{1}{12}$.

17. .34.

18. $\frac{2}{5}$.

19. $\frac{1}{169}$.

21. $\frac{1}{4}$.

22. $\frac{2}{11}$.

23. $\frac{1}{560}$, $\frac{1}{112}$, $\frac{1}{28}$.

EXERCISE 20.5, PAGE 297

1. $\frac{5}{16}$, $\frac{1}{2}$.

2. $\frac{3}{8}$, $\frac{11}{16}$.

3. $\frac{3}{16}$.

5. 34, $\frac{375}{34}$, 992.

6. $\frac{5}{72}$, $\frac{2}{27}$.

7. $\frac{121}{3456}$, $\frac{257}{6912}$.

9. $\frac{169}{16}$, 384, $\frac{211}{16}$, 384.

10. $\frac{15}{1024}$, $\frac{1}{64}$.

11. $\frac{48}{28}$, 561, $\frac{49}{28561}$.

13. $\frac{5}{16}$, $\frac{1}{2}$.

14. $\frac{125}{3888}$.

15. $\frac{1}{243}$.

17. $\frac{27}{64}$.

18. $\frac{144}{625}$.

19. $\frac{16}{243}$.

21. $\frac{12}{343}$.

22. $\frac{32}{81}$.

23. $\frac{1792}{15}$, 625.

EXERCISE 21.1, PAGE 304

1. $2x$.

2. $2x$.

3. $4x + 1$.

5. $3x^2 - 6x$.

6. $3x^2 + 2$.

7. $3x^2 - 2$.

9. 1.

10. 11.

11. -1.

13. -3.

14. -2.

15. -11.

17. 3.

18. 91.

19. 4.

22. $-\frac{1}{2}\sqrt{1-x}$.

25. $-1/x^2$.

27. $-2/(x-1)^2$.

21. $\frac{1}{2}\sqrt{x}$.

23. $x/\sqrt{x^2+1}$.

26. $-2/x^3$.

EXERCISE 21.2, PAGE 308

1. $3x^2+4x$.

3. $10x^4-12x^3+7$.

6. $1-1/x^2+4/x^3$.

9. 7.

11. 6.

14. 19.

17. 17.

19. -1.

22. $x=-2$, $x<-2$, $x>-2$.

25. $x=1, 2$, $1<x<2$, $x<1$ and $x>2$.

26. $x=-1, 3$, $-1<x<3$, $x<-1$ and $x>3$.

27. $x=-2, .5$, $-2<x<.5$, $x<-2$ and $x>.5$.

2. $4x^3-14x+2$.

5. $2x-3/x^2$.

7. $6x^2-\frac{1}{2}\sqrt{x}$.

10. 3.

13. 1.

15. 3.

18. 12.

21. $x=1$, $x<1$, $x>1$.

23. $x=1.5$, $x<1.5$, $x>1.5$.

EXERCISE 21.3, PAGE 313

1. 15, 18.

3. $-15, -12$.

6. Min for $x=.5$.

9. Max for $x=1$, min for $x=2$.

10. Max for $x=-1$, min for $x=3$.

11. Min for $x=.5$, max for $x=1$.

13. Max for $x=-2$, min for $x=1$.

14. Max for $x=-1$, min for $x=4$.

15. Max for $x=-.5$, min for $x=\frac{2}{3}$.

17. Max for $x=1$, min for $x=-3$.

18. Max for $x=2$, min for $x=-3$.

19. Max for $x=\frac{1}{3}$, min for $x=-1$.

2. 16, 16.

5. Min for $x=2$.

7. Max for $x=1.5$.

EXERCISE 21.4, PAGE 316

1. Each part is 6.

3. 6, 18.

6. $l=w$.

9. 3 inches.

11. $(1, 3)$.

13. 60 feet along the barn, other dimension 30 feet.

14. Width, 6 inches; height, 3 inches.

15. Width is 90 feet, length is 180 feet.

2. Each part is 8.

5. $l=w$.

7. $5\sqrt{2}$.

10. Base is 8 feet, altitude is 4 feet.

17. 300.

19. After 4 seconds, 256 feet.

22. Height $=$ radius.

18. 5 seconds.

21. Base $= 2$ altitude.

23. Width $= (1.5 + 1.5\sqrt{7})$ inches; length $= (2 + 2\sqrt{7})$ inches.

EXERCISE 21.5, PAGE 320

1. $9x^2 - 4x + 3$.

3. $5x^4 - 12x^2 - 6x + 6$.

6. $-17/(3x - 1)^2$.

7. $(6x^4 - 28x^3 - 4x^2 + 28x - 35)/(2x - 7)^2$.

9. $8(2x - 1)^3$.

2. $18x^5 - 12x^3 + 6x^2 - 2$.

5. $(x^2 + 4x + 3)/(x + 2)^2$.

10. $6(3x^2 - 4x + 7)^2(3x - 2)$.

11. $(x - 1)/\sqrt{x^2 - 2x + 3}$.

13. $2(3x - 1)(6x^3 - x^2 + 12x - 7)/(x^3 + 3x - 2)^{1/3}$.

14. $2x(2x^2 - 3)^2(8x^4 - 17x^2 + 21)/\sqrt{x^4 - 2x^2 + 3}$.

15. $6(2x - 1)^2(3x + 2)(5x + 1)$.

17. $(3x^2 - 5)^2(33x^2 + 54x + 5)/(2x + 3)^{3/2}$.

18. $2(x^2 - x)(22x^3 - 3x^2 - 6x + 1)/(8x - 1)^{3/2}$.

19. $2(2x^3 - 5x)(22x^3 - 18x^2 - 15x + 15)/(4x - 3)^{3/2}$.

21. No min and no max.

22. A min for $x = 0$, a max for $x = 2$.

23. A min for $x = 0$, a max for $x = -4$.

EXERCISE 21.6, PAGE 324

1. $x^3/3 + c$.

3. $-x^{-2}/2 + c$.

6. $-x^{-7} + c$.

9. $2x^5/5 - 5x^3/3 + c$.

2. $x^4/4 + c$.

5. $x^5 + c$.

7. $4x^3/3 + c$.

10. $x^5 - 2x^4 + c$.

11. $2x^3 - 7x + c$.

13. $\dfrac{x^2}{2} + \dfrac{2x^{3/2}}{3} + c$.

14. $\dfrac{4x^{3/2}}{3} + x^{-1} + c$.

15. $\dfrac{3x^{4/3}}{4} - \dfrac{3}{2}x^{2/3} + c$.

17. $(3x^2 - 4x)^4/4 + c$.

19. $-(x^4 - 2x^{-3})^{-1} + c$.

22. $(x^4 + 2x^2)^3/12 + c$.

25. $y = x^2 - x - 2$.

27. 208 feet.

18. $(x^3 - 5x)^3/3 + c$.

21. $(x^6 - x^4)^4/8 + c$.

23. $(7x^4 - 4x)^{1.5}/6 + c$.

26. $y = x^3 - 2x^2 + 1$.

\

Exercise 21.7, page 328

1. 14.	**2.** 19.5.
3. 28.	**5.** $\frac{62}{3}$.
6. $\frac{50}{3}$.	**7.** 4.5.
9. $86\frac{1}{4}$.	**10.** 31.5.
11. 192.	**13.** 15.
14. 12.5.	**15.** 2.
17. $\frac{125}{6}$.	**18.** $\frac{1}{6}$.
19. $\frac{256}{3}$.	**21.** $\frac{16}{3}$.
22. 36.	**23.** 48.

INDEX